SMART MIXES FOR TRANSBOUNDARY ENVIRONMENTAL HARM

This work offers a multidisciplinary approach to legal and policy instruments used to prevent and remedy global environmental challenges. It provides a theoretical overview of a variety of instruments, making distinctions between levels of governance (treaties, domestic law) and types of instruments (market-based instruments, regulation, and liability rules), and between government regulation and private or self-regulation. The book's central focus is an examination of the use of mixes between different types of regulatory and policy instruments and different levels of governance, notably in climate change, marine oil pollution, forestry and fisheries. The authors examine how, in practice, mixes of instruments have often been developed. This book should be read by anyone interested in understanding how interactions between different instruments affect the protection of environmental resources.

JUDITH VAN ERP is Professor of Public Institutions in the Faculty of Law, Economics and Governance at Utrecht University, the Netherlands, and Designated Research Chair in Utrecht University's strategic interdisciplinary theme of Institutions for Open Societies.

MICHAEL FAURE is Academic Director of the Maastricht European Institute for Transnational Legal Research and Professor of Comparative and International Environmental Law at Maastricht University. He is also Academic Director of the Ius Commune Research School and a half-time professor of Comparative Private Law and Economics at the Rotterdam Institute of Law and Economics of Erasmus University.

ANDRÉ NOLLKAEMPER is Dean and Professor of Public International Law at the University of Amsterdam. He is also an external legal advisor to the Minister of Foreign Affairs of the Netherlands, Member of the Permanent Court of Arbitration, Member of the Institut de Droit International Law, former President of the European Society of International Law and Member of the Royal Netherlands Academy of Arts and Sciences.

NIELS PHILIPSEN is Professor of Shifts in Private and Public Regulation at the Erasmus School of Law of Erasmus University Rotterdam, Vice-Director of the METRO research institute and Associate Professor in Law and Economics, Faculty of Law, at Maastricht University. Since November 2017, Philipsen has also been Adjunct Professor of the School of Law and Economics at the China University of Political Science and Law (CUPL) in Beijing.

Smart Mixes for Transboundary Environmental Harm

Edited by

JUDITH VAN ERP

Utrecht University

MICHAEL FAURE

Maastricht University, Erasmus University Rotterdam

ANDRÉ NOLLKAEMPER

University of Amsterdam

NIELS PHILIPSEN

Erasmus University Rotterdam, Maastricht University

CAMBRIDGE UNIVERSITY PRESS

CAMBRIDGE
UNIVERSITY PRESS

University Printing House, Cambridge CB2 8BS, United Kingdom

One Liberty Plaza, 20th Floor, New York, NY 10006, USA

477 Williamstown Road, Port Melbourne, VIC 3207, Australia

314-321, 3rd Floor, Plot 3, Splendor Forum, Jasola District Centre, New Delhi - 110025, India

103 Penang Road, #05-06/07, Visioncrest Commercial, Singapore 238467

Cambridge University Press is part of the University of Cambridge.

It furthers the University's mission by disseminating knowledge in the pursuit of
education, learning and research at the highest international levels of excellence.

www.cambridge.org
Information on this title: www.cambridge.org/9781108449526
DOI: 10.1017/9781108653183

© Cambridge University Press 2019

This publication is in copyright. Subject to statutory exception
and to the provisions of relevant collective licensing agreements,
no reproduction of any part may take place without the written
permission of Cambridge University Press.

First published 2019
First paperback edition 2022

A catalogue record for this publication is available from the British Library

ISBN 978-1-108-42838-5 Hardback
ISBN 978-1-108-44952-6 Paperback

Cambridge University Press has no responsibility for the persistence or
accuracy of URLs for external or third-party internet websites referred to in
this publication, and does not guarantee that any content on such websites is,
or will remain, accurate or appropriate.

Contents

Figures

Tables

Contributors

Richard Barnes, Professor, School of Law and Politics, University of Hull, United Kingdom.

Simon R. Bush, Professor and Chair of the Environmental Policy Group, Wageningen University, the Netherlands.

Judith van Erp, Professor of Public Institutions, Utrecht University, the Netherlands.

Michael Faure, Professor of International and Comparative Environmental Law, Maastricht University, and Professor of Private Comparative Law and Economics, Erasmus School of Law, Rotterdam, the Netherlands.

Lars Gulbrandsen, Professor and Deputy Director, Fridtjof Nansen Institute, Oslo, Norway.

Neil Gunningham, Professor, Australian National University, Canberra, Australia.

Andrew Jordan, Professor, Tyndall Centre for Climate Change Research, School of Environmental Sciences, University of East Anglia, United Kingdom.

Markos Karavias, Assistant Professor in International Law, Utrecht University School of Law, the Netherlands.

Jing Liu, Associate Professor, Research Institute of Environmental Law, Wuhan University, China.

Constance L. McDermott, Senior Fellow in Forest Governance and Leader of Ecosystems Governance Group, University of Oxford.

Mathias Müller, PhD Candidate, Maastricht European Institute of Transnational Legal Research, Maastricht University, the Netherlands.

André Nollkaemper, Dean and Professor of Public International Law, Faculty of Law, University of Amsterdam, the Netherlands.

Philipp Pattberg, Professor of Transnational Environmental Governance and Policy, Free University of Amsterdam, the Netherlands.

Marjan Peeters, Professor of Environmental Policy and Law, Maastricht University, the Netherlands.

Niels Philipsen, Professor of Shifts in Private and Public Regulation, Erasmus School of Law, Rotterdam, and Associate Professor in Law and Economics, Maastricht University, the Netherlands.

Linda Senden, Professor of EU Law, Utrecht University, the Netherlands.

Jan van Tatenhove, Professor of Marine Governance and MSP, Department of Planning, IFM – Center for Blue Governance, Aalborg University, Denmark.

Hui Wang, Former PhD Researcher, METRO, Maastricht University, the Netherlands.

Oscar Widerberg, Assistant Professor, Institute for Environmental Studies (IVM), Free University of Amsterdam, the Netherlands.

Rüdiger Wurzel, Professor of Comparative European Politics and Jean Monnet Chair in European Union Studies, University of Hull, United Kingdom.

Agnes Yeeting, Environmental Policy Group, Wageningen University, the Netherlands.

Anthony Zito, Professor of European Public Policy, University of Newcastle, United Kingdom.

Acknowledgments

This book is one of the results of a project subsidised by the Netherlands Royal Academy of Arts and Sciences on smart mixes in relation to transboundary environmental harm. The project was sponsored within the framework of the Academy project "Beyond borders" (Over grenzen). It consisted of a collaboration between the Erasmus School of Law in Rotterdam and the Faculty of Law of the University of Amsterdam, with additional contributions by members of the Faculty of Law of Maastricht University. The project work was largely executed by Dr. Markos Karavias and Dr. Jing Liu (two postdocs financed by the Academy), and supported by an expert team consisting of: Professor Michael Faure (Erasmus University Rotterdam and Maastricht University); Professor André Nollkaemper (University of Amsterdam); Professor Peter Mascini (Erasmus University Rotterdam); Professor Judith van Erp (Utrecht University); Professor Marjan Peeters (Maastricht University); and Professor Niels Philipsen (Erasmus University Rotterdam and Maastricht University).

The project team benefitted from the expertise of an advisory committee consisting of: Jan van den Broek (VNO-NCW); Professor Joyeeta Gupta (University of Amsterdam); Dr. Veerle Heyvaert (London School of Economics); Gerald Kok (Shell); Professor Catherine Redgwell (University of Oxford); Dr. Peter Sand (University of Munich); Professor Jaap Spier (Supreme Court of the Netherlands); Jasper Teulings (Greenpeace International); and Ludo Veuchelen (Erasmus University Rotterdam).

Within the framework of this research project, two workshops were organised: one, on smart mixes in relation to forest and climate governance, was held on 4–5 February 2015, and the other, on smart mixes in relation to fishery and oil pollution governance, was held on 7–8 October 2015; both workshops were held at the Facility of the Netherlands Royal Academy of Arts and Sciences in the Trippenhuis in Amsterdam.

As authors of this book, which largely builds on the mentioned research project, we are grateful to our colleagues, members of the expert team, to the members of the advisory committee, to the participants in the two mentioned workshops and especially to the Netherlands Royal Academy of Arts and Sciences for their invaluable input and support, without which this book could not have been published.

In addition, as editors of this book we are equally grateful to Marianne Breijer who, as project manager, supported the entire research project, and to Martine van Trigt of the University of Amsterdam. We are also grateful to Marina Jodogne (Maastricht University) for invaluable editorial help in harmonising footnotes and references in preparation of the publication of this volume.

Abbreviations

ABNJ	Areas Beyond National Jurisdiction
ACI	Airport Council International
ANSI	American National Standards Institute
Art.	Article
ASL	Asociaciones Sociales de Lugar
ASRC	Arctic Slope Regional Corporation
BASR	Basic Aviation Risk Standard
BAT	Best Available Technology
BCR	Binding Corporate Rules
BMSY	Biomass that would provide Maximum Sustainable Yield
BP	British Petroleum
BS	British Standards
CAB	Certification Assessment Body
CANSO	Civil Air Navigation Services Organization
CBD	Convention on Biological Diversity
CBFM	Community-Based Forest Management
CCAMLR	Commission for the Conservation of Antarctic Marine Living Resources
CDM	Clean Development Mechanism
CDS	Catch Documentation Scheme
CEC	Commission of the European Communities
CEDRE	Centre de documentation de recherches et d'expérimentations sur les pollutions accidentelles des eaux
CEN	European Committee for Standardization
CFV	Certificacion Forestall Voluntaria
CGMTA	Coast Guard and Maritime Transportation Act
CLC	Civil Liability Convention

CO_2	Carbon Dioxide
COFI	Committee of Fisheries
COLTO	Coalition of Legal Toothfish Operators
COPE	European Compensation Fund for Oil Pollution
CPET	Central Point of Expertise on Timber
CRISTAL	Contract Regarding an Interim Supplement to Tanker Liability
CSA	Canadian Standards Association
CSG	Coal Seam Gas
CSO	Civil Society Organization
CSR	Corporate Social Responsibility
DARRP	Damage Assessment, Remediation and Restoration Program
DGR	Dangerous Goods Regulations
EC	European Commission
EEA	European Environmental Agency
EEZ	Exclusive Economic Zone
EMAS	Environmental Management and Audit Scheme
EMS	Environmental Management Scheme
ENDS	Environmental Data Services
ENGO	Environmental Non-Governmental Organisation
EP	European Parliament
ESO	European Standardisation Organisation
ETS	Emissions Trading Scheme
EU	European Union
EUE	Ecologically Unequal Exchange
EU ETS	European Union Emissions Trading Scheme
EUR	Euro(s)
EUTR	EU Timber Regulation
FAD	Fish Aggregating Device
FAO	Food and Agriculture Organization of the United Nations
FLEGT	Forest Law Enforcement, Governance and Trade
FSA	Fisheries Stock Assessment
FSC	Forest Stewardship Council
FSF	Fisheries Survival Fund
FSF	Flight Safety Foundation
GHG	Greenhouse Gas
GSGSSI	Government of South Georgia and South-Sandwich Islands
HCR	Harvest Control Rule
HNA	Hyperlink Network Analysis
HSBC	Hongkong and Shanghai Banking Corporation
IATA	International Air Transport Association
IBAC	International Business Aviation Council

ICAO	International Civil Aviation Organization
ICAP	Interagency Committee for Aviation Policy
ICCAT	International Commission for the Conservation of Atlantic Tunas
IFALPA	International Federation of Air Line Pilots' Associations
IGO	International Governmental Organization
IMF	International Monetary Fund
IMO	International Maritime Organization
IO	International Organization
IOPC Fund	International Oil Pollution Compensation Fund
IOSA	IATA Operational Safety Audit
IOTC	Indian Ocean Tuna Commission
IRC	International Regime Complexes
IS-BAO	International Standard for Business Aircraft Operations
ISO	International Organisation for Standardisation
ITOPF	International Tanker Owners Pollution Federation
ITQ	Individual Transferable Quota
ITTO	International Tropical Timber Organization
IUU	Illegal, Unreported and Unregulated
LEI	Lembaga Ekolabel Indonesia
MARPOL	International Convention for the Prevention of Pollution from Ships
MSC	Marine Stewardship Council
MSC	FSG Marine Stewardship Council Fisheries Standard and Guidance
MSPEA	Maldives Seafood Processors and Exporters Association
NAO	Nenets Autonomous Okrug
NEPI	New Environmental Policy Instrument
NGO	Non-Governmental Organization
NOAA	National Oceanic and Atmospheric Administration
NOGEPA	Netherlands Oil and Gas Exploration and Production Association
NSMD	Non-State Market-Driven
ODS	Ozone-Depleting Substances
OECD	Organization of Economic Co-Operation and Development
OGP	Oil and Gas Producers
OJ	Official Journal
OPA	Oil Pollution Act
OPAGAC	Organisation of Associated Producers of Large Tuna Freezer Vessels
OPEGUI	Spanish Inshore Producers Organisations from Guipuzcoa
OPESCAYA	Spanish Inshore Producers Organisations from Biscay
OPRC	Oil Pollution Preparedness, Response and Cooperation
OSPAR	Offshore Industry from the Paris
PEFC	Programme for the Endorsement of Forest Certification

PI	Performance Indicator
PNA	Parties to the Nauru Agreement
POLMAR	Pollution Maritime
RBM	Rights-Based Measure
REDD+	Reducing Emissions from Deforestation and Forest Degradation
REEEP	Renewable Energy and Energy Efficiency Partnership
RFMO	Regional Fisheries Management Organisation
RFO	Regional Fisheries Body
SAE	Society of Automotive Engineers
SAR	Search and Rescue
SARPS	Standards and Recommended Practices
SCRS	Standing Committee on Research and Statistics
SDR	Special Drawing Right
SEM	Single European Market
SFI	Sustainable Forestry Initiative
SG	Scoring Guidepost
SGSSI	South Georgia and the South Sandwich Islands
SLO	Social License to Operate
SOLAS	International Convention for the Safety of Life at Sea
STCW	Standards of Training, Certification and Watchkeeping
STOPIA	Small Tankers Oil Pollution Indemnification Agreement
TAC	Total Allowable Catch
TBGI	Transnational Business Governance Interactions
TCO	Tierras Communitarias de Origen
TOPIA	Tanker Oil Pollution Indemnification Agreement
TOVALOP	Tanker Owners Voluntary Agreement concerning Liability for Oil Pollution
TPAC	Timber Procurement Assessment Committee
TPR	Transnational Private Regulation
TRCCC	Transnational Regime Complex for Climate Change
UK	United Kingdom
UN	United Nations
UNCLOS	United Nations Convention on the Law of the Sea
UNFCCC	United Nations Framework Convention on Climate Change
UNFSA	United Nations Fish Stocks Agreement
UN REDD	United Nations Programme on Reducing Emissions from Deforestation and Forest Degradation
UoA	Unit of Assessment
US	United States
USA	United States of America
USD	US Dollar
USOAP	Universal Safety Oversight Audit Program

VMS	Vessel Monitoring System
VPA	Voluntary Partnership Agreement
VPN	Virtual Policy Network
WCPFC	Western and Central Pacific Fisheries Commission
WCPO	Western and Central Pacific Ocean
WTO	World Trade Organisation
WWF	World Wide Fund for Nature

Conceptual Approaches to Smart Mixes

1

Introduction

The Concept of Smart Mixes for Transboundary Environmental Harm

*Judith van Erp, Michael Faure, Jing Liu, Markos Karavias,
André Nollkaemper and Niels Philipsen*

1.1 INTRODUCTION

The complex nature of transboundary environmental problems, such as global warming, ozone depletion, land degradation, oil pollution and biodiversity loss, and the risks associated with such problems, pose a fundamental challenge to policy makers worldwide, namely that of designing an effective global environmental governance system.

An important part of the quest for such a governance system, though one that has been recognised only relatively recently, consists of finding 'smart mixes' of regulatory instruments. We define this term in Section 1.2, but at this stage already note that the idea is that particular combinations of instruments may work better than others.

An example of such a smart mix constitutes the combination of safety regulation and civil liability in the US Oil Pollution Act (OPA). Normally, under the OPA the civil liability of tanker owners is limited to a financial cap. However, a responsible party can lose its right to limitation if the incident was caused by the violation of an applicable federal safety, construction or operating regulation. The construction of the civil liability regime therefore provides incentives for compliance with safety regulations intended to prevent oil spills.[1]

The emergence of the notion of smart mixes is one further stage in the development of modern environmental regulation. This development is intimately linked with the realisation that many environmental problems are transboundary in nature. Initially, such problems arose from the use of transboundary resources (such as rivers), or from the movement of pollutants across national boundaries. The advancement of science gradually generated concern over problems of a wider reach, namely global commons problems, such as the depletion of the ozone layer,

---

[1] See Chapter 13.

climate change, biodiversity loss or depletion of fish stocks. This has led to the expansion of the scope of environmental law from domestic to international to global.

Parallel to the expansion of the scope of environmental law to address transboundary and global problems, a wider variety of regulatory and governance actors and instruments has emerged. Traditional top-down command-and-control rules prohibiting or restricting environmentally harmful industrial activities[2] have been supplemented by a diverse spectrum of regulatory approaches. Gradually, starting in the mid-1980s a shift took place in the way of thinking about environmental law, both on the international and domestic plane. The ascent of a neoliberal thinking gave birth to the assumption that environmental problems, previously thought to require direct state intervention, could also be solved by (combinations of) deregulation, privatisation, voluntarism, outsourcing, and/or the use of market and suasive mechanisms.[3] This assumption has affected the character of environmental policy instruments, and a diversification of instruments has occurred, both at the domestic and international plane. This can be illustrated by the fact that in 1992 a call for the 'effective use of economic instruments and market and other incentives' was included in chapter 8 (C) of the UN's Agenda 21 for sustainability.[4]

Second, the emergence of global value chains, and other forms of increased connectivity – such as the current revolution in forms of information technology – have facilitated and stimulated private forms of regulation.[5] Thus, private actors (such as corporations, NGOs, regulatory intermediaries[6] and citizen initiatives[7]) and transnational networks have assumed key roles alongside states. A constellation of private environment-related instruments has emerged, such as standardisation instruments, certification/labelling schemes, transparency initiatives and corporate codes of conduct. The coexistence of the regulatory state, with its proliferating private or hybrid modes of regulation, has led to a pluralist environmental governance system.[8]

In sum, transnational environmental governance conjures an image of polycentricity. A diversity of international and domestic laws and regulations operate in parallel with market-based and suasive instruments and private standards promulgated by nonstate entities, and private actors operate alongside state actors and international organisations.

---

[2] Gunningham (2009).

[3] For a detailed classification of instruments, see Gunningham, Grabosky & Sinclair (1998), at 37–92.

[4] United Nations Conference on Environment & Development, Agenda 21, Rio de Janerio, Brazil, 3–14 June 1992, (Agenda 21), https://sustainabledevelopment.un.org/content/documents/Agenda21.pdf.

[5] Auld (2014).

[6] Abbott, Levi-Faur & Snidal (2017).

[7] Trevisanut (2014).

[8] See *inter alia* the contributions to Van Rooij, McAllister & Kagan (2010).

Despite – or perhaps because of – this polycentricity, the success of environmental governance remains modest, as evidenced by the state of the natural environment. Scholars have responded by engaging in empirical analyses of the effectiveness of a number of international environmental treaties, regimes,[9] and regulatory instruments.[10] While for a long time the effectiveness of the treaties and instruments was examined in isolation, scholarly approaches are increasingly responding to the increased pluriformity and complexity of the global regulatory landscape. A growing body of international law scholarship is shifting the focus from single regulatory instruments to more holistic analyses of interactions between regulatory institutions and between various levels of governance – international, state, local and within markets – as all of these contribute to the environmental outcome.[11] An important contribution is made by the scholarship on regime complexes, which studies complex and interwoven institutional landscapes consisting of nested, overlapping and parallel regimes.[12] This scholarship often views regime complexity as a source of ineffectiveness, as it finds that the interconnected and interdependent character of different regimes governing the same subject area generates a variety of problematic interactions, results in suboptimal outcomes, and creates a variety of structural opportunities for actors to strategically exploit regulatory diversity to further their self-interest. This raises the question of whether international governance interactions can also create positive incentives for environmental protection. For example, the rise of global value chains and other forms of international connectivity enables more stringent regimes to cast 'shadows of hierarchy' to less stringently regulated areas.[13]

This volume is induced by the quest for positive regulatory interactions and proposes that conceptions of 'smart regulatory mixes'[14] may enable an analytical way forward. The idea behind smart regulation is that various regulatory and governance instruments, both public and private and both international and local, can be combined into mixes of complementary instruments and actors, tailored to the specific needs of the situation. Such a 'smart mix' approach acknowledges that all environmental policy instruments taken separately (for example, liability rules,

---

[9] See Sand (1992); Young (1999).

[10] Early research tended to assess instruments independently, and, thus to conceptualize policy design as a zero-sum option. The question was often reduced to the superiority of one instrument to another in certain situations. See Howlett (2004), at 2–3; Woodside (1986), at 775–793. Later research incorporated the effectiveness of a combination of a variety of policy instruments. See *inter alia* Faure (2012).

[11] Winter (2011); Eberlein, Abbot, Black, Meidinger & Wood (2013).

[12] Raustiala & Victor (2004); Alter & Meunier (2009); Orsini, Morin & Young (2013).

[13] Borzel & Risse (2010, 2016).

[14] The concept of Smart Regulation was initially coined by Gunningham, Grabosky & Sinclair (1998).

taxation, emission trading or command-and-control regulation) have particular limitations,[15] thus justifying a need for a combination of instruments. A smart mix combines multiple instruments or programmes that interact; and can engage a wide circle of actors and networks.

The concept of 'smart regulation' does not necessarily mean that instrument mixes can easily be purposively designed. In situations of polycentric governance, combinations of institutions and actors emerge spontaneously and interact, often in unexpected and unintended ways, within governance networks. Governance arrangements are path-dependent, and their impact is context-specific and depends on the specific institutional, social, economic and environmental conditions. The idea that regulators could rationally and independently select and combine instruments out of a toolbox, or that regulatory mixes could be purposefully designed by a central actor, is mostly a fiction.

However, this does not mean that attempts to coordinate and orchestrate the interaction of instruments are entirely fruitless. Social change occurs through accident, evolution or intervention, and mostly through a combination of these three processes.[16] Although, as Goodin stated, 'institutions are often the product of intentional activities gone wrong', at least some form of intentionality almost always plays a role. Thus, studies oriented towards institutional design should acknowledge the multiplicity of designers and interactions of their intentions rather than advocating a Grand Design.[17]

This project is grounded in the conviction that a better understanding of instrument interactions can contribute to institutional design that is tailored to a specific situation where the need for environmental regulation arises. The contributions to this volume attempt to draw lessons from the experiences that have been gained with existing instrument mixes. These suggest that some instrument combinations are more effective than others; certain conditions are more beneficial to the emergence of smart mixes, and some actors have more effective strategies than others. Future instrument mixes can benefit from these lessons, whether they are purposively designed or incrementally shaped and reshaped.

This volume will look specifically into four key areas of environmental concern, namely deforestation, greenhouse gas emissions, overfishing and marine oil pollution. Of course, the selection of four areas as the testing ground of smart regulation evokes the question of context. The causes and drivers for each of the four threats to the environment vary highly, and therefore the appropriate strategy to address them will likely be context-specific. Other studies have also found that outcomes of regulatory instruments are influenced by the political and institutional context of

---

[15] See Faure (2014), at 690.
[16] Goodin (1996).
[17] Ibid.

specific countries[18] and the composition of the particular market[19] in which these instruments operate. This entails that conclusions about what constitutes a smart mix of instruments are necessarily context-specific. Therefore, this volume does not aim to identify 'the' optimal mix that would apply to all scenarios. Rather, it will seek to establish whether, in particular contexts, existing 'mixes' of forms of regulation and instruments in relation to the aforementioned four areas of concern have been 'smart' in addressing both the causes of environmental pollution and drivers for its prevention.

Identifying 'smart mixes' in a context-sensitive way has several benefits. First, the identification of such mixes may inspire a shift of research paradigm from the choice among regulatory strategies to the interaction between regulatory strategies.[20] Second, it may provide valuable insight into the design aspects of mixing forms of regulation and instruments that prove effective in a particular context. Finally, it may allow for a context-specific understanding of the role of nonstate entities within the framework of global environmental governance. Thus, the focus on smart mixes brings a complementary, yet distinctive, focus to the field of transnational environmental law and governance, and possibly also beyond that, to regulatory theory.

1.2 THE CONCEPT OF 'SMART MIXES'

The concepts of 'smart mixes' and 'smart regulation' were introduced by Gunningham and Grabosky in their seminal book in 1998,[21] and have been widely adopted since. 'Smart regulation' departs from a broad interpretation of 'regulation' that is not limited to state-based law[22] but also includes self- and co-regulation and a wide variety of other forms of social control exercised by business and NGOs.[23] 'Regulation' thus can take various forms: international treaties; domestic law; private standards; economic incentives; transparency and information disclosure; and procedural rights. 'Smart regulation' thus fits in the broader shift from 'government' to 'governance' in networks of states, businesses and civil society.[24]

Essential to smart regulation is the idea that the *combination* of regulatory instruments and actors is often more effective than a single instrument, and that instruments can be complementary. Since most instruments and actors have strengths and weaknesses in specific circumstances, combining instruments and regulatory actors into a mix allows them to take advantage of their strengths while

---

[18] Liu, Faure & Mascini (2017).
[19] Auld (2014).
[20] Cf. Eberlein et al. (2013).
[21] Gunningham, Grabosky & Sinclair (1998).
[22] Gunningham (2009).
[23] Gunningham & Sinclair (1999), at 49–76.
[24] Gunningham (2009); Howlett & Rayner (2004).

compensating for their weaknesses.[25] For example, command-and control regulation may be dependable and predictable, but also inflexible and inefficient; economic incentives, on the other hand, are generally flexible and efficient, but less dependable. Smart mixes combine instruments tailored to specific environmental goals and circumstances. They also can balance coercive and noncoercive regulatory techniques, and organise public regulation in such a way that it mobilises, facilitates and supports third-party regulation and informal social control.

Precisely because there is no one single instrument that can be considered as the silver bullet that would solve all environmental problems, smart regulation necessarily entails a search for smart mixes of instruments. The challenge for regulators and policy makers is thus to assess how the regulatory instruments and other governance initiatives regarding a certain environmental problem interact, and, where possible, to coordinate and orchestrate this interaction to stimulate a productive and compatible mix in a particular context.

A 'smart mix' does not necessarily include many instruments.[26] If too many instruments are included, there is a risk that the mixing of instruments simply results in a 'messy mix', rather than in a 'smart mix'.[27] Some instruments are even inherently incompatible and will turn out ineffective or even counterproductive when combined, such as command-and-control regulation imposing fixed performance levels on industry, in combination with economic instruments, such as tradable pollution rights. As performance standards limit choice, and tradable rights enable flexibility, their combined outcome will be at least suboptimal.[28] Other combinations, however, such as industry self-regulation backed up by command-and-control regulation, may be complementary.[29]

Smart combinations of instruments do not only appear between different policy instruments at the domestic level, but also between different levels of governance. One could, for example, imagine a combination of standard setting at the international level (for example, by a treaty aiming to reduce overfishing) with an implementation at the domestic level via regulatory standards, quotas and certification by the Marine Stewardship Council (MSC).[30] Therefore, one needs to examine how different forms of regulation and instruments incorporate, promote, limit or replace one another, also at different levels of governance.[31] To some extent, of course, most environmental treaties rely on domestic implementing measures; therefore, a combination of instruments is inherent in international regulation.

---

[25] Gunningham & Sinclair (1999).

[26] Ibid.

[27] See Peeters (2014), at 173–192.

[28] Gunningham & Sinclair (1999).

[29] Gunningham, Grabosky & Sinclair (1998).

[30] See Stokke (2012); Garcia, Rice & Charles (2014).

[31] See Stewart (2008), who indicates that the distinctive characteristics of international environmental regulation also affect the instrument choice in the international context.

However, this does not necessarily mean that all combinations of international and national (implementing) law and regulations are necessarily productive; and in that respect, the concept of smart mixes may have conceptual traction to help us identify which combinations do or do not work.

The concept of 'smart regulation' has been adopted by various states and supra-national authorities – though mostly without using that very term. One particularly influential adoption of a 'smart' approach is perhaps in the UN Guiding Principles on Business and Human Rights, implementing the UN's 'Protect, Respect and Remedy' Framework.[32] An express application can be found in the Canadian Smart Regulation initiative (2005).[33] 'Smart' and 'better' regulation also are often used in the context of deregulation in policy practice,[34] although this is far from the original purpose of these ideas.

In further clarifying the concept of smart mixes, we will identify four aspects of the concepts: the elements of the mix; forms of regulation; policy instruments; and the emergence of mixes.

1.2.1 *The Elements of the Mix*

A wide variety of mixes can be thought of, such as mixes of actors, levels of governance and institutional structures. As alluded to in Section 1.1, this volume will focus on (1) the forms of regulation – law versus private instruments, (2) the level of regulation and (3) the specific policy instruments. The three dimensions are interconnected. Thus, a specific policy instrument, whether command-and-control, market-based or informational, may be included in an international treaty, a domestic statute or a set of private standards. Of course, certain policy instruments, such as permits and environmental taxes, can only be adopted by States and thus will necessarily form part of the law. Many others, however, can be instituted either by States, private actors or both, such as certification and performance/process standards.

1.2.1.1 Forms of Regulation

Demarcating lines are often drawn between regulation by law (whether international or domestic), as the quintessential forms of public regulation, and private standards or guidelines, promulgated by corporations or NGOs. The latter category is sometimes called 'soft law' to distinguish it from formal legal rules, but that term is rather inaccurate as in reality, private regulation cannot be labelled as 'law', and it

---

[32] www.ohchr.org/Documents/Publications/GuidingPrinciplesBusinessHR_EN.pdf, also see Eijsbouts (2013).
[33] Hanebury (2006), at 33–63.
[34] Wood & Johannson (2009).

may also be far from 'soft'. Preferred-buyer agreements imposing private standards on suppliers in global value chains, reputational sanctions invoked by NGOs or media publicity or the threat of withdrawal of a certificate may exercise stronger influences on behaviour than formal public regulation. Especially in the area of transnational environmental problems, the capacity of states to regulate and enforce vis-à-vis transnationally operating actors may be limited, and nonstate monitoring and enforcement may have important added value.

Dichotomous conceptions of relations between international law, domestic law and private regulation make for blunt thinking about the modalities and actors involved in environmental governance. It appears that international law and private standards constitute two ends of the spectrum, between which a host of innovative and collaborative hybrid forms of regulation exists. In other words, public, private and hybrid forms of regulation interact on all levels of governance. This phenomenon has been examined in the governance literature under the heading of the 'layering of rules'.[35] Analyses of 'smart regulation', should take into account how law relates to other forms of regulation and governance, both public and private, and both domestic and international, to enable insight on the significance of the law in the wider context of environmental governance.

One particular question with regard to the relation between public and private forms of regulation, and between international and domestic law and institutions, concerns the role of state law. Whereas the state has traditionally been considered to have exclusive regulatory power, modern and pluralist forms of regulation have introduced many other nonstate, private, hybrid and supranational regulatory actors, as well as broad governance regimes, leading to the question if the state is just 'one actor among many' or still 'primus inter pares'.[36] How important is state law in a smart mix of regulatory initiatives? Does it exercise a 'shadow of hierarchy' and is that a necessary component in a smart regulation regime, or can 'governance without a state'[37] work, and under what circumstances? As previous research has shown, it is most likely that public and private regulation need to complement or even reinforce each other, as each has its own strengths and the one cannot replace the other.[38]

The transnational environmental governance literature has pointed out that the role of state law changes in a transnational governance setting from 'command and control' to orchestration and participation in governance networks.[39] This gives more prominence to the question, as formulated by Auld,[40] 'how the diversity of private governance processes interact with the diversity of intergovernmental

---

[35] On the concept of 'layering', see Bartley (2011), at 517–542.
[36] Gunningham (2009).
[37] Borzel & Risse (2010).
[38] Liu, Faure & Mascini (2017).
[39] Abbott & Snidal (2009), at 501–576.
[40] Auld (2014), at 250.

processes that are directly or indirectly affecting a particular social or environmental problem'. State law can be an important form of directive orchestration, by, for example, relaxing legal requirements for firms that adhere to transnational CSR schemes, by imposing requirements on standard setting arrangements and their monitors, or by threatening with mandatory regulation. Nevertheless, states often lack the authority, power, and administrative capacity necessary for directive orchestration, and more facilitative forms of orchestration are more likely to be successful in the transnational arena. These can range from subsidising NGOs to convening actors to providing knowledge and technical assistance to transnational standard-setting bodies. However, some authors question the ability of states to effectively steer and orchestrate at all (see Chapter 3).

1.2.1.2 Policy Instruments

Public, private and hybrid regulation can be further broken down to specific policy instruments. Among various ways to categorize instruments,[41] this volume will distinguish between substantive and procedural instruments and, within the category of substantive instruments, between command-and-control regulation, market-based economic instruments and suasive instruments.[42] The substantive instruments essentially target polluting behaviour that has direct environmental implications. The procedural instruments only indirectly impact the environment by establishing/supporting institutions or by targeting third parties, whose behaviour in turn influences the behaviour of polluters.[43]

---

[41] There are mainly two approaches of classification: the 'resource' and the 'continuum' approach. According to the former, the instruments are categorised according to the resources actors use in the governing process, such as nodality (information), authority, treasure and organisation. See Hood (1983). The latter approach ranges the instruments against some choices government/actors must make in the implementation process. See Dahl & Lindblom (1953). See further, Howlett (1991), at 2–4, in which the author argues that the first approach focuses on the differences between instruments and their technical aspects while the second focuses on the similarities and the contextual aspects.

[42] Typologies and classifications abound. Vedung differentiates between sticks, carrots and sermons, Bahr uses the typology of command-and-control, economic and suasive instruments, whereas Wurzel adopts the category of regulatory, market-based and suasive instruments. Cf. Vedung (1998), at 21–58; Bahr (2010); Wurzel, Zito & Jordan (2013). On the international level, Sand (2003) has referred to sticks, carrots and games. Bodansky (2010) refers to command-and-control measures, informational measures and market-based approaches. A similar approach is used by Sands & Peel (2012) and Stewart (2008). See also Wiener (1999).

[43] According to Howlett, substantive instruments directly affect 'the production and delivery of goods and services in society', including advice, training, licenses, grants, taxes, administration, public enterprises, etc. Procedural instruments are 'designed to indirectly affect outcomes through the manipulation of policy process', including information provision/withdrawal, treaties and commissions creation, interest group funding/creation, government reorganisation and so on. Howlett (2000), at 412–31.

TABLE 1.1 *Elements of a Mix*

		Public Regulation		Private/Hybrid Regulation
		International Law	Domestic Law	
Substantive Instruments	Command and Control	Process Standards ('driftnet fishing' ban) Product Standards ('double hull' tankers under MARPOL) Emission Standards (atmospheric emissions from aircraft)	Permits/Licences/ Performance/ Process-Related Standards Zoning/Planning Regulation Generally	Performance/ Process-Related Standards Planning
	Economic Instruments	Emission Trading Schemes International Law Rules on Liability Investment Incentives (under the CDM)	Environmental Taxes Emission Trading Schemes Subsidies/Public Procurement Policies Liability and Property Right– Based Instruments	Emission Trading
	Suasive Instruments		Public Voluntary Agreements Certification	Certification Environmental Management and Audit Schemes, Codes of Conduct CSR
Procedural Instruments		Access to Information/ Information Disclosure (cf. the obligation to conduct an EIA) Public Participation Access to Justice	Access to Information/ Information Disclosure Public Participation Access to Justice	Access to Information/ Information Disclosure Public Participation Access to Justice

We summarize the various options in Table 1.1, which shows that the substantive instruments can increasingly be found at all the different levels (international/ domestic/private or hybrid). For example, command-and-control types of instruments specifying required or prohibited conduct for particular regulated actors[44] can have their origins at the international level (for example, the phasing out of so-called single-hull tankers to prevent oil pollution), but also at the domestic level

---

[44] Stewart (2008), at 150.

(via permits, zoning and planning) but to some extent in private or hybrid regulation as well. The same is the case for economic instruments. Incentive systems that impose a price or opportunity cost on a unit of pollution[45] can, for example, be found in the international climate change regime, but also in international treaties governing liability rules. Taxation, emission trading and subsidies can also be found at the domestic level, and economic incentive systems are, to an important extent, also incorporated in private or hybrid regimes. This shows the enormous potential of mixes, both between different instruments and between different levels of governance.

1.2.2 *The Emergence of 'Mixes'*

When various instruments apply to the same environmental problem, they interact into a mix. In their seminal analytical framework on transnational business governance interactions, Eberlein et al. distinguish four forms of potential interaction between governance instruments: competition; coordination; cooptation and chaos. Competition between regulatory governance instruments can take the shape of competition for price (certification schemes); products (more or less stringently regulated); reputation; or participants. Competition, it has been argued, leads to a race to the top *or* bottom: both effects can occur depending on the circumstances and actors. Coordination can occur out of deliberate design and negotiations between governance actors,[46] but also through exchange and learning. Cooptation occurs when a governance actor or instrument increasingly dominates the other or when norms converge through meta-governance. Last, chaos is related to unpredictable and undirected interactions and regime complexity. These characterisations are not static, but dynamic, as the character of interactions is likely to change over time.

The focus on 'instruments' as a central concept in this volume brings along a specific interest in the dynamics and development of instruments, rather than a static perspective. Path dependencies are a source of both stability and dynamic change, as they generate advantages for some actors, processes and interests.[47] Instruments may evolve through processes of experimentation and learning,[48] as well as through exercise of power and contestation.[49] For example, as they develop, certification schemes can change from voluntary and suasive to more controlling and coercive as market pressures leave producers less and less choice over the course of time as the certificate gains significance. Thus, the actual working mechanism of an instrument is not given by nature, but depends on how it is constructed and enacted. This volume takes a specific interest in the pathways through which

---

[45] Ibid., at 151.
[46] Von Moltke (2011).
[47] Auld (2014).
[48] Overdevest & Zeitlin (2014).
[49] Auld (2014).

instruments and instrument mixes evolve. Chapter 7 on fishery, for example, shows that private certification institutions like the Marine Stewardship Council (MSC) have played a considerable role in influencing the management of transboundary fish resources in regional fisheries management organisations. However, the pathways through which MSC has attempted to create change very much differed depending upon, *inter alia*, the specific structure of the regional fisheries management organisation involved. The chapter shows that the pathways used (in this case by private organisations) to reach particular goals may therefore diverge in practice.

The prevailing focus in 'smart regulation' literature is on the *design* of smart regulatory mixes, as is also reflected in the formulation of 'design principles'.[50] Although examples exist of deliberately designed transnational environmental regulatory governance arrangements, such as the EU's FLEGT VPAs,[51] many regulatory mixes emerge spontaneously as regulatory instruments and actors interact and integrate in the course of their operation. More particularly in the area of transboundary regulation, a 'grand design' is not very likely, as this would require a powerful supranational actor or at least a high degree of consensus among participants on the best governance regime. These conditions are often absent in international settings: the 'orchestration deficit' is, in fact, considered the greatest limitation to transnational 'new governance',[52] as the lack of orchestration leads to suboptimal outcomes in the sense that private regulation is not optimally aligned to the public interest.

The optimism that smart regulation breathes about the potential of regulatory design has been subject to critique by authors who argue that smart-regulation literature fails to acknowledge political interests and institutional limitations.[53] For example, Von Moltke has argued that while a merger of various environmental regimes and treaties governing the same subject area – such as climate change, maritime pollution or air pollution – would have obvious gains in terms of coordination, the interests of organizations and stakeholders of each treaty inhibit the merger – or even the clustering – of instruments, and have made this politically impossible.[54] Similar resistance can be found with regard to public-private governance mixes: in an overview of several case studies of CSR throughout the world, business organisations are consistently found unwilling to engage with public regulators: instead, they prefer to keep private governance private. Businesses' opposition to smart mix regulation suggests that the metaphor of a partnership is seriously misguided and that, in reality, the relationship between public and private regulators is often 'dysfunctional'.[55] In addition to political opposition from

---

[50] Wood & Johannson (2009); Gunningham & Sinclair (1999).
[51] Trubek & Trubek (2006).
[52] Abbott & Snidal (2009).
[53] Baldwin & Black (2008); Hanebury (2006).
[54] Von Moltke (2011).
[55] Kinderman (2016).

businesses, institutional factors – such as limitations to regulatory competence; poor instrument fit with the existing regulatory style; and dependencies on other regulators – may limit the available instrument options. These constraints are likely to result in a much more limited 'menu' of instruments to choose from and combine than that imagined by smart regulation theory, and in some situations, it may even be impossible to achieve a smart mix.[56] Last, path dependencies play a role in the selection of instruments, their competition and survival, and their evolution.[57] Existing characteristics of markets, political systems, and other institutional factors may influence how mixes of regulatory instruments are composed, how actors perceive the costs and benefits of compliance and participation, and how effectively they function in their specific context.

The contributions to this volume take this critique into account by paying attention to the dynamics in the development of regulatory mixes, including path dependencies, institutional constraints, and the role of power. In general, this volume departs from the prevailing focus on the deliberate design or even manipulability of smart mixes in 'smart regulation' literature. Instead, it takes a more bottom-up approach by asking – quite irrespective of the question of design – whether, and in what respect, certain mixes that have emerged are smart, as well as how instruments play out in local circumstances; and what their unintended consequences are – both positive and negative. In other words, rather than looking for evidence from a positivist perspective, we take a constructivist perspective by studying regulatory mixes as social constructions, shaped through interactions in dynamic institutional structures and by the meaning they have for different actors in different circumstances.[58]

As indicated previously, this volume does not presume that mixes of instruments are the product of deliberate design. Instead, it asks how regulatory mixes have emerged, by whom and to what extent attempts at orchestration have been carried out and why they have been successful. Although instrument mixes may sometimes be the result of effective orchestration, they presumably are more often a question of experimentation or 'bricolage' and sometimes this may even simply be uncoordinated. In particular, the involvement of NGOs cannot always be controlled by state or international actors. This is not to say, however, that the outcomes and impacts of unplanned mixes cannot be 'smart' and effective.

1.3 EVALUATING THE 'SMARTNESS' OF THE MIX

The fundamental concept around which this volume revolves is that of the 'smartness' of the mix. Scholars have used a variety of criteria to evaluate environmental

---

[56] van Gossum, Arts & Verheyen (2010).
[57] Auld (2014); Auld, Renckens & Cashore (2015).
[58] Reus-Smit (2005).

governance or instruments. For example, economists usually focus on efficiency and effectiveness,[59] and political scientists and lawyers explore legitimacy, coherence, equity and unintended effect.[60] This volume chooses *effectiveness* as the primary criterion: we define the 'smartness' of a regulatory mix ultimately by its effectiveness, as it provides a direct evaluation of the practical influence of a mix.[61] Where necessary and relevant, other and related criteria will be used, such as coherence, efficiency, unintended effects, legitimacy and the adaptability of the instrument mixes.[62]

1.3.1 *The Criterion of Effectiveness*

The goal of this volume is to examine whether mixes of regulation and other instruments can contribute to the improvement of the quality of the environment. We thus view effectiveness of a mix of instruments primarily in terms of its impact, in terms of its contribution to the sustainable (rather than just temporal) reversal or alleviation of an environmental problem (*problem-solving effectiveness*). Therefore, the key and final criterion in evaluating the overall effectiveness of a mix is the degree to which it contributes to environmental problem solving.

This notion of effectiveness comprises elements of several common conceptualisations of effectiveness, in particular effectiveness in terms of output and outcomes.[63] The output of a regulatory arrangement refers to the extent of compliance with legal obligations (*legal effectiveness*). Outcome refers to the ability of a mix to induce states, private actors and individuals to modify their behaviour (*behavioural effectiveness*). Problem solving introduces a more dynamic perspective, in which effectiveness is not limited to predetermined goals but also includes overcoming negative side effects or new problems that have emerged as a result of an instrument mix. Obviously, outputs, outcomes and impacts are closely related. They may constitute 'three distinctive steps in a causal *chain* of events, where one serves as a starting point for analysing the subsequent stages'.[64]

In an ideal scenario, assessing the effectiveness of instrument mixes would require a point of reference against which observed outputs, outcomes and impacts can be compared.[65] Obviously, however, transnational environmental problems and their governance mechanisms are entirely unsuitable for the application of the social scientific 'gold standard' of randomized controlled trials. For some environmental

---

[59] Richards (2000), at 222.
[60] Cf. Howlett & Rayner (2007), at 1–18; Meidinger (2002); Oikonomou et al. (2014). The adaptiveness, accountability and legitimacy of governance structure have received intense attention under the Earth System Governance Project. See Biermann et al. (2009).
[61] Cf. Mitchell (2008).
[62] Mintz (2014).
[63] Mintz (2013).
[64] Underdal (2001).
[65] Mitchell (2008), at 897.

problems, data are available that enable the measurement of developments in environmental quality, such as air quality or levels of (de)forestation. However, the reliability of these data is not always obvious, particularly in nondemocratic regimes or remote and inaccessible ecosystems. Also, it is not always possible to establish a causal relation between a regulatory regime and its effects, such as reduction of environmental harm. Comparing different situations with varying regulatory mixes is often impossible given the complex and context-specific nature of environmental problems and arrangements.[66] In addition, changes in environmental quality are long-term processes, and it is hardly possible to isolate the contribution of concrete instruments and instrument mixes from other factors contributing to improvements or deterioration of the environment, as well as to assess the resilience of positive outcomes over time. The effectiveness of regulatory instruments is also a dynamic question: certification schemes, for example, may have high take-up upon initiation, but face the challenges of exit and free riding over time.[67] On the other hand, competition between instruments can lead to ratcheting up of standards, mutual learning and convergence.[68] All in all, any assessment of effectiveness therefore usually contains many provisional claims, caveats, limitations and contextual restrictions, and caution should be exercised when claiming 'success'.

Despite the popularity of the concept of 'smart regulation' in environmental policy and beyond, surprisingly few studies have provided concrete examples and analyses of smart mixes in terms of effectiveness. However, several case studies of smart mixes have provided examples of effective and less effective instrument mixes. Van Erp and Huisman[69] analysed how the regulation of disposal of electronic waste moved from a legal prohibition of exporting hazardous waste directed towards the single target group of exporters, to a mix of preventative interventions directed towards a variety of partners in the supply chain of electronic waste. Supranational regulation in the shape of EU directives – obliging manufacturers to substitute harmful and toxic material with safer alternatives – was complemented with self-regulation through an industrial code of conduct. In addition, and uncoordinated by public regulators, Greenpeace undertook the initiative towards naming and shaming manufacturers with suboptimal disposal policies, which is likely to have stimulated corporate adherence to legal and industry norms.

Another case study (by Howlett and Rayner), this one of Canadian shellfish aquaculture,[70] reveals that economic incentives are not particularly suitable for industries with large amounts of Small and Medium-Sized Enterprises (SMEs), while these are important players in the move towards more sustainable fishery. SMEs do not always have the capacity and communicative structure to receive,

---

[66] Liu, Faure & Mascini (2017).
[67] Auld (2014).
[68] Eberlein et al. (2013).
[69] Van Erp & Huisman (2010), at 579–590.
[70] Howlett & Rayner (2003).

process and respond to information about economic incentives. Also, industry–government partnerships require the participation of industry associations, and thus depend on their effective organization, as SMEs are too small in themselves to organize partnerships. Smart regulation is often directed to including large organized private actors and supranational bodies, but fails to involve small businesses and citizens, who are important in realizing sustainability and preventing environmental damage on a day-to-day basis.[71]

This volume will add to these early studies by providing case studies and analyses in the areas of fishery, forestry, climate change and oil pollution. We will analyse various mixes between instruments in these areas (either international and domestic and/or between various policy instruments) and will assess to what extent these mixes can be considered 'smart' in terms of effectiveness.

1.3.2 *Coherence, Efficiency, Unintended Effects, Legitimacy and Adaptability as Criteria for Smart Mixes*

While the goal of this volume is to examine whether mixes of instruments can contribute to the improvement of the quality of the environment, many of the contributions to this volume demonstrate that effectiveness is closely intertwined with other criteria such as coherence, efficiency, legitimacy, the absence of unintended negative effects and adaptability. Where such criteria are related to effectiveness, the contributions in this volume also use these criteria and their contribution to the effectiveness of instrument mixes.

The coherence of an instrument mix is a particularly relevant issue when studying combinations of instruments. The parallel operation of two or more forms of regulation or instruments gives rise to problems of coordination. They may complement or antagonize each other, which would make their combined effects deviate from the aggregate of effects that result when these instruments are used individually.[72] In this light, coherence assumes a significant role in securing the 'smartness' of the mix. This is particularly the case with instruments targeting the same activities and actors. For example, both public regulation and forest certification regulate the forest management activities of forest owners and operators. Both include specific standards regarding riparian buffer zone, biodiversity protection, forest tenure and so on. Whether the multiple layers of standards are conflicting or consistent with each other influences the behaviour of forest owners and hence the extent to which the problem solving occurs.

Using coherence as an evaluation criterion directs the attention to the relation between policy instruments. With regard to this relation, scholarship on international regime complexes (IRC) may provide useful insights. Like smart mixes,

---

[71] Wood & Johannson (2008).
[72] Simões, Huppes & Seixas (2005).

regime complexes are combinations of different relations. Relations between regimes can have a 'nested' character when one institution is hierarchically superior over others, or they can be overlapping, when different regimes in different issue areas possess authority over the same behaviour without being mutually exclusive or hierarchical.[73] Keohane and Victor have added the concept of 'loosely coupled' regime complexes being positioned in between highly fragmented and highly integrated regime complexes, and being neither nested nor overlapping.[74] Transnational climate change, for example, is highly fragmented, with multiple regimes regulating different subtopics around the globe, and involving a large variety of actors.[75]

The IRC literature has highlighted several types of costs that fragmentation can have. Overlapping regimes may generate inconsistencies between them, and induce competition between regimes and forum shopping, which may undermine regulation. However, and perhaps surprisingly, fragmentation can also have benefits: it may allow fine-tuning to specific contexts; by allowing for benchmarking, it makes learning and quality improvement possible; and by offering flexibility, fragmentation makes it possible to modify arrangements in response to changing conditions. These benefits are particularly likely to emerge with some form of orchestration within the network.[76] Last, the IRC literature draws attention to dynamics in relations between regimes, as they can move from chaotic to more coherent over time. These insights may be useful in the analysis of smart mixes.

Like the measurement of effectiveness, assessing the degree of coherence, as well as its costs and benefits, is not straightforward. For example, the competing interests between forestry and agriculture in forest areas are important drivers for deforestation. The interaction of forest regulation and agriculture policies hence has an important effect in addressing deforestation. However, since they target different activities of forest owners/operators and farmers, the coherence of these instruments cannot be judged in a straightforward way.

Other criteria may also be relevant in examining instrument mixes and their effectiveness, such as efficiency, legitimacy, unintended effects and the adaptability of the governance system. The costs generated by regimes and instruments can also be far from trivial and may influence the success or choice of instrument mixes.[77]

Legitimacy and accountability are crucial for private regimes since, unlike the state, which has sovereign authority according to constitutional law, private regimes rely on the voluntary acceptance and uptake by private regulatees and the public.[78] The importance of legitimacy, however, is not at all restricted to private instruments,

---

[73] Alter & Meunier (2009), at 13–24.
[74] Keohane & Victor (2011), at 7–23.
[75] Abbott (2012), at 571–590.
[76] Ibid.
[77] Ibid.
[78] Bernstein & Cashore (2007).

particularly in the international area where regulatory competition and the existence of multiple governance instruments may enable venue shopping. Even with international treaties, legitimacy-related elements such as capacity building, financial assistance or other mechanisms to help countries comply with the agreement when it is clear that they are unable to comply may enhance compliance.[79]

One last factor that needs to be considered is the adaptability and flexibility of the mix. Regulatory strategies are of course context-specific and path dependent, and for that reason mixes should also have the capacity to adapt to ever-changing circumstances, and new understandings of social and environmental problems.[80]

1.4 OUTLOOK

This volume brings together a variety of theoretical analyses and empirical case studies of smart instrument mixes to address transboundary environmental harm in a multilevel governance setting. This introductory chapter is followed by Chapters 2, 3, and 4, which discuss the theoretical background of smart mixes and together comprise Part I of the volume. In Chapter 2, Linda Senden discusses how relations between public and private actors in hybrid regulatory regimes can develop into complementary relations. She stresses the importance of the law for ensuring a smart public-private mix, as a smart mix requires constitutional guarantees to ensure not only its output but also input legitimacy. In Chapter 3, Philipp Pattberg and Oscar Widerberg draw upon complexity theory to argue that governance arrangements need to be studied in the broad and complex interaction network in which they operate, and support this argumentation with illustrations from global climate governance. In Chapter 4, Rüdiger Wurzel, Anthony Zito and Andrew Jordan provide a descriptive analysis of the emergence of smart mixes in various EU states, thus demonstrating the variation in adoption of smart mixes between countries.

Part II of this volume addresses case studies of the two environmental issues – forestry and fishery – in which private governance is relatively strong. In these areas, certification plays an important governing role. Forestry and fishery are also characterised by a divide between the global North and South, with the North often imposing extraterritorial rules on the South. This part of the volume comprises three case studies of fishery governance – Chapter 5 (Richard Barnes); Chapter 6 (Markos Karavias); and Chapter 7 (Agnes Yeeting and Simon Bush) – more precisely, of certification in combinations with state law and policy – and two case studies of forest governance – Chapter 8 (Jing Liu) and Chapter 9 (Constance McDermott). Finally, Chapter 10 (Lars Gulbrandsen) draws upon both forest and fisheries certification to address the engagement of public actors with certification programmes.

---

[79] Gupta (2011).
[80] Auld (2014).

Part III of this volume presents case studies on the global problem of climate change and its most important cause; oil extraction and pollution. In some of these cases, state and international regulation form the main ingredient of the instrument mix. In Chapter 11, Neil Gunningham extends the smart instrument mix concept to the analysis of 'deep green' initiatives with the aim of more fundamental transformation such as the divestment movement. The fossil fuel divestment movement is an example of a successful instrument mix without involvement of states. In Chapter 12, Marjan Peeters and Matthias Müller address the role of EU information disclosure obligations to enable citizen and NGO control of greenhouse gas emission reduction schemes. In Chapter 13, Michael Faure and Hui Wang present a case study of Marine Oil Pollution, and in Chapter 14 Jan van Tatenhove analyses mixes of governance arrangements in offshore oil production.

In Part IV of the book, The final chapter (Chapter 15) draws the case studies together and presents conclusions.

REFERENCES

Abbott, K. 2012. 'The Transnational Regime Complex for Climate Change'. *Environment and Planning C: Government and Policy* 30, 571–590.

Abbott, K., Levi-Faur, D. & Snidal, D. 2017. 'Introducing Regulatory Intermediaries'. *The Annals of the American Academy of Political and Social Science* 670(1), 6–13.

Abbott, K. & Snidal, D. 2009. 'Strengthening International Regulation through Transnational New Governance: Overcoming the Orchestration Deficit'. *Vanderbilt Journal of Transnational Law* 42, 501–576.

Alter, K. & Meunier S. 2009. 'The Politics of International Regime Complexity'. *Perspectives on Politics* 7, 13–24.

Auld, G. 2014. *Constructing Private Governance: The Rise and Evolution of Forest, Coffee, and Fisheries Certification*. New Haven, CT, Yale University Press.

Auld, G., Renckens, S. & Cashore, B. 2015. 'Transnational Private Governance between the Logics of Empowerment and Control'. *Regulation & Governance* 9(2), 108–124.

Bahr, H. 2010. *The Politics of Means and Ends*. Abingdon, Routledge.

Baldwin, R. & Black, J. 2008. 'Really Responsive Regulation'. *The Modern Law Review* 71(1), 59–94.

Bartley, T. 2011. 'Transnational Governance as the Layering of Rules'. *Theoretical Inquiries in Law* 12, 517–542.

Bernstein, S. & Cashore, B. 2007. 'Can Non-state Global Governance Be Legitimate? An Analytical Framework'. *Regulation & Governance* 1(4), 347–371.

Biermann, F., Betsill, M., Gupta, J., Kani, N., Lebel, L., Liverman, D., Schroeder, H. & Siebenhüner, B. 2009. *Earth System Governance: People, Places and the Planet, Science and Implementation Plan of the Earth System Governance Project*. Bonn, The Earth System Governance Project.

Bodansky, D. 2010. *The Art and Craft of International Environmental Law*. Cambridge, MA, Harvard University Press.

Börzel, T. A. & Risse, T. 2010. 'Governance without a State: Can it Work?'. *Regulation & Governance* 4(2), 113–134.

Börzel, T.A. & Risse, T. 2016. 'Dysfunctional State Institutions, Trust, and Governance in Areas of Limited Statehood'. *Regulation & Governance* 10(2), 149–160.

Dahl, R. & Lindblom, C. 1953. *Politics, Economics and Welfare*. Chicago, IL, University of Chicago Press.

Eberlein, B., Abbott, K., Black, J., Meidinger, E. & Wood, S. 2013. 'Transnational Business Governance Interactions: Conceptualization and Framework for Analysis'. *Regulation & Governance* 8, 1–21.

Eijsbouts. J. 2013. 'Guest Editorial'. *The Dovenschmidt Quarterly* 4, 165–167.

Faure, M. G. 2012. 'Effectiveness of Environmental Law: What Does the Evidence Tell Us?'. *William & Mary Environmental Law and Policy Review* 36(2), 293–336.

   2014. 'The Complementary Roles of Liability, Regulation and Insurance in Safety Management: Theory and Practice'. *Journal of Risk Research* 17(6), 689–707.

Garcia, S. M., Rice, J. & Charles, A. 2014. *Governance of Marine Fisheries and Biodiversity Conservation: Interaction and Coevolution*. Hoboken, NJ, John Wiley & Sons.

Goodin, R. 1996. *The Theory of Institutional Design*. Cambridge, Cambridge University Press.

Gunningham, N. 2009. 'Environmental Law, Regulation and Governance: Shifting Architectures'. *Journal of Environmental Law* 21(2), 179–212.

Gunningham, N., Grabosky, P. & Sinclair, D. 1998. *Smart Regulation: Designing Environmental Policy*. New York, Oxford University Press.

Gunningham, N. & Sinclair, D. 1999. 'Regulatory Pluralism: Designing Policy Mixes for Environmental Protection'. *Law & Policy* 21(1), 49–76.

Gupta, Y. 2011. 'Regulatory Competition and Developing Countries and the Challenge for Compliance Push and Pull Measures'. In Winter, G. (ed.), *Multilevel Governance of Global Environmental Change. Perspectives from Science, Sociology and the Law*. Cambridge, Cambridge University Press, 455–469.

Hanebury, J. 2006. 'Smart Regulation – Rhetoric or Reality?'. *Alberta Law Review* 44(1), 33–63.

Hood, C. 1983. *The Tools of Government*. London, Palgrave McMillan.

Howlett, M. 1991. 'Policy Instruments, Policy Styles and Policy Implementation: National Approaches to Theories of Instrument Choice'. *Policy Studies Journal* 19(2), 1–21.

   2000. 'Managing the "Hollow State": Procedural Policy Instruments and Modern Governance'. *Canadian Public Administration* 43, 412–431.

   2004. 'Beyond Good and Evil in Policy Implementation: Instrument Mixes, Implementation Styles, and Second Generation Theories of Policy Instrument Choice'. *Policy and Society* 23(2), 1–17.

Howlett, M. & Rayner, J. 2004. '(Not so) "Smart Regulation"? Canadian Shellfish Aquaculture Policy and the Evolution of Instrument Choice for Industrial Development'. *Marine Policy* 28(2), 171–184.

   2007. 'Designing Principles for Policy Mixes: Cohesion and Coherence in "New Governance Arrangements"'. *Policy and Society* 26, 1–18.

Keohane, R. & Victor, D. 2011. 'The Regime Complex for Climate Change'. *Perspectives on Politics* 9, 7–23.

Kinderman, D. 2016. 'Time for a Reality Check: Is Business Willing to Support a Smart Mix of Complementary Regulation in Private Governance?'. *Policy and Society* 35, 29–42.

Liu, J., Faure, M. & Mascini, P. 2017. *Environmental Governance of Common-Pool Resources*. London, Routledge.

Meidinger, E. 2002. 'The New Environmental Law: Forest Certification'. *Buffalo Environmental Law Journal* 10, 214–303.

Mintz, J. A. 2013. 'Assessing National Environmental Enforcement: Some Lessons from the United States Experience'. *The Georgetown International Environmental Law Review* 26, 1–12.

2014. 'Measuring Environmental Enforcement Success: The Elusive Search for Objectivity'. *Environmental Law Reporter* 44(9), 10751–10756.

Mitchell, R. 2008. 'Compliance Theory: Compliance, Effectiveness and Behavior Change in International Environmental Law'. In Bodansky, D., Brunnée, J. & Hey, E. (eds.), *Oxford Handbook of International Environmental Law*. Oxford, Oxford University Press, 893–921.

Oikonomou, V. et al. 2014. *Understanding Policy Contexts and Stakeholder Behaviour for Consistent and Coherent Environmental Policies: A Synthesis of Results from the APRAISE Project.*

Orsini, A., Morin, J. & Young, O. 2013. 'Regime Complexes: A Buzz, a Boom, or a Boost for Global Governance?'. *Global Governance: A Review of Multilateralism and International Organizations* 19(1), 27–39.

Overdevest, C. & Zeitlin, J. 2014. 'Assembling an Experimentalist Regime. Transnational Governance Interactions in the Forest Sector'. *Regulation & Governance* 8(1), 22–48.

Peeters, M. 2014. 'Instrument Mix or Instrument Mess? The Administrative Complexity of the EU Legislative Package for Climate Change'. In Peeters, M. G. W. M. & Uylenburg, R. (eds.), *EU Environmental Legislation. Legal Perspectives on Regulatory Strategies.* Cheltenham, Edward Elgar, 173–192.

Raustiala, K. & Victor, D. 2004. 'The Regime Complex for Plant Genetic Resources'. *International Organization* 58(2), 277–309.

Reus-Smit, C. 2005. 'Constructivism'. In Burchill, S., Linklater, A., Devetak, R., Donnelly, J., Paterson, M., Reus-Smit, C. & True, J., *Theories of International Relations*, 3rd edn. Basingstoke, Palgrave, 188–212.

Richards, K. 2000. 'Framing Environmental Policy Instrument Choice'. *Duke Environmental Law & Policy Forum* 10, 221–285.

Sand, P. H. (ed.). 1992. *The Effectiveness of International Environmental Agreements: A Survey of Existing Legal Instruments.* Cambridge, Cambridge University Press.

Sands, P. & Peel, J. 2012. *Principles of International Environmental Law.* Cambridge, Cambridge University Press.

Simões, S., Huppes, G. & Seixas, J. 2015. A Tangled Web: Assessing Overlaps between Energy and Environmental Policy Instruments in Place along Electricity Systems. *Environmental Policy and Governance* 25(6), 439–458. https://onlinelibrary.wiley.com/doi/abs/10.1002/eet.1691.

Stewart, R. B. 2008. 'Instrument Choice'. In Bodansky, D., Brunnée, J. & Hey, E. (eds.), *Oxford Handbook of International Environmental Law*. Oxford, Oxford University Press, 147–181.

Stokke, O. S. 2012. 'International Fisheries Politics: from Sustainability to Precaution'. In Andresen, S., Lerum Boasson, E. & Hønneland. G. (eds.), *International Environmental Agreements: An Introduction.* London/New York, Routledge, 97–116.

Trevisanut, S. 2014. 'The Role of Private Actors in Offshore Energy: Shifting Models of Participation'. *International Journal of Marine and Coastal Law* 29(4), 645–665.

Trubek, D. & Trubek, L. 2006. 'New Governance & Legal Regulation: Complementarity, Rivalry and Transformation'. *Columbia Journal of European Law* 13, 539–564.

Underdal, A. 2001. 'One Question, Two Answers'. In Miles, L., Andresen, S., Carlin, E., Skjaerseth, J., Underdal, A. & Wettestad, J. (eds.), *Environmental Regime Effectiveness: Confronting Theory with Evidence.* Cambridge, MA, The MIT Press, 3–45.

United Nations Conference on Environment & Development, Agenda 21, Rio de Janeiro, Brazil, 3–14 June 1992, (Agenda 21), https://sustainabledevelopment.un.org/content/docu ments/Agenda21.pdf.

Van Erp, J. & Huisman, W. 2010. 'Smart Regulation and Enforcement of the Illegal Disposal of Electronic Waste'. *Criminology & Public Policy* 9(3), 579–590.

Van Gossum, P., Arts, B. & Verheyen, K. 2010. 'From "Smart Regulation" to "Regulatory Arrangements"'. *Policy and Science* 43, 245–261.

Van Rooij, B., McAllister, L. & Kagan, R. A. (eds.). 2010. 'Pollution Law Enforcement in Emerging Markets'. *Law & Policy* (special issue) 32(1), 1–13.

Vedung, E. 1998. 'Policy Instruments: Typologies and Theories'. In Bemelmans-Videc, M.-L., Rist, R. & Vedung, E. (eds.), *Carrots, Sticks and Sermons: Policy Instruments and their Evaluation*. New Brunswick, NJ, Transaction Publishers, 21–58.

Von Moltke, K., 2011. 'On Clustering International Environmental Agreements'. In Winter, G., (ed.), *Multilevel Governance of Global Environmental Change. Perspectives from Science, Sociology and the Law*. Cambridge, Cambridge University Press, 409–429.

Wiener, J. 1999. 'Gobal Environmental Regulation: Instrument Choice in Legal Context'. *Yale Law Journal* 108, 677–800.

Winter, G. (ed.). 2011. *Multilevel Governance of Global Environmental Change: Perspectives from Science, Sociology and the Law*. Cambridge, Cambridge University Press.

Wood, S. & Johannson, L. 2009. 'Six Principles for Integrating Non-governance Environmental Standards into Smart Regulation'. *Osgoode Hall Law Journal* 46(2), 345–395.

Woodside, K. 1986. 'Policy Instruments and the Study of Public Policy'. *Canadian Journal of Political Science* 19, 775–793.

Wurzel, R. K. W., Zito, A. R. & Jordan, A. J. 2013. *Environmental Governance in Europe: A Comparative Analysis of New Environmental Policy Instruments*. Cheltenham, Edward Elgar.

Young, O. (ed.). 1999. *The Effectiveness of International Environmental Regimes*. Cambridge, MA, The MIT Press.

2

'Smart' Public-Private Complementarities in the Transnational Regulatory and Enforcement Space

Linda Senden

2.1 INTRODUCTION

Chapter 1 determines that the main focus of this volume is geared towards establishing whether existing mixes of forms of regulation and instruments have been 'smart' in addressing the causes of environmental pollution and drivers for its prevention in different contexts. 'Smart' has primarily been taken to mean whether a mix of instruments has effectively contributed to the reversal or alleviation of environmental pollution. Underlying this smart mixes notion is the presumption that the *combination* of regulatory instruments and actors is often more effective than a single instrument and that instruments are thus *complementary*. As Gunningham and Sinclair have argued, most instruments and actors have strengths and weaknesses in specific circumstances. Combining regulatory instruments and actors into a mix, then, allows us to take advantage of their strengths while compensating for their weaknesses.[1] The challenge for regulators and policy makers has thus been said 'to select and combine those instruments that form a productive and compatible mix',[2] considering that this focus also brings a number of benefits in terms of considering the interaction rather than the choice of regulatory strategies; of concentrating on design aspects thereof; and of nuancing the role of nonstate entities within the framework of global (environmental) governance.

Given this presumption of complementarity underlying the smart mixes approach, an important question is not only when one can actually speak of complementarity, but also what factors are relevant in turning this into a 'smart' complementarity. This chapter will address these questions by first fleshing out the meaning of complementarity somewhat further from a regulation theory perspective (Section 2.2). It will then elaborate on this complementarity by zooming in more

---

[1] Gunningham & Sinclair (1999), as represented in Chapter 1.
[2] See Chapter 1, Section 1.1.

specifically on the regulatory role of private actors in relation to public regulators; what is the nature of this interaction and what forms does it take? In dealing with these questions, this contribution relies also on insights gathered from case studies that have been performed on transnational private regulatory regimes (TPR) in domains other than environmental policy, including, in particular, data protection, civil aviation and private security companies.[3] It will be argued that it is often the case that some form of public-private complementarity can indeed be identified, taking shape along the public-private axis, the private-public axis or a mixed form thereof, the latter often entailing some form of institutional cooperation between public and private actors (Section 2.3). The chapter will then proceed with considering what makes such a hybrid public-private regulatory regime 'smart' in terms of effectiveness and what other context-specific elements – beyond the regime in question – may be of relevance with a view to realising an effective smart mix. This also includes a consideration of the relevance that needs to be given to ensuring the legitimacy of such regimes. In connection with this, it will be seen that it is vital to also take account of the role of the law and the ways in which it may impact – and also impose limits on – the shaping of a smart mix. In most smart, instrumental and optimal choice approaches, however, the law still appears to be a rather neglected dimension (Section 2.4). The chapter will end with some concluding remarks (Section 2.5).

2.2 PUBLIC-PRIVATE COMPLEMENTARITY AS A FUNCTIONAL GOAL

The terminology and conceptual framing of the scholarly debate on instrument selection varies quite a bit, ranging in particular from instrumental choice/selection[4] to optimal policy mix, optimal instrumental mix,[5] responsive regulation[6] and smart regulation.[7] This framing can be seen as indicative of the underlying nature or goal of the various approaches; some of these are merely explanatory, offering underlying reasons for why certain choices occur or identifying aspects that have a bearing on instrumental choice, while others are clearly normatively tainted, offering a perspective on what would be the ideal or optimal instrumental choice and mix.

---

[3]   HiiL-project on Transnational Private Regulatory Regimes. Constitutional Foundations and Governance Design (http://www.hiil.org/project/corporate-laissez-faire-or-public-interference-effective-regulation-of-cross-border-activities), its overall main findings represented in: Cafaggi (2014). This project included eleven case studies in the domains of consumer protection, financial services and fundamental rights. This contribution builds mostly upon the findings of the fundamental rights case studies, which included a consideration of transnational private regulatory regimes in the domains of data protection, civil aviation and private security companies. See Senden (2013).

[4]   See for example Linder & Peters (1989). For a more general discussion of such theories, see Howlett (2004).

[5]   For an account of various optimal mix approaches, see for example Wiener & Richman (2010).

[6]   Ayres & Braithwaite (1994); Baldwin & Black (2008).

[7]   Gunningham, Grabosky & Sinclair (1998); and Gunningham, Grabosky & Sinclair (2017).

Howlett and Rayner have argued that 'smart regulation takes its place in a larger literature about the transition from a more hands-on, interventionist style of government to "governing at a distance", "governance" and the widely employed metaphor of governments achieving public purposes by "steering" complex networks of public and private actors rather than directing an expensive and possibly ineffective bureaucracy'.[8] The public and private levels of regulation are thus seen as complementary to one another, in the sense that they improve each other's qualities'. The smart regulation approach thus underscores the importance of designing policies that employ a mix of instruments, taking account of and responding to the specific context and features of the policy sector in question. So, it is not simply a choice between regulation and markets; smart regulation is about a policy mix that considers the fullest range of possible instruments. Another core element of smart regulation, however, which the previous quote reflects, is that, whenever possible, within this mix less interventionist instruments are favoured instruments. This includes motivational, informative and incentive-based instruments as well as various forms of co-regulation and self-regulation by industry. This leads us to a third core element, namely that this possibility depends very much on the effectiveness of the proposed [mix of] instruments; or, governance is deemed smart 'when it is conducive to timely and effective collective problem-solving under conditions of high problem complexity and contextual uncertainty and volatility'.[9] This effective problem-solving capacity may thus require the choice of more interventionist instruments in a particular context or situation.

Another scholarly line of thought on instrument choice puts more emphasis on responsiveness, rather than on effectiveness. Most prominently, Ayres and Braithwaite developed the theory of responsive regulation in the early 1990s.[10] Central to this theory is that regulators should not immediately turn to law enforcement solutions to solve certain problems, but that they should first focus on considering a range of approaches that support capacity-building. Problems of concern to regulators should thus first be addressed through developing or reinforcing a pyramid of support that is capable of expanding strengths to deal with a certain problem. Only when this fails to realize the desired results should the regulator then move up a pyramid of sanctions. The presumption is that regulation should always start at the base of the pyramid and that sanctions only come into play when dialogue fails. This presumption applies regardless of the seriousness of the problem and the risks at stake, meaning that dialogue and the lowest form of intervention should be tried first and only when there are compelling reasons, when there is failure, should one move up the ladder. Baldwin, Cave and Lodge have since built upon this pyramid idea by connecting regulation and enforcement strategies more directly. Their approach

---

[8] Howlett & Rayner (2004), at 173.
[9] NWO (2013), at 10.
[10] Ayres & Braithwaite (1992).

thus reflects a preference for the use of self-regulation and then moving towards enforced self-regulation before following that with forms of command-and-control regulation coupled with limited or more stringent sanctioning.

While the responsive regulation and enforcement approaches are thus very strongly focused on the lowest form of intervention, seemingly as a goal in itself, smart regulation puts effective problem-solving centre stage, as well as the mix of instruments this may require. Smart-regulation and responsive-regulation approaches thus differ from one another as regards the preference to be given to less interventionist forms of regulation and enforcement; whereas in the responsive-regulation approach this preference is at its very heart, in the smart-regulation approach the idea of the mix of instruments and its effectiveness to realise the set policy goal is put centre stage, and the choice of the less interventionist instrument is functional to this aim. Smart regulation also underscores the need to search for 'next generation' instruments to meet this challenge, so it is also more about regulatory innovation. As Van Gossum et al. have stated, smart regulation is about ascending a 'dynamic instrumental pyramid' to the extent necessary to achieve policy goals and to maximize opportunities for win-win outcomes.[11] As such, this underscores an understanding of smart regulation in a rather technical, nonideological sense by seeking, first and foremost, efficiency in the delivery of public policies. This connects with economic and political science approaches that emphasise regulatory quality as a means of expressing the extent to which regulation is successful in realizing certain policies, e.g. stable institutions, the development of the private sector, fair market conditions etc.[12]

Yet, within such a smart-mixes approach, due account should be given to a range of elements that will have a bearing on the problem-solving capacity of an instrumental mix. These include its capability to deal with unexpected events, developments taking place over time in domestic society as well as globally, the presence of other actors who adjust their behaviour in response to policies and each other's actions, and progressive information.[13] Moreover, policies have to be devised while there are profound uncertainties about the future – whether these uncertainties are due to unknown, or unforeseen, vulnerabilities or due to assumptions that fail to hold – for actors taking actions that undermine the utility of the set policy, or there being exogenous events that fundamentally change the conditions under which the policy must operate.[14] So, importantly, as Walker, Rahman and Cave have held, 'policies should be adaptive; devised not to be optimal for a best estimate future, but robust across a range of plausible futures'.[15] This can be said to also hold true for the instruments relied upon to shape such policies. Smart regulation thus requires

---

[11] Van Gossum, Arts & Verheyen (2009).
[12] Cf. Voermans (2017).
[13] Walker, Rahman & Cave (2001).
[14] Ibid., at 283.
[15] Ibid.

the instrumental mix to be effective not just here and now, but also to be robust in the sense of having future-proof capacity, and to demonstrate, where needed, adaptability and flexibility to deal with changing circumstances with a view to realising its overarching problem-solving goal. In addition, however, regulation will be truly smart only when those affected by it consider it to be legitimate and follow up on it. As will be seen, public-private complementarity in regulation and enforcement regimes may contribute to these various elements.

2.3 THE NATURE OF PUBLIC-PRIVATE INTERACTION

Shifting the focus from the state as the main and exclusive public regulator to a smart-mixes approach implies taking a much wider and open stance towards the question of how certain socioeconomic or societal problems need to be dealt with, as well as by whom and by what measures, instruments and procedures. How may the regulatory and enforcement space be used and filled in particular areas – not only by public regulators and enforcers but also by private ones – at the international, regional and national levels? What interaction is taking place and how can we explain it in terms of complementarity? This, then, will be the focus of this section: in it we will consider how transnational private regulation (TPR) (Section 2.3.1) interacts with public regulation (Section 2.3.2) and other private regimes (Section 2.3.3) before concluding with some observations on how public-private interaction impacts on or transforms the public-private distinction (Section 2.3.4).

2.3.1 *Transnational Private Regulation*

In earlier work, the concept of transnational private regulation has been said to capture 'the idea of governance regimes which take the form of "coalitions of non-state actors which codify, monitor, and in some cases certify firms' compliance with labor, environmental, human rights or other standards of accountability". They are transnational, rather than international, in the sense that their effects cross borders, but are not constituted through the cooperation of states as reflected in treaties. They are non-state (or private, as we prefer) in the sense that key actors in such regimes include both civil society or nongovernmental organisations (NGOs) and firms (both individually and in associations)'.[16] TPR is cast in private law forms and instruments such as regulatory contracts, codes of conduct, guidelines and internal regulations, all of which are characterised by voluntariness; in addition, TPR has its own governance, regulatory and enforcement processes, which are subject to only limited judicial review.

---

[16] Scott, Cafaggi & Senden (2011).

It must be stressed, though, that TPR is a heterogeneous phenomenon that is very context-dependent. It may thus emerge within the framework of a multilevel, multiactor regulatory and enforcement system featuring numerous international, regional and national – both public and private – regimes, such as in the area of civil aviation. In other areas such as data protection, public international law regimes may still be absent while there are various regional (including EU) and national public law regimes as well as private corporate rule systems. One can also identify different drivers behind TPR, such as the harmonization of (technical) standards, the need for regulating market entry, cost reduction, risk management, safeguarding fundamental rights and the monitoring of compliance.[17] As such, TPR may also occur at different stages of the policy cycle; rule-making, implementation, monitoring and enforcement. So, when using the concept of a transnational private regulatory regime, it must be understood that the meaning of 'regulatory' goes beyond rule-making, and may refer to any of these stages. Very importantly also, while 'private' in this concept denotes the fact that the role of private actors is key in such regimes, it must be understood at the same time that most TPR regimes are of a hybrid public-private nature and therefore interact with public regulation and regulators in one way or another.

The question that interests us here is: Can these different forms of interaction be explained in terms of a mutually reinforcing interaction that contributes to achieving the set policy goals and the solution of identified problems, or are they functioning rather as an alternative to public regulation or in a competitive way? Or, what may this combination of regulatory instruments and actors that is advocated by a smart-mixes approach actually entail; what forms can 'complementarity' take? This enquiry could start from different interaction mechanisms: comparison (and benchmarking); collaboration; coercion; conceptual interaction; cognitive interaction; and also competition.[18] Here, however, we will not look through these specific lenses, but rather start from the different levels at which interactions occur: the vertical interaction level (between public and private actors/regulation) and the horizontal interaction level (between private actors/regulation). In doing so, our main focus will be on vertical complementarity, identifying the formal and informal connections that may exist between public and private actors and regulation at the various stages of the policy cycle, so we are considering not only rule-making, but also its implementation, monitoring and enforcement. In doing so, manifestations of the aforementioned mechanisms will come to the fore as well. The case studies on TPR that I will be drawing on concern primarily 1) Binding Corporate Rules (BCR) in the area of data protection, 2) a number of private regulatory regimes that have been established to secure the compliance of private security companies with fundamental rights, 3) several private regulatory regimes to secure the safety of civil

---

[17] Cafaggi (2014).
[18] On such mechanisms, see e.g. Eberlein et al. (2012) and Gulbrandsen (2014).

aviation[19] and 4) the technical standardization regime within the context of EU internal market legislation. The analysis of these case studies has revealed that public and private regulators complement each other in many ways on the vertical and horizontal levels and that this complementarity can take different forms, depending on the prevailing institutional, market and legal landscapes in a certain domain or policy area.[20]

2.3.2 *Vertical Complementarity*

The analysis of these case studies reveals that there are different dynamics at play in the regulatory space between private and public organisations and that we can distinguish between three different situations of vertical complementarity, depending on who takes the lead in the initiation or development of a regulatory regime. These are: the public regulator bringing in the private regulator; the private regulator bringing in the public regulator; or the case of a mixed public-private initiative in which neither of them is clearly in the driver's seat.

Looking first more closely at the *public-to-private axis*, this refers in particular to cases where complementarity comes about by way of public regulators involving private actors or regulation in public regulatory processes or by setting rules for private regulation through co-regulation or conditioned self-regulation mechanisms. Co-regulation is a term that is used to refer to a 'whole spectrum of regulatory set-ups between the two extremes of pure self-regulation and pure state regulation'.[21] Again, this involvement may occur at the various stages of the policy cycle. The nature of the public involvement may also differ, from being formally mandated to being merely informally supported. The intensity of public involvement may also vary, from setting out a detailed legal or sanctioning framework to (much) lighter forms of public involvement.[22]

The corporate rules that may be adopted in the area of EU data protection provide an interesting example of conditioned self-regulation. The newly adopted General Data Protection Regulation[23] does not so much mandate the adoption of such rules as it allows for their lawful establishment and even for their binding nature. It thus defines 'binding corporate rules' (BCR) as 'personal data protection

---

[19] As ensued from the HiiL project that was mentioned in Footnote 3 in this chapter. Furthermore, account will be taken of recent research in the area of technical standardisation in EU internal market law.

[20] Cf. also Cafaggi (2014).

[21] Heremans (2012), at 81–82. See also: Maxwell (2014), at 67.

[22] See in more detail, Senden et al. (2015), at 35–36, available at: https://ec.europa.eu/digital-single-market/sites/digital-agenda/files/dae-library/mapping_self-and_co-regulation_in_the_eu_context_0.pdf.

[23] Regulation EU 2016/679 of the European Parliament and Council on the protection of natural persons with regard to the processing of personal data and on the free movement of such data, and repealing Directive 95/46/EC, OJ 2016, L 119/1 which will apply as from 25 May 2018.

policies which are adhered to by a controller or processor established on the territory of a Member State for transfers or a set of transfers of personal data to a controller or processor in one or more third countries within a group of undertakings, or group of enterprises engaged in a joint economic activity'.[24] In particular, the Regulation makes it legally possible for a controller or processor to transfer personal data to a third country or an international organisation on the basis of BCR, as it considers that BCR may provide appropriate safeguards and effective legal remedies for data subjects as long as these comply with the conditions set out for this in the Regulation. The Regulation's Article 47 thus prescribes the binding nature of BCR, the substantive and legal protection requirements they need to fulfil, the procedures for their approval by national data protection authorities and mechanisms for securing compliance and enforcement. So, while the adoption of BCR is voluntary in itself, the process of their adoption, their compliance and effects is regulated top-down by EU law, in effect transforming them into legally binding rules.

Yet, public regulators may also formally delegate regulatory powers to private bodies or confer a regulatory mandate upon them. This occurs, for instance, in the area of technical standard-setting – private regulators being made responsible for providing technical standards that specify the principles defined by the public organisation. Exemplary in this respect is the case of the 'New Approach' to technical harmonisation and standards within the framework of EU internal market legislation.[25] As a result of its introduction, harmonisation in EU directives can be limited to establishing essential safety requirements, the development of technical specifications being left to specific private European standardisation organisations (ESOs).[26] While such specifications are formally of a nonbinding, voluntary nature, Member States are obliged to presume that products 'manufactured in conformity with harmonized standards … conform to the "essential requirements" established by the Directive' and thus to allow them to have market access.[27] Harmonised standards are those standards that the European Commission has mandated,[28] and which are to be adopted following a specific procedure which entails the Commission's assessment of whether the harmonised standards comply with its initial request and the requirements contained in the corresponding EU harmonisation legislation. If so, they will be officially published. The EU also contributes to the financing of these standards to ensure that they are developed and revised in support of the

---

[24] GDPR, Article 4(20).

[25] This account draws on Senden (2017).

[26] More elaborately, Joerges, Schepel & Vos (1999), at 7.

[27] Council Resolution of 7 May 1985 on a new approach, OJ 1985, C136/1. See e.g. Pelkmans (1987) and Joerges, Schepel & Vos (1999) and the latest revision of its legal framework: Regulation 1025/2012, OJ 2012, L 316/12.

[28] The number of which has now increased to some 20 percent of all European standards. See Commission Communication 'A strategic vision for European standards: Moving forward to enhance and accelerate the sustainable growth of the European economy by 2020', COM 2011/311 final.

objectives, legislation and policies of the Union.[29] While such a harmonised standard is formally still a piece of private regulation adopted by a private organisation, it is at the same time a top-down highly conditioned, supervised and recognised form of private regulation. Recently, this has led the Court of Justice of the EU to conclude that these harmonised standards 'form part of EU law' and that they are 'necessary implementation measures' of an act of EU law.[30] Harmonised standards must also be considered as having legal effects, regardless of the fact that they are only of a voluntary nature, as compliance with the Directive's essential requirements may be evidenced by other means as well.[31]

Another example in the transnational domain concerns the interaction in regulating safety in civil aviation between ICAO (the public International Civil Aviation Organization, established in 1944) and IATA (the private International Air Transport Association, established in 1945).[32] While ICAO has set so-called Standards and Recommended Practices (SARPS) in different fields and developed a global aviation safety programme and an audit and oversight mechanism (USOAP), IATA has also set numerous standards as well as the IATA Operational Safety Audit (IOSA). In the development of these regulatory regimes, we can identify different types of dynamics. ICAO thus incited the establishment of the multi-stakeholder International Safety Strategy Group, including IATA, its stated aim being to move beyond an adversarial relationship between government and industry and to provide a common frame of reference for all stakeholders involved; states, regulatory authorities, airline and airport operators, aircraft manufacturers, pilot associations, safety organisations and air traffic control service providers.[33] As such, it was to develop a global safety strategy which was concluded in the form of the Global Aviation Safety Roadmap. ICAO then aligned its own global aviation safety programme to this Roadmap. When it comes specifically to the relations of ICAO with technical standardisation bodies, it is stated in one of its resolutions[34] that 'in the development of SARPs, procedures and guidance material, ICAO should utilize, to the maximum extent appropriate and subject to the adequacy of a verification and validation process, the work of other recognized standards-making organizations. Where deemed appropriate by the Council, material developed by these other standards-making organizations should be referenced in ICAO documentation'. There is thus no question of mandating private organisations in this regard, but there is clearly the possibility of incorporating such standards into own regulations. This appears just one form of the

---

[29] See recitals 38 and 39 of the Regulation and Articles 15–19.

[30] Case C-613/14, Elliott, ECLI:EU:C :2016:821, paras. 40–43.

[31] Ibid., paras. 42–43.

[32] Now with 280 airline members, representing 83 percent of total air traffic. See www.iata.org/Pages/default.aspx, last accessed 12 February 2018.

[33] Global Aviation Safety Roadmap, at 4; see https://flightsafety.org/files/roadmap1.pdf, last accessed 12 February 2016.

[34] Appendix A to Assembly Resolution A35–14.

public endorsement of private regulation. As regards IATA's Dangerous Goods Regulation, one can thus note ICAO's administrative approval thereof, as it recognised it as the field guide for the application of its own rules in this field.[35] One can even witness the legislative endorsement or codification of private regulation, such as in the case of the EU Regulation[36] and the CITES Convention referring to the IATA Live Animals Regulations as the applicable legal standard to follow for the transport of live animals by aircraft. Another impact we can see concerns the reference in (domestic and EU) law to the IOSA audit standards that IATA has set, as a compliance criterion that aircraft have to meet before being allowed to enter (domestic and EU) airspace.

These examples show the different ways in which the public regulator may decide to bring private regulation into its own realm and therewith even decide to give it public law effects or effects akin to that. But the dynamics also go the other way round. The *private-to-public axis* thus refers to cases where complementarity comes about as a result of the private regulator itself taking the lead in determining the core standards and requirements of the regulatory regime and bringing in the public regulator or regulation in one way or another in this process. Here the area of civil aviation is again highly illustrative when it comes to the many forms complementarity can take. IATA has clearly been a front-runner in the development of certain safety standards, such as the already-mentioned regulations for the transport of dangerous goods and of live animals. Especially when it comes to the development of the dangerous goods regulations over time, the interaction between IATA and ICAO shows an interesting picture. One could say that initially private regulation in this area emerged for gap-filling reasons, as ICAO was not taking any action and that next there was some regulatory competition between IATA and ICAO in this field, while then moving to a situation of regulatory cooperation and coordination, before ICAO approved IATA's rules as the applicable field guide. In a similar vein, the International Business Aviation Council (IBAC) has engaged in standard-setting because of the need for sector-specific standards, the International Standard for Business Aircraft Operations (IS-BAO) being an example of this. IBAC worked closely with ICAO to achieve international standardisation under the ICAO umbrella and the standards thus incorporated in the IS-BAO were already contained in ICAO standards but were partially rewritten in consultation with IBAC before being incorporated into the IS-BAO. This reflects the fact that the IBAC operates in a manner in which public and private regulatory processes influence and nudge one another on a mutual basis. IATA has also collaborated on a voluntary basis with ICAO to produce a Guide on Fatigue Risk Management Systems, which is linked to the ICAO SARPs on fatigue management. As such, it aims first and foremost to

---

[35] Senden (2013).
[36] Council Regulation 1/2005 on the protection of animals during transport and related operations, OJ 2005, L3/1, Annex I, chapter 2, point 4.1.

provide air operators with information as to how a fatigue risk management system can be put in place that would ensure an appropriate implementation of the public law SARPs. So, here IATA is also providing more implementing types of measures of public law standards. Furthermore, the approximately 900 IOSA audit standards and recommended practices have been derived from all relevant ICAO standards, as well as from regulations of the European Aviation Safety Authority.

Yet, it must also be noted that complementarity originating in the private-to-public axis may have a different bearing, in the sense that private regulation emerges because of the gaps left by the public regulator and without the public regulator or regulation being involved at all in the private regulatory process. The area of private security companies is a more general example of this. In such cases the private regulation complements the public regulatory framework because it has been incited by the very absence of public standards, thus serving the purpose of gap filling. Sometimes, public law standards are then incorporated in the private regulatory regime or reference is made to public law standards. The latter has occurred, for instance, in the case of the Voluntary Principles which have been referred to in TPR in the field of private security companies. So, in these cases there is no question of public law setting top-down rules that need to be complied with in private regulation with a view to securing market access or any other public policy goal, but of public law standards having 'filtered through' to the private regulatory regime on a voluntary basis via incorporation, reference or as a source of inspiration.

The *mixed public-private axis* concerns in particular those cases where the initiative for a regulatory regime cannot be clearly located with the public or private regulator, but where this can rather be seen as a joint public-private initiative. Sometimes such an initiative is geared not so much towards the development of new standards or rules as it is towards establishing institutional cooperation for a better implementation and compliance with both public and private standards. These initiatives are geared towards enhancing consistent and integrated approaches, reducing overlap and the duplication of efforts. Such complementarity gets its shape in particular through bringing about more structural and institutionalised forms of cooperation through the conclusion of memoranda of cooperation and understanding and through having observer status or participating in another way in each other's committees, working groups etc. The area of safety in civil aviation again provides interesting examples of this, including the Memoranda between ICAO and IATA in relation to training programmes for young professionals; for the sharing of information, also in relation to their respective audit programmes (IOSA and USOAP); and the setting of joint standards for assessing accident rates.[37]

---

[37] Assessing the accident rate in global civil aviation requires, amongst other things, common taxonomies and definitions for what actually constitutes an aviation accident and in incident reporting systems.

2.3.3 *Horizontal Complementarity*

One can also distinguish between different forms of horizontal complementarity. These concern a reference by private regulators to other private standards or even the adoption, incorporation or recognition of other private standards, as well as the joint development of private standards. When it comes to a reference to other private standards, the BASR standard developed by the Flight Safety Foundation, which refers to IATA's Dangerous Goods Regulation, can be mentioned. The adoption or incorporation of other private standards goes a step further, with the IOSA standard of IATA – which gleaned materials from several standardisation bodies, including ISO, SAE and ANSI – providing an example. The already mentioned IS-BAO standard has been granted official recognition by the European Committee for Standardization (CEN) in the form of a CEN Workshop Agreement. The Global Aviation Safety Roadmap presents an example of the joint development of standards; as mentioned previously, it was developed jointly by the Industry Safety Strategy Group, which comprises a number of international private organisations and companies: IATA, Airbus, Boeing, ACI, CANSO, the FSF and IFALPA.

Such horizontal complementarity seems to be more present and also likely to occur in areas where private regulation does not come about in a top-down regulated framework. Private regulation in the areas of private security companies and, most visibly, civil aviation has developed in a more bottom-up way and both areas show horizontal complementarity, albeit to varying degrees. One explanation for this may be that safety regulation in this area has developed over a very long period of time, and as a result is highly regulated both on the public and private level, and in both normative and technical terms. This has also added to the need for alignment between private regulators and regulation. Furthermore, the private standard-setting within the area of civil aviation takes place within the framework of private organisations that are of a permanent and representative nature. As such, they not only have a certain institutional stability but also have gained a certain reputation over time. In comparison, in the area of private security companies such frameworks are (still) very much lacking and the private standard-setting activity itself is still at a very embryonic level.

2.3.4 *Transforming the Public-Private Distinction*

Concluding this section, it can be noted that, in terms of actors, processes, instruments and effects, TPR clearly differs from public hard law regulation. Yet, we have also seen that while the distinction between public and private actors remains relevant, the distinction between public and private instruments may become blurred; while private standards and rule-making subscribe to a voluntary approach

in a way that is similar in character to soft public law,[38] meaning that in principle
they will not have legally binding force in and of themselves, we have seen that they
can certainly acquire legal effects, depending on their relationship with, and
embedment in, public law frameworks. The distinction between a public and
private norm may thus lie foremost in the public or private nature of the adopting
organisation, but no longer so much in the substance and legal effect of the norm
itself. Whenever the public regulator decides to mandate the adoption of a private
norm, refers, confirms or endorses it in its legislation, uses it as a yardstick in auditing
and compliance procedures or in administrative approval procedures and so on, the
dividing line between a public and a private norm becomes fluid. Not so surprisingly, this also evokes the question of whether or not such private regulation actually
qualifies as law.[39] Yet, in other cases the mutual public and private regulatory impact
is limited to being a source of inspiration or voluntary referencing, or to mere
cooperation in standard-setting. We therefore conclude that public-private complementarity does not eliminate, but rather transforms, the public-private distinction.

2.4 ASSESSING AND SHAPING PUBLIC-PRIVATE COMPLEMENTARITY

The analysis in the previous Section 2.3.4 has already touched upon the relevance of
the policy context for the development of the public-private relationship in regulation and enforcement. In this section, I will be reflecting somewhat more in depth
on this context-dependency in the case studies, connecting this to the question of
when, and in what ways, complementarity may actually contribute to the effectiveness of smart mixes (Section 2.3.1). Then other, more general contextual factors that
may impact the shaping of effective smart mixes will be considered (Section 2.3.2),
while finally looking at how the legitimacy concern and the law may constitute such
a factor in and of itself (Section 2.3.3).

2.4.1 *The Contribution of Public-Private Complementarity to an*
Effective Smart Mix

Taking a formal or procedural approach to effectiveness would mean that compliance with the set rules would already be considered as being effective.[40] Yet, what
interests us is what the added value of public-private complementarity from a
substantive perspective can be. Regulation would then be considered effective only
when it had led to the achievement of the set policy goals, such as the alleviation or

---

[38] Cafaggi (2014).
[39] See Senden (2017) for a discussion of this as regards harmonised standards in EU internal
   market legislation and the mentioned Elliot case.
[40] Cf. the framing thereof as legal effectiveness in Chapter 1.

reversal of an environmental problem. An important question, then, is: What elements need to be part of a complementary public-private regulatory approach in order for it to be capable of providing such problem-solving effectiveness? Mechanisms ensuring compliance with the set rules may actually be a precondition for this, based on the presumption that the rules put into place could indeed contribute to solving the problem in question. So, in that sense, formal or legal effectiveness may also be a precondition for ensuring substantive or problem-solving effectiveness, even if this will not suffice in itself.

The following assessment elements have been identified in earlier research as being relevant to ensuring that private regulation can realise the policy goals for which it strives.[41] The first two elements relate to the extent to which private and public actors, substance-wise, are on the same page: Do private interests align with public policy objectives, and is there industry commitment and capacity to realise the set goals? Three other elements point to formal or procedural effectiveness mechanisms: what are the means used to render the regulation effective (what is the design/governance of the TPR and according to what impact/performance indicators are its effects measured); is there any form of government pressure and oversight; and are credible sanctioning policies provided for? In themselves, these elements already show that there is a presumption that a certain level of public-private complementarity of a TPR regime will add to ensuring its effectiveness. As such, they can also be considered to be key issues that need to be addressed with a view to designing smart regulatory mixes. In the following text we will reflect on how these elements may have come to the fore in the development of TPR in the areas considered in this chapter – data protection, civil aviation and private security companies – by looking deeper into the drivers behind these TPR regimes and into why a certain interaction with the public regulation level has occurred.

In the area of data protection, the importance that is given to data protection as a fundamental right in the European context has led to a strong role for the European legislature, which has established a regulatory framework for this particular purpose. Thus, the starting point here has been the existing public regulatory framework, and it is only because European legislation has fallen short in adequately dealing with the problem of data protection beyond EU borders that BCR are seen as an adequate tool to fill gaps in this protection and, in a way, to complement public regulation.[42] Under the new EU Regulation, BCR are seen as an adequate basis for the transfer of data, but a strong public regulatory framework is still considered necessary to ensuring a high level of protection, and BCR should be given shape within the boundaries set by this public law framework. This highly conditioned self-regulation presumes great alignment between EU policy and corporate goals, as

---

[41] Within the framework of the HiiL-project mentioned in Footnote 3, this chapter.
[42] Moerel (2012).

well as significant corporate investment in complying with the public regulatory framework. It also goes along with stronger public oversight mechanisms, as national data protection supervisors have to ensure compliance.

In the area of private security companies, there is an at least equal, if not higher, need for fundamental rights protection because people's lives may actually be at stake. The main problem here lies in the fact that private security companies engage in security activities that traditionally were within the exclusive competence of public authorities, such as the police force. While fundamental rights conventions and rules do apply to such public bodies, they do not in and of themselves apply to private bodies like private security companies. Moreover, so far no specific international public regulatory framework has been developed with a view to regulating private security companies from the human rights perspective. So, the focus of the various TPR regimes in this field has essentially been on providing a regulatory framework in the first place, by 'translating' public human rights standards to this new private sphere and tailoring them to the activities of private security companies. As such, these regimes are primarily geared towards filling the gap that public regulation leaves. While one could thus consider the private regimes in this domain to complement existing public rules, which are limited in their application to public bodies, one must also observe here that the level of vertical complementarity is still limited. Industry commitment and capacity, as well as the alignment of the public and private interests of these regimes, are still problematic; in addition, government pressure and oversight are limited for the reasons previously explained. This is also reflected in the overall effectiveness of these regimes, which is still limited, and the free rider problem, or the way in which private security companies that are caught infringing on fundamental rights can rather easily disappear and start a new business elsewhere.

A similar interest is at stake in the area of civil aviation, as ensuring aviation safety involves an extensive duty of care for all public and private parties concerned; if this is not taken seriously enough, the consequences for passengers can be grave. For airlines and other responsible actors, it is difficult to cover up negligent behaviour by simply disappearing and starting a new business as a private security company can. One could say that because of the fewer possibilities of free rider behaviour, and the potentially high implications for reputation and business in case of accidents and incidents, there is a huge incentive to work towards the same goal with other relevant actors – both public and private. The private and public interest in securing safe air traffic are therefore very much aligned, and one can also say that there is a high industry commitment to this. In that sense, it comes as no surprise that this sector has already been quite highly regulated for many years. So in contrast to the area of private security companies, the area of civil aviation demonstrates a multitude of public and private regulatory actors on all levels (national, regional and global) that have been engaged in dealing with and regulating the manifold safety and security aspects of civil aviation since its very inception. This very multitude of actors can also be seen as an explanatory factor for the more diversified interaction between the public and private levels. An important distinction with the areas of data protection and private security companies is that civil aviation not only has a longer history of national regulation, but also has one leading international public law organisation in the field – ICAO – that has been engaged in setting global safety and security standards since the 1950s. Many private organisations, including IATA, have been established out of a desire to influence the standard-making process within that public organisation. As we have already seen, this desire – but also the desire for one's own private regulation and enforcement in this sector – has been incited by the fragmentation, gaps (e.g. de-icing standards) and (low) level (DGR) of public standard-setting, the need for sector-specific standards such as for business aircraft (IS-BAO) and the deficiencies in public compliance/enforcement which, for example, led IATA to set up its own auditing system (IOSA). Another benefit of private regulation is considered to be that it is geared directly towards private, industry actors and is not dependent upon implementation by states. This has provided a platform for not only for developing more *ad hoc* and structural forms of both vertical and horizontal complementarity, but also for building vertical institutional cooperation, which has enhanced inter alia the development of complementary safety strategies and the exchange of information. The alignment of public and private audit standards, and their application in ensuring compliance with the set safety standards, is also a very strong feature of this public-private complementarity. The drivers for actively engaging in building such a public-private partnership are stronger when there is a huge common interest on the part of both the private sector and public government, such as where there is a joint duty of care and high reputational risk at stake. Looking at the ever-decreasing accident rate in the ever-increasing air traffic, this complementarity can be considered to have contributed to the set policy goal of ensuring a high level of safety in civil aviation.[43]

On a final note, empirical research has also shown that transnational private standards are often stricter than public standards. Public regulatory regimes are considered to establish minimum mandatory standards; private regulatory regimes often adopt stricter standards or focus more on implementation and compliance monitoring. Such a relationship implies that the public standard, where it exists, constitutes the common basis, which may lead to regulatory competition between private actors proposing stricter standards as has been the case, for instance, in food safety regimes. Public standards define a floor and private standards go beyond, adding requirements or calling for more rigorous compliance programmes.[44] Codes of conduct and guidelines often include rules that explicitly demand compliance

---

[43] See https://aviation-safety.net/statistics/, last accessed 12 February 2018.
[44] Cafaggi (2014).

with international and domestic laws. Such TPR topping up of public standards may also be considered to add to the effectiveness of the overall regulatory framework.

2.4.2 Contextual Factors Impacting on Shaping Effective Public-Private Complementarity

Accepting or adopting public-private complementarity as a means or strategy to create an effective smart mix with a view to the realisation of certain policy goals is just one side of the coin. The other side of the coin is that such a proactive design policy has to take due account of a number of contextual factors that will impact on and/or limit such a policy. While it is thus submitted in the literature that 'different instruments involve varying degrees of effectiveness, efficiency, equity, legitimacy, and partisan support' and that 'some instruments will be more effective in carrying out a policy in some contexts than another',[45] it is also considered that 'policy instruments are rarely selected on the basis of their implementability and effectiveness. Different policy fields tend to show preferences for their own "favourite" types of policy instruments and use these repeatedly regardless of their actual contribution to problem solving'.[46] It is thus necessary to look somewhat more into the specific factors or variables that have been identified as having a bearing on instrumental choice. These can be located at the macro, meso and micro levels of analysis.

2.4.2.1 The Macro Level

The macro level refers to the relevance of context at a higher, more general level, going beyond the individual realm (the micro level) and the specific policy domain (the meso level). It rather concerns factors which determine societal structures, processes and problems and their interrelationships. Such factors include, first of all, time and place; these are key elements in determining what may be seen as core values of a state or other type of community, both at a given point in time and in a specific place. For example, while the efficient use of financial resources may be highly valued in times of economic crisis and budgetary constraints, this may be considered less important in prosperous times. Likewise, differences in place may refer to a different perception or assessment of the importance of certain values and the legitimacy of certain instruments because of different cultural norms and institutional or political arrangements (e.g. liberal democracies and more conservative regimes) in different locations and regions of the world.[47] Values like efficiency and

---

[45] Howlett (2004), at 5–6, with reference to Salamon & Lund (1989).
[46] Bressers & O'Toole Jr (1998), at 214.
[47] Howlett (2004), at 5–6.

legitimacy may thus be differently balanced because of such differences. Zooming in further on place as a relevant factor, there is the importance of the general economic, social and political context thereof, in which some combinations of instruments may work well in one context but not in another. Political culture, national policy styles, social divides, organisational structures of the decision-maker, networks and agencies are all issues that are considered to have an important impact on both instrument selection and policy implementation. Preferences of state decision-makers, and the nature of the constraints within which they operate, are thus understood to be highly relevant determinants for the choice of policy instruments.[48] This also includes political calculations, which may vary from electoral advantage, policy learning, past precedents and the ideological preferences of state and societal actors.[49] Howlett also found on the basis of existing studies of instrument choice that these take 'great care to observe sectoral nuances in instrument choice but still indicate that there are different national propensities to favour certain instruments in "normal" circumstances – a clear indication of the existence of national policy styles'.[50] Also important is that the preservation of the balance of power has been emphasised in the literature as a relevant choice determining factor. Majone has thus argued that, because of the balance of power, the selection of different policy instruments has little impact on a policy's success. Instruments that could change this balance would probably not be selected to begin with, and using them in the implementation process would also occur only insofar as the balance of power allowed for this. In other words, the selected instruments would never pose a serious threat to the existing balance of power.[51]

2.4.2.2 The Meso Level

At the meso level, the specific policy/sectoral context is a determining factor. Importantly, instrument preferences are thus considered to be linked to relatively long-term aspects of the policy-making context. Howlett concludes in this regard that instruments are primarily chosen on the basis of the empirical situation on the ground in the sector or issue area concerned, in conjunction with the nature of the constraints and capacities of the applicable political regime. Since state capacities and societal targets, as factors that affect implementation styles, are rather long lasting, such styles and instrument choices can be expected to change infrequently.[52] Elements that have been said to have a bearing on instrumental choice at this level are 1) the clarity and unambiguous nature of the policy goals that are

---

[48] Howlett (2004), with reference to Bressers & O'Toole (1998).
[49] Howlett (1991), at 15.
[50] Ibid.
[51] Majone (1998), at 221.
[52] Howlett (2004), at 9.

set,[53] 2) costs and benefits and the extent to which these are known (information gaps) and 3) aspects of the social setting in a sector or policy domain, such as the network of actors involved or the nature of the networked relationship. Bressers and O'Toole have thus argued that 'in general, the more an instrument's characteristics help to maintain the existing features of the network, the more likely it is to be selected during the policy formation process' and that there is a process that results in instrument determination, rather than a particular actor who 'chooses'.[54] The analysis in Section 2.3 has also shown that it matters how such public/private networks may have developed over time, and the role the law may play in framing not only substantive rules in the field at issue but also the network cooperation. Another relevant element is considered to be the adaptive and responsive nature of policy instruments, given also the specific features and complexities of the policy/ sectoral context and the uncertainties and risks involved.

2.4.2.3 The Micro Level

The micro level concerns the relevance of the personal context for instrumental choice. Individual perceptions and subjective values – key elements in instrumental choice – interact with organisational and systemic factors. Two relevant issues, therefore, are 1) how actors inside and outside governments view instruments and 2) who is involved in making choices about instruments. What criteria do those actors use to judge the suitability of instruments for addressing policy problems?[55] Clearly, behavioural aspects come into play here, giving rise to the important empirical question of whether decision-makers tend to choose the same instruments regardless of the problem at issue, or whether they select different instruments to match the given situation.[56] As Linder and Peters argue, favouring the same instruments across problem contexts suggests a strong link to the decision-maker's attributes and setting while, 'conversely, if choice varies systematically by problem situation, the setting then would exert its influence on choice not through how instruments are viewed but by the way problems are structured'.[57]

2.4.3 The Law As a Factor Impacting on Effective Public-Private Complementarity

In general, little attention is given to the role of the law in the various instrumental-choice/smart-mixes approaches as discussed in Section 2.1. Yet, the instrumental

---

[53] Ibid., at 5.
[54] Bressers & O'Toole (1998), at 220.
[55] Howlett (1991), at 15.
[56] Linder & Peters (1989), at 37–38.
[57] Ibid.

function of the law – if we see it as a tool to be used to steer towards certain goals – squares well with such approaches, as do important legal principles such as proportionality and subsidiarity and good governance principles such as consistency and coherence. At the same time, these approaches rather fail to take into account limits to instrumental choice ensuing from the guarantee function of the law, which concerns the realization and protection of certain values by the law. Such limits may thus concern not only fundamental and human rights and compliance with competition law rules, but also key elements of sound governance,[58] which include the democratic balancing of interests and the participatory nature of decision-making; a broad, diverse membership; balanced funding sources; a functional and legal separation of powers in standard-setting, monitoring and enforcement functions; and a high level of accountability and transparency in the activity of the regulatory regime in order to avoid regulatory capture by certain private actors. More specifically, while private regulation can be seen as the expression of contractual freedom and private autonomy, there is not only the risk that the self-interest of private actors may prevail in the regulatory, monitoring and enforcement process, but also that it may have *de iure* or *de facto* binding effects on third parties, without this relying on an electoral mandate or a decision-making process that ensures democratic legitimacy or accountability. This risk is higher in a monopolistic market, where it is not possible for the parties concerned to follow other rules.

For these reasons, it is therefore insufficient for a smart-mixes approach to achieve legitimacy solely on the basis of output, or how effectively it achieves the desired policy goals; it must also ensure procedural and input legitimacy. If private actors involved are given any normative power akin to that of public authorities, this exercise of power should meet constitutional requirements to ensure that all of the interests of all concerned stakeholders are taken into account. This becomes even more important when private regulation is actually difficult to distinguish from public regulation in its nature and effects.[59] Any evaluation of the effectiveness of TPR regimes thus has to incorporate legitimacy values as well as preconditions for effectiveness.[60] Otherwise, these regimes risk not achieving sufficient trust and credibility amongst those who are affected by, and who must apply, the set rules.[61]

Often the law plays a role in ensuring such values and imposing limits at the three levels identified in Sections 2.3.1, 2.3.2 and 2.3.3. At the systemic, macro level, one

---

[58] Cf. also the limits contained in instrument choice in the former EU Inter-Institutional Agreement on Better Lawmaking, OJ C 321, 31.12.2003, p. 1–5 and the current Better Regulation Toolbox of the European Commission, in particular its Tool 18 on the choice of policy instruments; accessible via https://ec.europa.eu/info/files/better-regulation-toolbox-18_en, last accessed 12 February 2018.

[59] Cf. also Curtin & Senden (2011).

[60] Cf. Cafaggi (2014) and Senden (2013).

[61] See e.g. Six & Verhoest (2017).

can thus witness within Western democracies – as these have evolved in the post-World War II era – an increasing emphasis on fundamental rights protection in international, regional (e.g. EU) and national legal frameworks. In addition, however, market organisation principles such as those contained in national and regional competition laws, consumer protection laws and environmental laws are gaining in relevance in instrument choice and in framing public-private complementarity. At the meso level, relating to the specific policy contexts, similar legal constraints ensuing from general principles of law but also from a particular body of national, regional and/or international public regulation that may have been put in place to deal with a certain problem may impose themselves. The legislator may also have already devised a certain enforcement strategy that ties the hands of supervision and enforcing agencies. In this respect, it has also already been seen that the law may play a very important role in shaping the interaction and complementarity between the public and the private regulatory and enforcement level, not only by setting a legal framework and conditions for the use of TPR but also by giving it certain effects. As seen, the public endorsement of private regulation may thus take the form of ex post facto administrative approval, policy alignment or legislative endorsement, as well as incorporation in public standards or in monitoring/compliance processes. Such forms of complementarity may actually in themselves add to both the legitimacy and effectiveness of TPR, as they demonstrate an acceptance of the exercise of regulatory authority by the private regulator and therewith a certain level of trust in the private body as a source of regulation on the part of the public regulator or enforcer.[62]

In the context of the EU, the applicable legal framework of the various policy sectors is also a highly relevant factor, as the EU is constrained by the principle of conferred powers. The EU may only take regulatory action if it is so empowered by the European Treaties, and the powers attributed to the EU also differ in various policy domains in the sense that they can be of an exclusive or shared legislative nature or merely of a supporting and policy-coordinating nature.[63] Clearly, this also determines to a significant degree the instruments by which it may operate, not only in the internal EU sphere but also in the external sphere, e.g. to tackle global problems like climate change.[64] The attributed powers have to be exercised in conformity with the legal basis provisions in the Treaties from which they ensue, which also underscores the relevance of how the actors involved in the decision-making process actually interpret these competencies and determine how best to

---

[62] Van der Voort (2017).

[63] See Articles 5(1) and (2) Treaty on European Union and Articles 2–6 of the Treaty on the Functioning of the EU.

[64] It must also be noted that the set of EU instruments is itself already limited as it is very much constructed as an 'integration through law' approach and, for instance, the EU has limited economic/financial instruments at its disposal to steer towards certain policy goals.

implement them. But here one can also refer to the relevance of observation by Majone, quoted in Section 2.4.2.1, that the instrumental choices that will be made will not seriously upset the balance of power which actually underlies the framing of competences in the Treaties.

2.5 CONCLUDING REMARKS

The focus of this chapter has been on considering what vertical, public-private complementarity entails and what factors impact on its 'smartness'. Considering examples of different policy domains and sectors, we have seen that there may be many different connections – both formal and informal – between public and private actors and regulation at the various stages of the policy cycle, including rule-making, implementation, monitoring and enforcement. Therefore, these may come about as a result of rather top-down conditioned co-regulatory processes, but also in a more bottom-up, interactive dynamic process of coordination, cooperation and voluntary confirmation, or even the codification of private regulation in public regulatory frameworks. Such dynamics clearly reveal a certain alignment of private and public interests, which has been seen to contribute to the realisation of a smart mix and therewith to effective problem-solving and the realisation of the set policy goal. This effectiveness has been put centre stage in many theoretical smart-mix approaches. Yet, this chapter has also sought to demonstrate that the achievement of a truly effective public-private smart mix requires consideration of a broader array of factors, which go beyond the specific features of the policy domain or sector at issue in the here and now (the meso level). These concern not only individual, behavioural aspects and patterns that have a bearing on instrumental choice (the micro level), but also systemic (i.e. political, economic and social) features of a community or state – including its adherence to certain values and principles (the macro level) – which may vary both in time and place.

At all of these levels, the law comes in as an important factor for ensuring a smart public-private mix, both as an instrumental steering mechanism in itself and as a substantive yardstick by providing constitutional guarantees. A smart mix requires constitutional guarantees with a view to ensuring not only its output legitimacy but also its input legitimacy. Output legitimacy in the sense of problem-solving effectiveness and the realization of policy goals – such as consumer protection and climate protection – may be enhanced by ensuring that the interests of all stakeholders are considered and all parties are involved in the decision-making process, thereby enhancing their support for the set rules. As such, the move towards better solutions for prevailing problems from a smart-mixes perspective must include a broad understanding of legitimacy as an indispensable variable and thus also more consideration of the role of the law; in its guarantee function it constitutes an important factor in conditioning the behaviour and actions of those involved in the regulatory and enforcement process on the basis of core values that underlie any particular system at a given point in time and that are expressed in the law as fundamental rights and principles. This is the case regardless of whether a smart-mixes approach comes about as a result of intentional design or as a result of some dynamic process. In the latter case, it may need adjustment or improvement from a legitimacy perspective to actually enhance the effectiveness of a smart mixes approach.

REFERENCES

Ayres, I. & Braithwaite, J. 1992. *Responsive Regulation. Transcending the Deregulation Debate*. Oxford, Oxford University Press.

   1994. *Responsive Regulation. Transcending the Deregulation Debate*. Oxford, Oxford University Press.

Baldwin, R. & Black, J. 2008. 'Really Responsive Regulation'. *Modern Law Review* 71(1), 59–94.

Bressers, H. Th. A. & O'Toole, L. J., Jr. 1998. 'The Selection of Policy Instruments: A Network-Based Perspective'. *Journal of Public Policy* 18(3), 213–239.

Cafaggi, F. 2014. 'A Comparative Study Analysis of Transnational Private Regulation: Legitimacy, Quality, Effectiveness and Enforcement', EUI Working Papers, 2014/12015 http://cadmus.eui.eu/bitstream/handle/1814/33591/LAW_2014_15.pdf?sequence=1/handle/ 1814/16526, www.eesc.europa.eu/resources/docs/a-comparative-analysis-of-transnational-private-regulation-fcafaggi_12062014.pdf.

Curtin, D. & Senden, L. A. J. 2011. 'Public Accountability of Transnational Private Regulation; Chimera or Reality?'. *Journal of Law and Society* 38(1), 163–188.

Eberlein, B., Abbott, K. W., Black, J. & Meidinger, E. 2012. 'Transnational Business Governance Interactions: Conceptualization and Framework for Analysis', *Comparative Research in Law & Political Economy*, Research Paper No. 29, 2012, available at: http://digitalcommons.osgoode.yorku.ca/clpe/33.

Gulbrandsen, L. 2014. 'Dynamic Governance Interactions: Evolutionary Effects of State Responses to Non-State Certification Programs'. *Regulation & Governance* 8 (1), 74–92.

Gunningham, N. & Sinclair, D. 1999. 'Regulatory Pluralism: Designing Policy Mixes for Environmental Protection'. *Law & Policy* 21(1), 49–76.

Gunningham, N., Grabosky, P. & Sinclair, D. 'Smart Regulation'. 2017. In Drahos, P. (ed.), *Regulatory Theory, Foundations and Applications*, Canberra, Australian National University Press, 133–149.

   1998. *Smart Regulation, Designing Environmental Policy*. Oxford socio-legal studies. Oxford, Oxford University Press.

Heremans, T. 2012. *Professional Services in the EU Internal Market – Quality Regulation and Self-Regulation*. Oxford, Hart Publishing.

Howlett, M. 1991. 'Policy Instruments, Policy Styles, and Policy Implementation: National Approaches to Theories of Instrument Choice'. *Policy Studies Journal* 19(2), 1–21.

   2004. 'Beyond Good and Evil in Policy Implementation: Instrument Mixes, Implementation Styles, and Second Generation Theories of Policy Instrument Choice'. *Policy & Society: Journal of public, foreign and global policy* 23(2), 1–17.

Howlett, M. & Rayner, J. 2004. '(Not so) "Smart Regulation"? Canadian Shellfish Aquaculture Policy and the Evolution of Instrument Choice for Industrial Development'. *Marine Policy* 28, 171–184.

Joerges, Ch., Schepel, H. & Vos, E. 1999. 'The Law's Problems with the Involvement of Non-Governmental Actors in Europe's Legislative Processes: The Case of Standardization under the "New Approach"', *EU Working Paper LAW*, No. 99/9.

Linder, S. H. & Peters, B. G. 1989. 'Instruments of Government: Perceptions and Contexts'. *Journal of Public Policy* 9(1), 35–58.

Majone, G. 1976. 'Choice among Policy Instruments for Pollution Control'. *Policy Analysis* 2, 589–613

Maxwell, W. J. 2014. 'Global Privacy Governance: A Comparison of Regulatory Models in the US and Europe, and the Emergence of Accountability as a Global Norm'. In Dartiguepeyrou, C. (ed.), *Cahier de Prospective. The Future of Privacy*. Paris, Foundation Telecom, 63–70.

Moerel, L. 2012. *Binding Corporate Rules, Corporate Self-Regulation of Global Data Transfers*. Oxford, Oxford University Press.

NWO. 2013. *Smart Governance. Part of the Social Infrastructure Agenda*, The Hague, NWO.

Pelkmans, J. 1987. 'The New Approach to Technical Harmonization and Standardization'. *Journal of Common Market Studies* 25(3), 249–269.

Salamon, L. M. & Lund, M. S. 1989. *Beyond Privatization: The Tools of Government Action*. Washington, D.C., Urban Institute Press.

Scott, C., Cafaggi, F. & Senden, L. 2011. 'Conceptual and Constitutional Challenges of Transnational Private Regulation'. *Journal of Law and Society* 38(1), 1–19.

Senden, L. A. J. 2013. *Transnational private regulation in the area of human rights*, Unitreport, available at: http://renforce.rebo.uu.nl/bouwsteenprojecten/private-actoren/.

2017. 'The Constitutional Fit of European Standardisation Put to the Test'. *Legal Issues of Economic Integration* 44(4), 337–352.

Senden, L. A. J., Kica, E., Hiemstra, M. & Klinger, K. 2015. *Mapping Self- and Co-regulation Approaches in the EU-context*, explorative study for the European Commission, DG Connect, available at: https://ec.europa.eu/digital-single-market/sites/digital-agenda/files/dae-library/mapping_self-and_co-regulation_in_the_eu_context_0.pdf.

Six, F. & Verhoest, K. 2017. 'Trust in Regulatory Regimes: Scoping the Field'. In Six, F. & Verhoest, K. (eds.), *Trust in Regulatory Regimes*. Cheltenham, Edward Elgar, 1–36.

Van der Voort, H. 2017. 'Trust and Cooperation over the Public-Private Divide: An Empirical Study on Trust Evolving in Co-regulation'. In Six, F. & Verhoest, K. (eds.), *Trust in Regulatory Regimes*. Cheltenham, Edward Elgar, 181–204.

Van Gossum, P., Arts, B. & Verheyen, K. 2012. '"Smart regulation": Can policy instrument design solve forest policy aims of forest expansion and sustainability in Flanders and the Netherlands'. *Forest Policy and Economics* 16, 23–34, available at: http://dx.doi.org/10.1016/j.forpol.2009.08.010.

Voermans, W. 'Legislation and Regulation'. 2017. In Karpen, U. & Xanthaki, H. (eds.), *Legislation in Europe. A Comprehensive Guide to Scholars and Practitioners*. Portland, Hart Publishing, 17–32.

Walker, W. E., Rahman, S. A. & Cave, J. 2001. 'Adaptive Policies, Policy Analysis, and Policy-Making'. *European Journal of Operational Research* 128, 282–289.

Wiener, J. B. & Richman, B. D. 2010. 'Mechanism Choice'. In Farber, D. A & O'Connell, A. J. (eds.), *Research Handbook on Public Choice and Public Law*. Northampton, Edward Elgar, 363–396.

3

Smart Mixes and the Challenge of Complexity

Lessons from Global Climate Governance

*Philipp Pattberg and Oscar Widerberg**

3.1 INTRODUCTION

Addressing global environmental challenges in the twenty-first century, from bio-diversity loss to climate change, requires more than intergovernmental agreements. As has been widely observed, command-and-control instruments have been complemented by a variety of new governance instruments across different issue areas and sectors.[1] The resulting institutional architecture of global governance has been described using concepts such as 'fragmegration',[2] 'assemblages',[3] 'bricolage'[4] or 'new medievalism'[5] to make sense of the seeming complexity of world politics. At the level of concrete institutional arrangements and governance instruments, scholars have noted (international) 'regime complexity',[6] defined as 'the presence of nested, partially overlapping, and parallel international regimes that are not hierarchically ordered'.[7] As a consequence of increased institutional density and overlap, authors have theorised the 'fragmentation of global governance architectures'[8] and the emergence of related 'regime complexes'.[9] However, while complexity is often assumed and hypothesised, few authors have attempted to conceptualise and measure complexity from the perspective of complexity theory. This seems particularly relevant when taking into account the idea of 'smart mixes' and 'smart

* The authors also acknowledge financial support from the Netherlands Organization for Scientific Research (CONNECT project, grant number 016.125.330).
1 Pattberg and Stripple (2008); Jordan et al. (2018).
2 Rosenau (1990).
3 DeLanda (2006); Sassen (2006).
4 Mittelman (2013).
5 Friedrichs (2001).
6 Drezner (2008); Hafner-Burton, Kahler & Montgomery (2009); Orsini, Morin & Young (2013).
7 Alter & Meunier (2009).
8 Biermann et al. (2009).
9 Raustiala & Victor (2004); Keohane & Victor (2011).

49

regulation'. These concepts describe an ideal situation in which we can combine various regulatory and governance instruments, both public and private and both international and local, into sophisticated mixes of complementary instruments and actors, tailored to the specific needs of the situation. However, does rational mixing and orchestrating work in situations where the parts of the system (the regulations, policies, institutions) interact in a non-linear, non-additive way to produce unexpected outcomes?

In this chapter, we analyse in what respects the governance architecture of climate change has properties of a complex system. We argue that potential emergent behaviour of complex climate governance, i.e. unforeseen impacts and unintended consequences, might stand in the way of attempts to better manage and orchestrate the many parallel initiatives that have sprung up in recent years next to the United National Framework Convention on Climate Change (UNFCCC). In other words, we critically engage with the concept of smart mixes from a complexity perspective. Rather than arguing that the idea of smart mixes and complex systems do not go together, we think it more useful to explore how a more experimental mode of governance might be utilised to improve the 'smartness' of existing institutional mixes at the global level. This chapter is organised as follows: in Section 3.2, we provide a short introduction to complexity theory and the notion of complex systems. We also operationalise the concept of a complex system for researching the global climate change governance architecture. Section 3.3 then provides an empirical illustration of our claim that the climate change governance architecture has some properties of complex systems. Finally, in Section 3.4 we conclude with outlining some lessons learned and developing a research program on governance complexity and smart mixes for the future.

3.2 COMPLEXITY THEORY, COMPLEX SYSTEMS AND HOW TO STUDY CLIMATE CHANGE GOVERNANCE

This section introduces basic ideas of complexity theory, elaborates further on the properties of complex systems and suggests a way to test whether global climate change governance has properties of a complex system.

3.2.1 *Complexity Theory and Complex Systems*

Next to the widely applied term *complexity science*, scholars use *complexity theory* to refer to a number of assumptions about and perspectives on complexity and complex systems. It is important to note that no general theory of complexity has so far emerged that would be able to explain all classes of complex phenomena, such as hurricanes, financial crises, cities, organisational ecologies and regime complexes. Complexity theory should therefore be understood less as a unified explanatory theory and more as an ontologically founded framework of analysis.

In this chapter, we employ complexity theory to analyse those phenomena that arise from and are visible in complex systems. But what are complex systems? A straightforward way to understand them (if such a thing exists in the context of complexity) is to say that the whole is greater than the sum of its parts. As a result of systemic interactions (to quote Kavalski), 'alterations occur whose outcomes are wholly unexpected and nearly impossible to predict'. [10] A complex system approach considers actions of agents (e.g. organisations/members to governance institutions) that produce macro-level phenomena by aggregation. Sometimes complex systems create complexity. For example, the global financial system is a complex system that has parts that are predictable while other parts have emergent properties, i.e. properties that do not fully derive from the simple sum of all individual parts. Bicycles and pendulums are examples of simple systems; the immune system and ecosystems are examples of complex systems.

We follow Byrne and Callaghan[11] and Bousquet and Curtis[12] in suggesting that most theorists of complexity agree upon the following concepts as key ingredients to complexity theory: non-linearity, emergence, self-organisation and open systems.

In linear systems, when the value of a causal element changes, we can predict the change in the value of the dependent element. The changes in the latter are proportional to the changes in the former. As Byrne and Callaghan note, 'Linearity is foundational to "Newtonian" science by which we mean scientific accounts in which we can describe a current state in terms of values of parameters and have a covering law, a universal/nomothetic specification, which describes how the state will change if values in the parameters change'. [13] Non-linear systems do not satisfy the superposition principle by which outputs are proportional to inputs. In other words, in non-linear systems, effects emerge that are disproportionate to the changes in the input and thereby might be qualified as 'surprising' and hard-to-predict.

What follows from non-linear dynamics is that often we can speak of emergent properties and phenomena that are qualitatively different from those of the individual units/agents that are aggregated (think of the difference between water molecules/water and brain cells/consciousness). Emergent properties are defined as the 'intricate intertwining or interconnectivity of elements within a system, and between a system and its environment'.[14] A good example is a group of commuters competing for space on a road and causing a traffic jam. While the individual commuter is motivated by a simplistic goal, to get home as soon as possible, the aggregate phenomenon, a traffic jam, is hard to predict, evolve and manage. It represents, in other words, an emergent phenomenon.[15]

---

[10] Kavalski (2007), at 437.
[11] Byrne & Callaghan (2013), at 8.
[12] Bousquet & Curtis (2011).
[13] Byrne & Callaghan (2013), at 18.
[14] Mitleton-Kelly (2000).
[15] See Johnson (2009).

Different from the observed emergent phenomenon, self-organisation is a process 'by which the autonomous interaction of individual entities results in the bottom-up emergence of complex systems'.[16] That means that emergent phenomena do not result from the operation of an 'invisible hand', central agency or intelligent designer. Applied to the field of governance studies, this implies that central steering of entire governance systems is difficult to observe in reality.

Finally, complex systems are usually open systems, i.e. they have porous borders and exchange information and energy with their environments. It is, however, an empirical question how open they are.[17] The assumption of openness contradicts most analysts of political systems, who posit that these are closed: 'disturbances are temporary and the system tends to return to equilibrium'.[18]According to Walzian Structural Realism, for example, the international system does not change in its fundamental features as anarchy and sovereignty construct the conditions of a closed system. Similar assumptions about closed systems can be found in Luhmann's Modern Systems Theory.

In sum, complexity theory assumes that non-linear relationships will dominate complex systems, that equilibrium is not automatically a preferred system state, that self-organisation and emergence are defining properties of complex systems and that, consequently, reductionist approaches are ill-suited for analysing many real-world phenomena.[19]

3.2.2 *How to Study Climate Governance as a Complex System*

To study global climate governance from a complexity perspective, we need appropriate methodologies. Social network analysis provides a particularly suitable approach because non-linearity is a key characteristic of complex systems, and networks are a perfect embodiment of non-linearity. In the words of Capra:

> The first and most obvious property of any network is its non-linearity — it goes in all directions. Thus the relationships in a network pattern are non-linear relationships. In particular, an influence, or message, may travel along a cyclical path, which may become a feedback loop. The concept of feedback is intimately connected with the network pattern. [20]

What follows from this is that network theories and relational ontologies[21] will feature prominently as an analytical tool to unravel complex systems.[22] While

---

[16] Bousquet & Curtis (2011), at 47.
[17] Singer (1971), at 13.
[18] Harrison & Singer (2006), at 47.
[19] Kavalski (2007); see also Rosenau (2003).
[20] Capra (1996), at 82.
[21] On the notions of relational ontologies and relational thinking in sociology, see Emirbayer (1997).
[22] Watts & Strogatz (1998).

networks as a specific mode of organisation (opposed to markets and hierarchies) have been recognised in International Relations and global governance scholarship for a while,[23] network analysis as a formal method of inquiry has been applied less frequently. This is rather regrettable, as network analysis allows for fine-grained but robust measurements of structure (e.g. interactions among institutions in the climate change regime complex). Network analysis is grounded in three principles that make it an ideal approach within complexity science:[24] first, nodes (i.e. agents) are behaviourally interdependent; second, ties between nodes can be channels for resource exchange (material and non-material); and third, repeated and persistent patterns of interaction among nodes create structures that exert influence on the behaviour of agents. We will utilise these insights in our application of network analysis to the climate change governance architecture in Section 3.3.

But how can we empirically verify whether the institutional structure visible in the climate governance architecture[25] constitutes a complex system? Following Page,[26] we focus on four properties of complex systems with the potential to generate complexity. First, complex systems consist of diverse entities; second, the entities interact in an interaction structure; and third, the behaviour of the entities in the system is interdependent (meaning that they influence each other). We add the criterion of open system as a fourth property, arguing that climate change as a governance system should be open to its environment for our assumption about climate change as a complex system to hold. Adaptation and learning are sometimes mentioned as an additional characteristic; however, due to data limitations, we do not analyse learning effects in this chapter.

In sum, we have argued in this section that complexity theory provides a potentially innovative perspective on global governance in that it allows to study governance systems (aggregations of regulations, institutions, rules, norms, decision-making procedures) as complex systems. Complexity – with its key attributes non-linearity, emergence, self-organisation and open systems – is a possible state of complex systems. Section 3.3 illustrates our claim that the climate change governance architecture,[27] also referred to as the regime complex for climate change,[28] has attributes of a complex system and should therefore analytically be treated as such.

---

[23] An example is Keck and Sikkink's concept of transnational activist networks; see Keck & Sikking (1998).

[24] Hafner-Burton, Kahler & Montgomery (2009), at 562.

[25] See governance triangle in Figure 3.1.

[26] Page (2015), at 24–25.

[27] Biermann et al. (2009).

[28] Keohane & Victor (2011).

3.3 ANALYSING GLOBAL CLIMATE GOVERNANCE AS A COMPLEX SYSTEM

The complexity of complex systems derives from the relationships among constituent parts, not from the parts themselves. In other words, complex systems are complex because of the interactions of their components and not because of additive effects of all parts. It is therefore not sufficient to map all governance institutions in the climate change regime complex in order to deduce outcomes; in fact, interactions among constituent parts, including feedback loops and non-linearity, result in system-wide, emergent properties. The implications for the idea of 'smart mixes' is potentially far-reaching. If the interaction of the many policies, governance arrangements and institutions operating within an issue area (as possibly between issue areas as well) produces unintended and unforeseeable effects, this might severely limit the ability of actors to engage in a rational design of broader governance architectures.

This section presents an empirical illustration of global governance as a complex system using climate change as an issue area for the case, where increasing institutional density has long been observed and analysed under headings such as 'interplay'[29] and 'fragmentation'.[30] Still, research has only recently started to map and partially measure the overall institutional structure of global governance, i.e. the clusters of norms, principles, regimes and other institutions generally referred to as the *governance architecture* of an issue area. In the words of Biermann and colleagues: 'Most research on global governance has focused either on theoretical accounts of the overall phenomenon or on empirical studies of distinct institutions to solve particular governance challenges'. [31] An important step towards analysing the macro-level of governance institutions in world politics was Raustiala and Victor's conceptualisation of *regime complexes* as 'an array of partially overlapping and non-hierarchical institutions governing a particular issue-area'. [32] While firmly rooted in a state-based, international ontology, their interest was in understanding institutionalisation beyond single regimes and clearly demarcated legal boundaries. Keohane and Victor[33] applied the regime complex conceptualization to climate governance, also with a focus on international cooperation.[34]

---

[29] Oberthür & Stokke (2011).

[30] Biermann et al. (2009).

[31] Biermann et al. (2009), at 14.

[32] Raustiala & Victor (2004), at 279.

[33] Keohane & Victor (2011).

[34] For an alternative conceptualisation, see Orsini, Morin & Young (2013), at 29: 'a network of three or more international regimes that relate to a common subject matter, exhibit overlapping membership, and generate substantive, normative, or operative interactions recognized as potentially problematic whether or not they are managed effectively'.

Building on, *inter alia*, Young's early work on institutional linkages,[35] the regime complex literature problematises and nuances our understanding of structure in global governance, situating regimes on a continuum, ranging from fully integrated institutions regulated via top-down authority, to nested regimes (semi-hierarchical), and collections of loosely coupled institutions (regime complexes) to fragmented institutional structures that lack coordination and linkages among constituent parts. Consequently, scholars have suggested that regime complexes are inherently complex and therefore justify speaking of *regime complexity*.[36] However, system-wide complexity of governance architectures has been more often assumed than measured. Improving the mapping of institutional diversity might provide further empirical insights into the structure of global governance, in particular understanding its characteristics as a complex system.

When mapping the wide array of institutional forms in global climate governance, Abbott employs a 'governance triangle'.[37] Institutions are situated in the governance triangle depending on the identity of their constituent actors – State, Firm or Civil Society Organisation (CSO) – or combinations thereof.[38] The placement of an institution is determined by judging each actor group's approximate 'share' in the governance of the scheme: in principle, the *State* category includes individual states and collections of states or international organisations (IOs) along with public bodies below the level of central states, e.g. cities and regions. Similarly, the *Firm* category includes individual business firms, groups of firms and industry associations; and ultimately, the *CSO* category includes individual CSOs as well as CSO coalitions and networks. All three actor groups are defined broadly, so that between them they encompass virtually all possible participants in transnational governance. The triangle is further divided into seven *zones*, which represent the major combinations of actor *types*. Institutions in the vertex zones (1–3) are dominated by a single type of actor; those in the quadrilateral zones (4–6) involve two types of actors; and those in the central zone (7) involve actors of all three types. Additionally, the two dashed horizontal lines divide the triangle into three 'tiers', defined by the nature of government involvement – *state-led* (public institutions are dominant), *private-led* (Firms and CSO are dominant), and *hybrid* (government bodies share governance with firms and/or CSO in public-private partnerships).

Figure 3.1 shows an updated and altered version of Abbott and Snidal's governance triangle applied to global climate governance, including a total of eighty-nine international and transnational governance institutions (updated as of January 2017). Criteria for inclusion in our dataset (1) have been international or transnational, (2) display intentionality to steer the behaviour of their members, (3) explicitly mention

---

[35] Young (1996).
[36] Orsini, Morin & Young (2013).
[37] See Abbott (2012); Abbott & Snidal (2009); Abbott & Snidal (2010).
[38] Abbott & Snidal (2009).

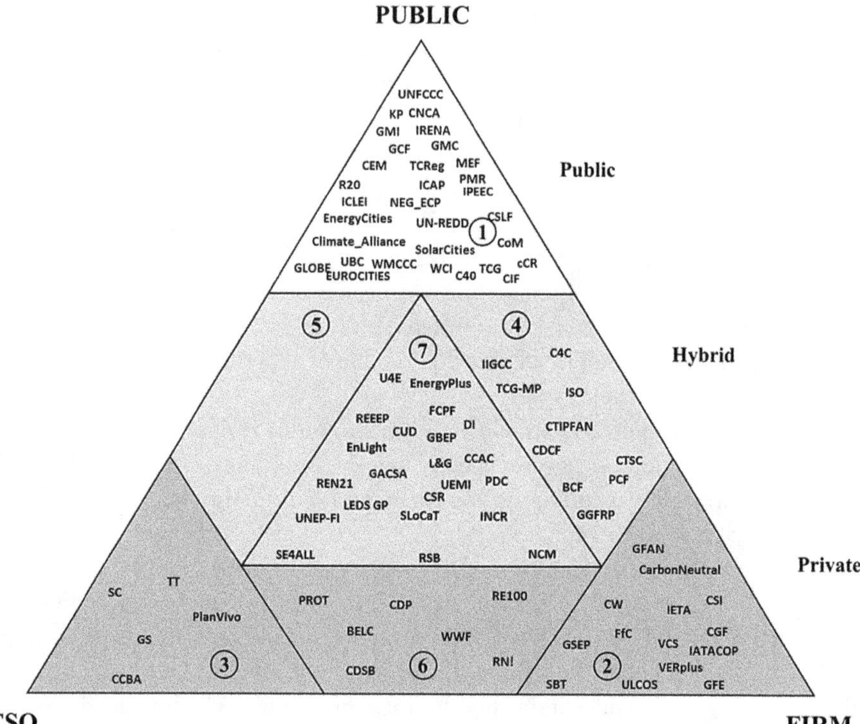

FIGURE 3.1: Global climate governance triangle
(Widerberg, Pattberg, and Kristensen 2016; based on Abbott and Snidal 2009; Abbott 2012).

a common governance goal, and (4) have identifiable governance functions (for further information on selection process, see the technical report by Widerberg, Pattberg, and Kristensen).[39] In the following text, each of the four properties discussed in Section 3.2.2 is analysed using data from the CONNECT-project as outlined previously.

3.3.1 *Diversity*

Complex systems consist of diverse entities (or agents in the parlance of agent-based modelling). Analysing a complex system thus requires a description of this diversity. Mapping the climate regime complex using the heuristic of a governance triangle (see Figure 3.1) reveals seven different zones of interaction determined by the type of actors involved: public actors (e.g. national or local governments); for-profit actors (e.g. corporations); and non-profit actors (e.g. non-governmental organisations). Within these three different broad actor types, we find both diversity (distinct types)

[39] Widerberg, Pattberg & Kristensen (2016).

and variation (distinctions within a type).[40] In terms of member types, the CONNECT-dataset identifies about 10,750 different organizations with different degree of governing functions in the institutions. These can further be divided across six categories: cities (81 per cent), companies (14 per cent), NGOs (3 per cent) and States (2 per cent). The distribution of members across institutions also differs widely, from institutions with one member to those with over 6,000 members. With an average of 142 members, a median of 25 members per institution, and a standard-deviation of 670, however, we observe that while the vast majority of institutions have fewer than 100 members, their size varies considerably. Moreover, institutions in our dataset perform a wide range of functions across a diverse set of themes. We have identified twelve themes: Carbon pricing and trading, CCS, Climate finance, Energy access, Energy efficiency, Forest mitigation (general), MRV, Non-CO_2 GHGs, Renewable energy, Sectoral (e.g. steel, transport, aviation and lighting) and Urban climate action. This finding corroborates previous research by (for example) Bulkeley and colleagues,[41] who study sixty transnational climate govern-ance institutions and identify twenty-one different climate related themes. Hence, we can observe large diversity across the institutions. Zooming in on individual institutions, we also find large variation. For instance, consider city size in the Covenant of Mayors, which, with over 6,000 members, is by far the largest insti-tution in the CONNECT dataset. The distribution ranges from villages such as Isuerre, located in northern Spain, with a population of fewer than 40, to cities such as London, with its population of nearly 8 million. Moreover, the median popula-tion of 5,476 in cities that are members of the Covenant of Mayors, and the standard-deviation of over 170,000 (indicating a highly positively skewed distribution) are both similar to the distribution of size of institutions in terms of members.

In sum, the global climate governance regime complex consists of a highly diverse set of institutions and organisations in terms of type, size and thematic focus. For example, even considering only type and theme of institutions generates eighty-four possible institutional types,[42] indicating that the system fulfils the attribute of 'diver-sity'. Interestingly, the distributions in terms of size in each of the preceding cases are positively skewed. According to Page,[43] these patterns are common in complex systems and allow for testing several explanatory models for this observation. For instance, perhaps we can explain the characteristics of the Covenant of Mayors by a 'preferential attachment' model – suggesting that popular nodes in a network tend to become disproportionally more central than less popular nodes – creating scale-free networks characterised by power-laws.[44] Hence, simply by describing the diversity in the global climate regime complex, and placing it in a complex systems framework,

---

[40] Page (2015), at 35.
[41] Bulkeley et al. (2014).
[42] 12 themes x 7 zones.
[43] Page (2015), at 34–35.
[44] Barabási & Albert (1999).

we allow for new hypotheses to be generated around the structure of global governance by borrowing concepts from complexity theory.

3.3.2 *Interconnectedness*

Emergent properties of a complex system can only be understood when analysing the interactions among the system's diverse entities. Entities (agents) interact with each other in a networked structure, allowing for non-linear and recursive behaviour to appear. Hence, to understand interaction in complex systems, we need to describe the structure in which the interaction occurs. Starting with the eighty-nine institutions presented in the previously mentioned governance triangle (see Figure 3.1), we have identified two different types of interlinkages – shared members and hyperlinks between homepages – as proxies for interaction.

A membership network is created by connecting those institutions that share members.[45] For example, if the Netherlands is a member to the UNFCCC and the Renewable Energy and Energy Efficiency Partnership (REEEP), then a link is created between the UNFCCC and REEEP. Figure 3.2 depicts the full membership network representing eighty institutions since nine institutions do not share members with other institutions. The nodes are coloured by zone (see also governance triangle in Figure 3.1).

The membership network illustrates the intricate structure of the climate governance complex, as well as how the institutions are by no means 'governance islands' working in isolation from each other, but are in fact quite densely connected across zones in the governance triangle. A particularly dense cluster appears around those institutions with public members, primarily states, such as the UNFCCC, CIF and IRENA. The membership network also shows signs of 'small-worldliness'[46] with a high clustering coefficient and small average path length compared to random graphs.[47] Small-world network topologies can have a range of implications on the system such as the speed of contagion in the network, something which could become important when trying to spread ideas, know-how and information.

The second network connects institutions via hyperlinks. Nearly all institutions in the CONNECT dataset have dedicated homepages providing different types of information, including embedded hyperlinks referring to other institutions and organisations. By linking in this cyberspace, institutions and organisations reveal existing or desired virtual connections. For instance, if an institution links to the UNFCCC homepage on its site, it may consider information on the UNFCCC homepage relevant for its own members and participants. Hyperlink Network

---

[45] See also Widerberg (2016).

[46] Watts & Strogatz (1998).

[47] The statistics for the random graph were produced taking averages of a 1000 Erdos-Renyi model-runs with the same node and edge count as the observed graph. Small-worldliness is defined as $C_{actual} / C_{random} > 1$ and $L_{actual} / L_{random} \approx 1$.

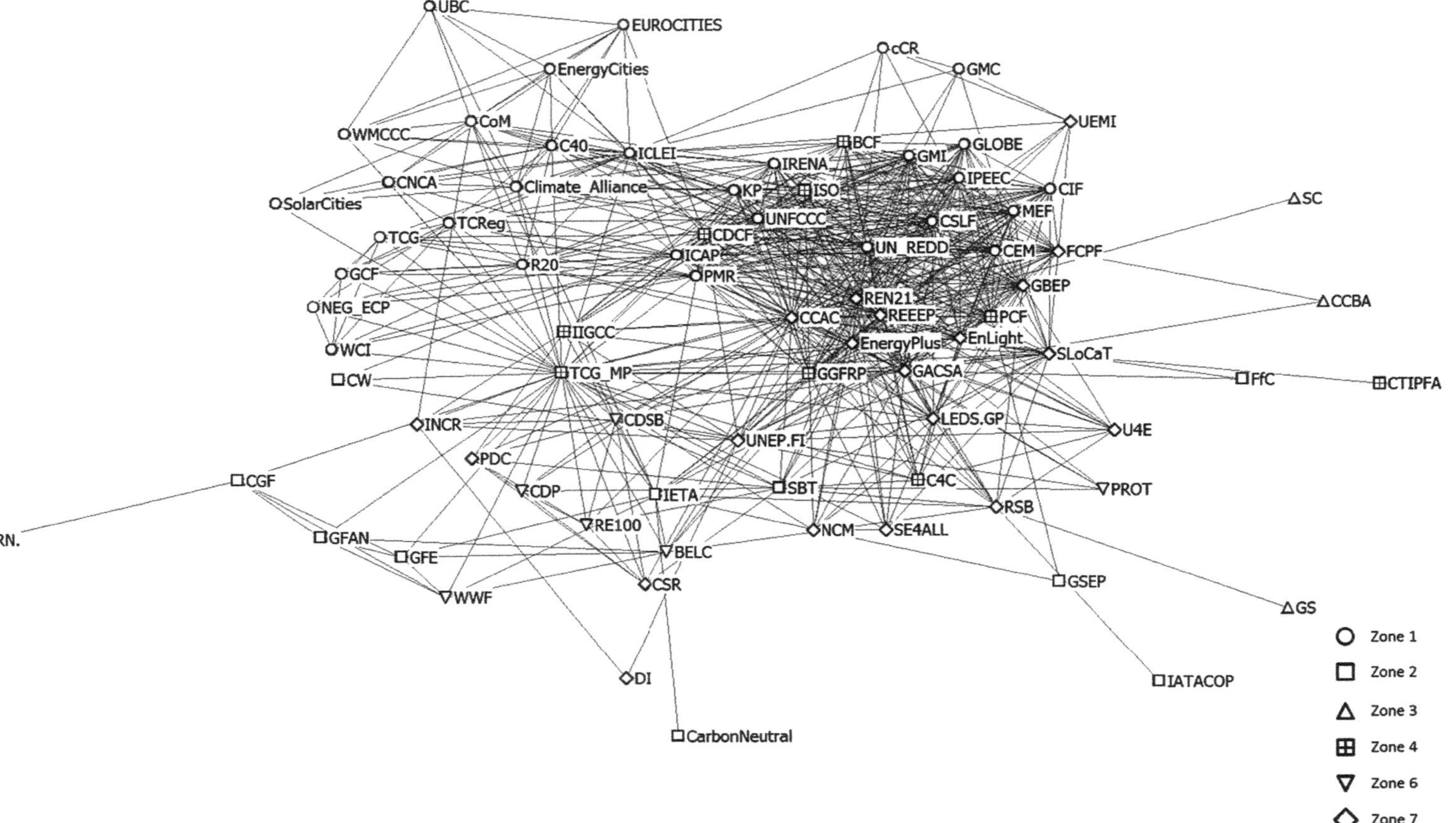

FIGURE 3.2: 1-mode membership network of institutions
(*source:* CONNECT)

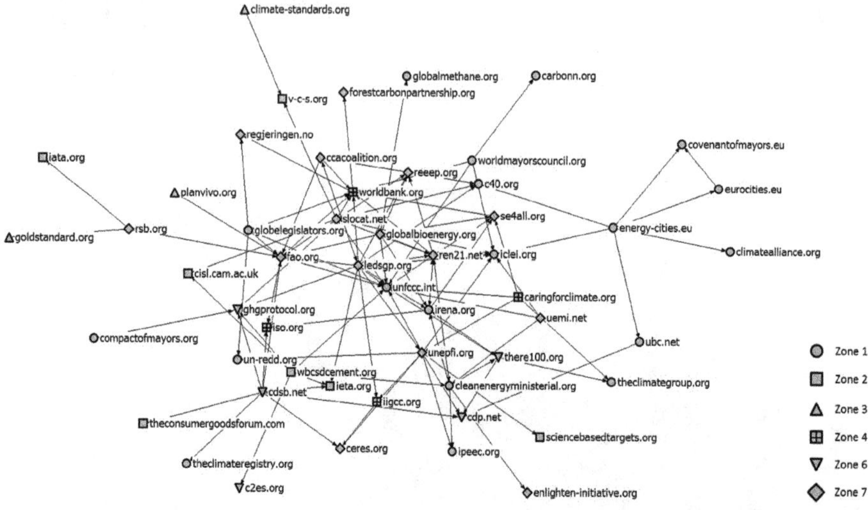

FIGURE 3.3: 1-mode hyperlink network of 53 institutions
(*source*: CONNECT)

Analysis (HNA)[48] and connecting institutions and organisations by homepages creates what have been referred to as Virtual Policy Networks (VPNs),[49] which are defined as web-based issue networks.[50] VPNs reflect the purposeful behaviour of one institution's attempt to publish information online and linking it to other institutions.[51]

In the context of complexity and 'interconnectedness', we are interested in the degree to which institutions connect to each other in cyberspace. To this end, an 'inter-actor' analysis has been conducted using Issue Crawler,[52] which requires a node to connect to another node from the list of starting points in order to be admitted to the network. As starting points, the homepages of the eighty-nine institutions in the CONNECT dataset have been used and tested at different depth, i.e. at how many iterations the harvesting of hyperlinks are carried out.[53] At the most conservative depth (1), only one iteration is performed collecting all the links at the sites of the starting homepages and checked for inter-actorness.

---

[48] Park (2003).

[49] McNutt (2010).

[50] McNutt (2012); McNutt & Pal (2011).

[51] McNutt (2012).

[52] Rogers (2010).

[53] Issuecrawler's instruction page provides a short explanation of crawl depth: 'The pages fetched from the starting point URLs are considered to be depth 0. The pages fetched from URL links from those pages are considered to be depth 1. In general, the pages found from URL links on a page of depth N are considered to be depth N+1. If you set a depth of 2, then no pages of depth 2 will be fetched. Only pages of depth 0 and 1 will be fetched (i.e. two levels of depth)' (www .govcom.org/Issuecrawler_instructions.htm).

The hyperlink network reaffirms the results in the membership and thematic networks of a highly dense and interlinked structure.[54] It suggests that institutions are actively engaging in trying to link with other institutions in the network, strengthening the observation that contemporary global climate governance is exercised in an intricate interaction structure.

The meaning behind linking homepages could be manifold. Homepages are places for institutions to send information to potential participants and users, and opportunities to show support or cooperation with other institutions. For instance, patterns in inward linking suggest that the system favours a few central hubs in terms of information and authority. Patterns in outward linking suggest that some institutions are trying to increase their own prestige and relevance by associating with certain institutions.

3.3.3 *Behavioural (Inter)dependence*

In networks, agents depend on the flows of resources, knowledge and norms. For instance, Kim[55] shows how nearly 750 multilateral environmental agreements connect in an increasingly interconnected dynamic network by referring to each other in the negotiated text. His findings suggest that the global body of international environmental law, in the absence of hierarchical coordination, has been self-organising into a 'mature' complex system.[56] Similarly, Green[57] has reported similar findings when linking private and public carbon accounting standards, revealing a hierarchical pattern in private accounting standards 'anchoring' to public international carbon accounting standards. Her findings show how agents create emergent behaviour in a system where different standards both compete and cooperate to carve out a niche.[58] The very concept of 'networked governance', contrasted against hierarchical or market governance, relies on these flows between agents and emergent interdependencies.

Within the CONNECT project, we have identified the central institutions that hold key positions in the network and therefore influence the flow of resources, knowledge and norms, thus creating forms of dependence. Figures 3.4 and 3.5 show two different measurements for identifying central institutions in the climate governance regime complex.

Figure 3.4 shows the centrality of institutions (highlighted in black) in terms of degree. Degree centrality is an indicator for institutions that have the most 'friends'. For flow of resources, however, an alternative measurement for centrality is more interesting, namely 'betweenness'. Betweenness centrality is an indicator for

---

[54] See Figure 3.3.
[55] Kim (2013).
[56] Ibid.
[57] Green (2013).
[58] Abbott, Green & Keohane (2015).

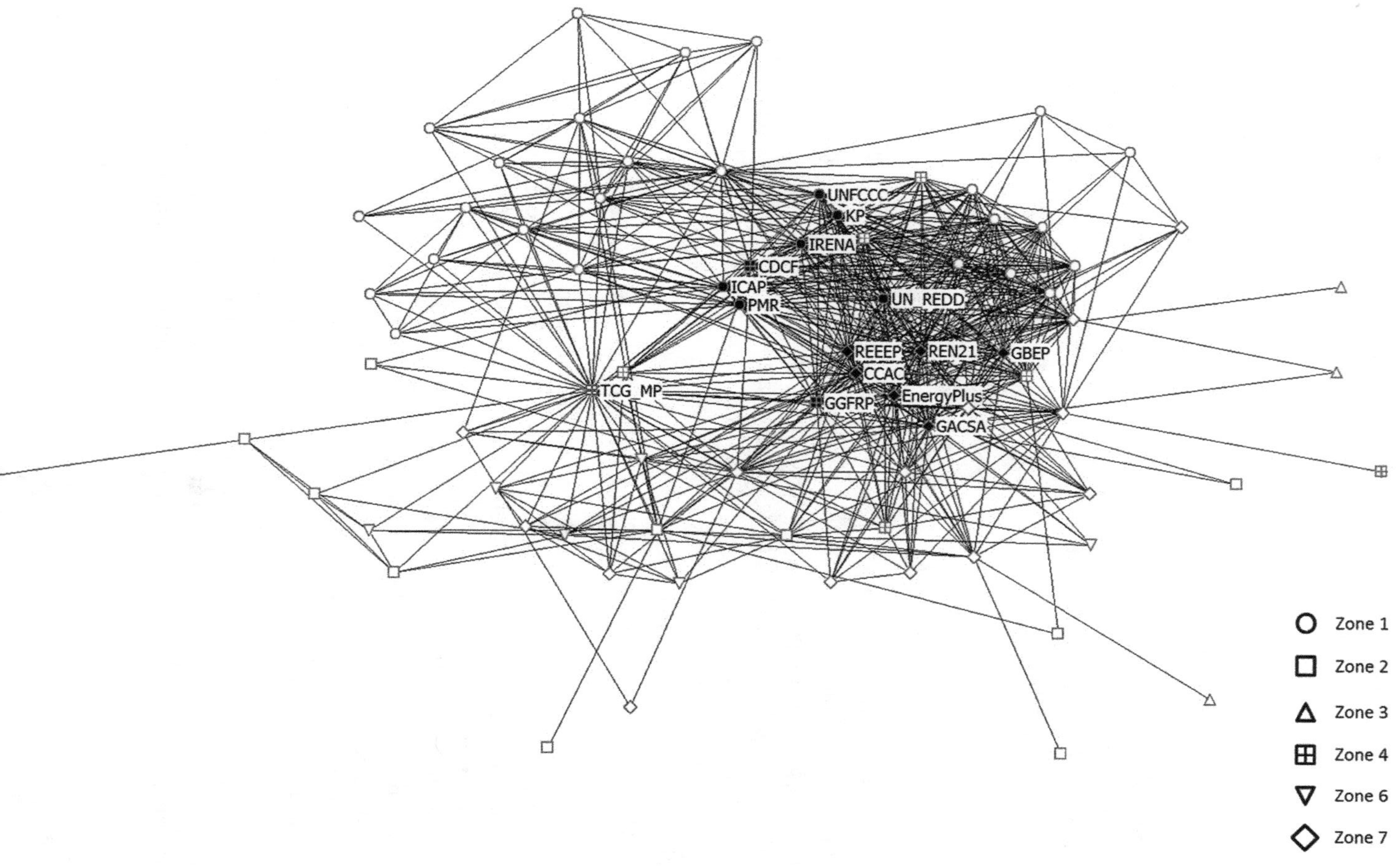

FIGURE 3.4: 1-mode membership network, 15 institutions with highest degree centrality in black
(*source:* CONNECT)

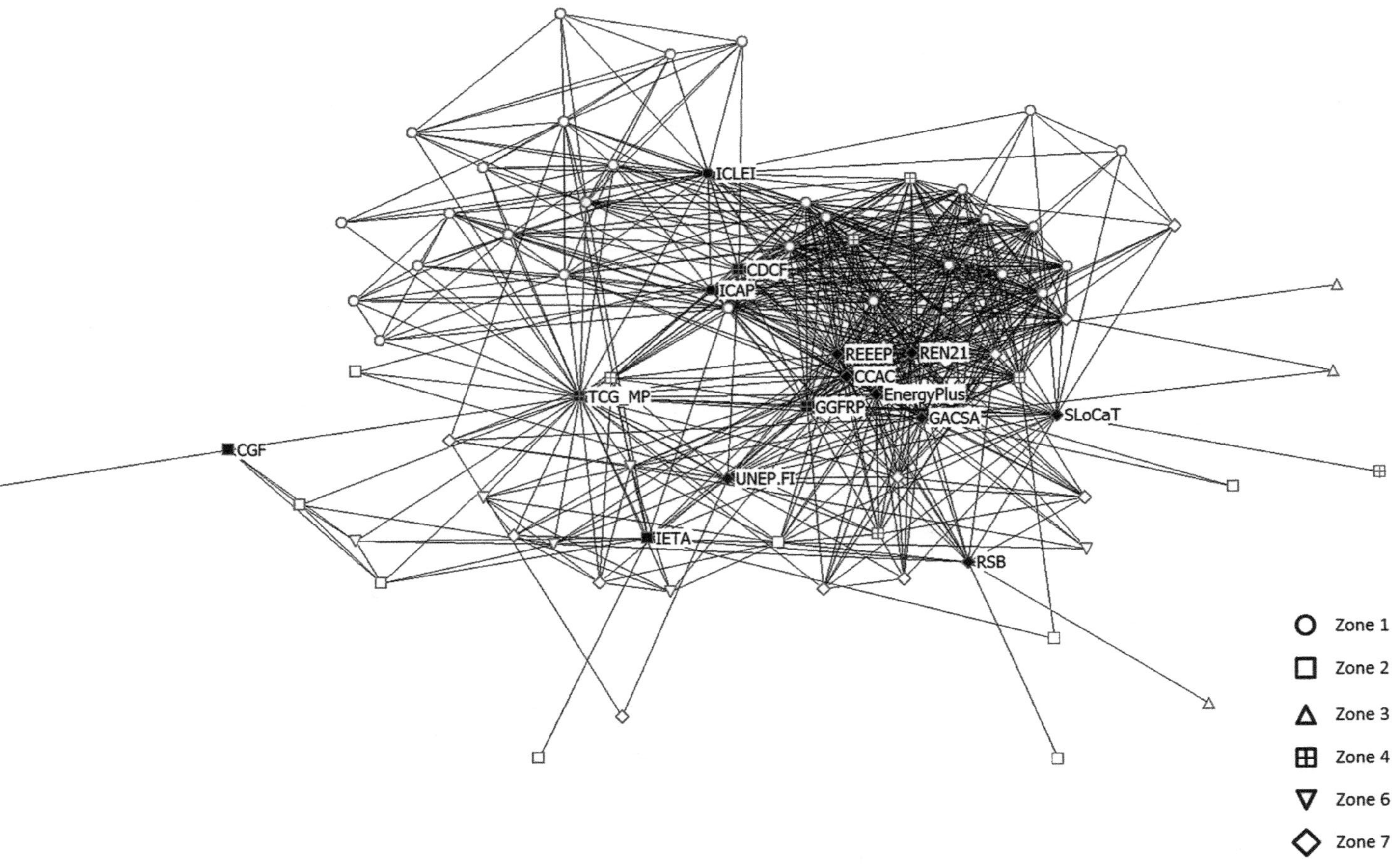

FIGURE 3.5: 1-mode membership network, 15 institutions with highest betweenness centrality in black
(*source:* CONNECT)

institutions that are located between other institutions, having the possibility to act as gate-keepers for steering different flows. The behaviour of these institutions is thus in theory extremely important for ensuring efficient transfers of whatever that travels through the network. In Figure 3.5, the ten nodes with the highest betweenness centrality are highlighted in black. While there is some overlap between institutions with a high degree centrality and a high betweenness centrality, we find some nodes that only become important as potential gate-keepers.

The exercise in testing different measurement for centrality shows how the structural position of institutions becomes important when thinking of global governance in terms of complex systems and networks. In Figure 3.5, for example, we find that the institutions with the highest betweenness centrality are not international institutions such as the UNFCCC, but are rather those such as ICAP and REEEP, hybrid institutions that connect international and transnational institutions and public and private actors. These nodes are thus in theory important gate-keepers for the flow of resources, information and norms between – in particular – state and non-state actors.

3.3.4 *Openness*

As the fourth and final criterion for evaluating whether or not global climate change governance constitutes a complex system, we employ the idea of openness, meaning that a complex system should be interacting with its systemic environment (i.e. all other systems around it). Different from our approaches in Sections 3.3.2–3.3.4, openness cannot easily be measured in a quantitative way. Rather, we use qualitative data to illustrate that climate change indeed is an open system. Our main argument here is that the climate governance system is routinely interacting with other governance systems, exchanging resources, information and actors. Examples include governance in other policy fields that interacts with the climate change domain, such as the Montreal Protocol on Substances that deplete the Ozone Layer (as some Ozone-Depleting Substances (ODS) are also greenhouse gases), and climate change governance impacting on other policy domains, such as the UN REDD process and its relation to forest governance. A recent study on the institutional nexus between global biodiversity governance and other environmental policy fields[59] has shown that, for example, out of 385 institutions governing climate change, agriculture, fisheries and forests, 108 institutions also have governance functions related to biodiversity. Similar studies on the Sustainable Development Goals have shown that the seventeen individual goals are connected by both bio-physical links and institutional interactions.[60] In sum, the climate change governance architecture is not only rich with internal

---

[59] Pattberg, Kristensen & Widerberg (2017).
[60] Boas, Biermann & Kanie (2016); Nilsson, Griggs & Visbeck (2016).

interactions and behavioural linkages (Sections 3.3.1–3.3.3) but also with linkages to other policy fields, both environmental and non-environmental.

3.4 IMPLICATIONS FOR SMART MIXES AND THE WAY FORWARD

This chapter has argued that learning from complexity theory offers new insights into the system of global environmental governance. In fact, we have attempted to demonstrate that some global governance systems, in this case the global climate governance architecture, are complex systems and can therefore be analysed from a complexity perspective.

Concluding that the current climate change governance architecture displays properties of a complex system has at least three important implications for the idea of 'smart mixes'. First, following from the insight that in complex systems the whole is greater than the sum of its parts, studying individual governance arrangements and ideal-types is insufficient for understanding the actual behaviour of the evolving regime complex. Instead, actors and interactions must be studied taking their broader environment, context and position within an interaction network into account. In this chapter, we have illustrated how mapping a governance system using network-based approaches could help research further in positioning events in their broader contexts.

Second, since non-linear behaviour is the rule rather than the exception, complex systems need different evaluation criteria than simple or complicated systems. Using concepts from complexity theory to open up the 'black box' between input and impacts of interventions in a policy field – for instance, by using complex program theory[61] – could improve our understanding of observed social change or how to more effectively reach governance goals.[62] For example, conventional linear thinking on how to bring about change is challenged if we understand outcomes as emergent rather than being planned, or the functioning of feedback loops in social systems to create large changes with small means.

And third, and related to the previous point, a complex systems perspective calls into question overly positive expectations about orchestration and the ability of actors to influence the system in a linear-causal way. While individual cooperative initiatives in the climate change policy field might well be considered as 'orchestrated', the overall system is not the result of rational planning. Feedbacks and unintended consequences (both positive and negative) undermine the linear-consequential ontology of orchestration. From this vantage point, the question of smart mixes is a relevant *ex ante* evaluation but less important in the design or planning phase.

Finally, we see in particular three areas for future research. First, efforts should be made to better measure and explain varying degrees of complexity across various

---

[61] Rogers (2008).
[62] Stame (2004).

issue areas and policy domains in a comparative perspective. Understanding governance systems as complex systems allows for a unified theoretical perspective on seemingly disparate phenomena in world politics, from migration to climate change, the financial system and sustainable development. A second challenge and research opportunity lies in modelling complex governance systems. The guiding question here is whether we can foresee (within certain margins of uncertainty) the evolution of complex governance systems. For example, would we rather expect increased synergies and integration, or fragmentation and welfare losses? This means that we make a step closer to modelling smart mixes as an outcome of complex systems. And third, we should make more efforts to accommodate failure into our theorising. Complexity theory could be used to better understand governance failures as well as adaptive behaviour of governance systems. From this perspective, designing smart mixes is not so much rational planning as constant experimenting, failing, adapting and learning. We hope that a complexity perspective can help to balance overly optimistic and technical approaches towards planning policy interventions.

REFERENCES

Abbott, K. W. 2012. 'The Transnational Regime Complex for Climate Change'. *Environment & Planning C: Government & Policy* 30(4), 571–590.

Abbott, K. W. & Snidal, D. 2009. 'Strengthening International Regulation Through Transnational New Governance: Overcoming the Orchestration Deficit'. *Vanderbilt Journal of Transnational Law* 42(2), 501–578.

2010. 'International Regulation without International Government: Improving International Organization Performance Through Orchestration'. *The Review of International Organizations* 5(3), 314–344.

Abbott, K. W., Green, J. & Keohane, R. O. 2016. 'Organizational Ecology and Organizational Diversity in Global Governance'. *International Organization* 70(2), 247–277, doi:10.1017/S0020818315000338.

Alter, K. J. & Meunier, S. 2009. 'The Politics of International Regime Complexity'. *Perspectives on Politics* 7(1), 13–24.

Barabási, A. -L. & Albert, R. 1999. 'Emergence of Scaling in Random Networks'. *Science* 286 (5439), 509–512.

Biermann, F., Pattberg, Ph., Van Asselt, H. & Zelli, F. 2009. 'The Fragmentation of Global Governance Architectures: A Framework for Analysis'. *Global Environmental Politics* 9(4), 14–40, doi:10.1162/glep.2009.9.4.14.

Boas, I., Biermann, F., & Kanie, N. 2016. 'Cross-Sectoral Strategies in Global Sustainability Governance: Towards a Nexus Approach'. *International Environmental Agreements: Politics, Law and Economics* 16(3), 449–464, doi: 10.1007/s10784-016-9321-1.

Bousquet, A. & Curtis, S. 2011. 'Beyond Models and Metaphors: Complexity Theory, Systems Thinking and International Relations'. *Cambridge Review of International Affairs* 24(1), 43–62.

Bulkeley, H., Andonova, L., Betsill, M. M., Compagnon, D., Hale, Th., Hoffmann, M. J., Newell, P., Paterson, M., Roger, C. & VanDeveer, S. D. 2014. *Transnational Climate Change Governance.* New York, Cambridge University Press.

Byrne, D. & Callaghan, G. 2013. *Complexity Theory and the Social Sciences: The State of the Art*. Abingdon, Routledge. Available at: https://books.google.nl/books?hl=en&lr=&id=mXC_AAAAQBAJ&oi=fnd&pg=PP1&dq=Byrne+and+Callaghan+2014+complexity+theory&ots=PNk1H-pLEf&sig=oICNWDfYjIE2gBm-EjOd7FkTIA4.

Capra, F. 1996. *The Web of Life: A New Scientific Understanding of Living Systems*. New York, Anchor.

DeLanda, M. 2006. *A New Philosophy of Society: Assemblage Theory and Social Complexity*. London, A&C Black. Available at: https://books.google.nl/books?hl=en&lr=&id=64TUAwAAQBAJ&oi=fnd&pg=PP1&dq=delanda+assemblages&ots=9AolgY2R4q&sig=WO3_YWFp1mRW9oUHDFAaEcOxRdQ.

Drezner, D. W. 2008. *All Politics Is Global*. Cambridge, Cambridge University Press. Available at: http://journals.cambridge.org/production/action/cjoGetFulltext?fulltextid=5466536.

Emirbayer, M. 1997. 'Manifesto for a Relational Sociology'. *American Journal of Sociology* 103 (2), 281–317, doi:10.1086/231209.

Friedrichs, J. 2001. 'The Meaning of New Medievalism'. *European Journal of International Relations* 7(4), 475–501.

Green, J. F. 2013. 'Order out of Chaos: Public and Private Rules for Managing Carbon'. *Global Environmental Politics* 13(2), 1–25.

Hafner-Burton, E. M., Kahler, M. & Montgomery, A. H. 2009. 'Network Analysis for International Relations'. *International Organization* 63(3), 559–592.

Harrison, N. E. & Singer, D. J. 2006. 'Complexity is More than System Theory'. In Harrison, N. E. (ed), *Complexity in World Politics. Concepts and Methods of a New Paradigm*. Albany, NY, State University of New York Press, 25–42.

Johnson, N. 2009. *Simply Complexity: A Clear Guide to Complexity Theory*. Oxford, Oneworld Publications. Available at: https://books.google.nl/books?hl=en&lr=&id=REFbbA15cX8C&oi=fnd&pg=PT8&dq=Simply+Complexity.+A+Clear+Guide+to+Complexity+Theory.+London:+Oneworld+Publications.&ots=fuE5B4Z-2r&sig=jX6FYan6J8ILmZdHxuUtmo7prZU.

Jordan, A., Van Asselt, H., Huitema, D. & Forster, J. (eds.). 2018. *Governing Climate Change: Polycentricity in Action*. Cambridge, Cambridge University Press.

Kavalski, E. 2007. 'The Fifth Debate and the Emergence of Complex International Relations Theory: Notes on the Application of Complexity Theory to the Study of International Life'. *Cambridge Review of International Affairs* 20(3), 435–454.

Keck, M. E. & Sikkink, K. 1998. *Activists beyond Borders: Advocacy Networks in International Politics*. Cambridge, Cambridge University Press.

Keohane, R. O. & Victor, D. G. 2011. 'The Regime Complex for Climate Change'. *Perspectives on Politics* 9(1), 7–23.

Kim, R. 2013. 'The Emergent Network Structure of the Multilateral Environmental Agreement System'. *Global Environmental Change* 23, 980–991.

McNutt, K. 2010. 'Virtual Policy Networks: Where All Roads Lead to Rome'. *Canadian Journal of Political Science* 43(4), 915–935.

2012. 'Climate Change Subsystem Structure and Change: Network Mapping, Density and Centrality'. *Canadian Political Science Review* 6(1), 15–50.

McNutt, K. & Pal, L. A. 2011. '"Modernizing Government": Mapping Global Public Policy Networks'. *Governance* 24(3), 439–467, doi:10.1111/j.1468-0491.2011.01532.x.

Mitleton-Kelly, E. 2000. *Complexity: Partial Support for BPR?*. Dordrecht, Springer. Available at: http://link.springer.com/chapter/10.1007/978-1-4471-0457-5_3.

Mittelman, J. H. 2013. 'Global Bricolage: Emerging Market Powers and Polycentric Governance'. *Third World Quarterly* 34(1), 23–37.

Nilsson, M., Griggs, D. & Visbeck, M. 2016. 'Map the Interactions between Sustainable Development Goals'. *Nature* 534(7607), 320–322.

Oberthür, S. & Stokke, O. S. 2011. *Managing Institutional Complexity: Regime Interplay and Global Environmental Change*. Cambridge, MA, MIT Press.

Orsini, A., Morin, J. -F. & Young, O. R. 2013. 'Regime Complexes: A Buzz, a Boom, or a Boost for Global Governance?'. *Global Governance: A Review of Multilateralism and International Organizations* 19(1), 27–39.

Page, S. E. 2015. 'What Sociologists Should Know About Complexity'. *Annual Review of Sociology* 41, 21–41.

Park, H. W. 2003. 'Hyperlink Network Analysis: A New Method for the Study of Social Structure on the Web'. *Connections* 25(1), 49–61.

Pattberg, Ph. & Stripple, J. 2008. 'Beyond the Public and Private Divide: Remapping Transnational Climate Governance in the 21st Century'. *International Environmental Agreements: Politics, Law and Economics* 8(4), 367–388.

Pattberg, Ph., Kristensen, K. & Widerberg, O. 2017. 'Beyond Biodiversity. Exploring the institutional landscape of governing for biodiversity', Amsterdam, IVM, R-17/06.

Raustiala, K. & Victor, D. G. 2004. 'The Regime Complex for Plant Genetic Resources'. *International Organization* 52(2), 277–309.

Rogers, P. J. 2008. 'Using Programme Theory to Evaluate Complicated and Complex Aspects of Interventions'. *Evaluation* 14(1), 29–48.

Rogers, R. 2010. 'Mapping Public Web Space with the Issuecrawler'. In Brossaud, C. & Reber, B. (eds.), *Digital Cognitive Technologies: Epistemology and the Knowledge Economy*. Hoboken, NJ, Wiley, 89–99.

Rosenau, J. N. 1990. *Turbulence in World Politics: A Theory of Change and Continuity*. Princeton, NJ, Princeton University Press.

2003. *Distant Proximities: Dynamics beyond Globalization*. Princeton, NJ, Princeton University Press. Available at: https://books.google.nl/books?hl=en&lr=&id=grYHrK5MS RYC&oi=fnd&pg=PR9&dq=rosenau+2003&ots=_QpyUr4is4&sig=Nd7ilkIkMywatFsal GWNrTMQicc.

Sassen, S. 2006. *Territory, Authority, Rights: From Medieval to Global Assemblages*, Vol. 7. Cambridge, Cambridge University Press. Available at: http://journals.cambridge.org/pro duction/action/cjoGetFulltext?fulltextid=1039036.

Singer, D. J. 1971. *A General Systems Taxonomy for Political Science*. New York, General Learning Press.

Stame, N. 2004. 'Theory-Based Evaluation and Types of Complexity'. *Evaluation* 10(1), 58–76.

Watts, D. J. & Strogatz, S. H. 1998. 'Collective Dynamics of "Small-World" Networks'. *Nature* 393(6684), 440–442.

Widerberg, O. 2016. 'Mapping Institutional Complexity in the Anthropocene: A Network Approach'. In Pattberg, Ph. & Zelli, F. (eds.), *Environmental Politics and Governance in the Anthropocene: Institutions and Legitimacy in a Complex World*. London, Routledge, 81–102.

Widerberg, O., Pattberg, Ph. & Kristensen, K. 2016. 'Mapping the Institutional Architecture of Global Climate Change Governance - V.2.', Technical Paper, Amsterdam, Institute for Environmental Studies (IVM).

Young, O. R. 1996. 'Institutional Linkages in International Society: Polar Perspectives'. *Global Governance* 2, 1–24.

4

Smart (and Not-So-Smart) Mixes of New Environmental Policy Instruments

Rüdiger Wurzel, Anthony Zito and Andrew Jordan

4.1 INTRODUCTION

When environmental policy emerged as a distinctive policy field in the early 1970s, traditional 'command-and-control' regulation quickly became the dominant environmental policy instrument in most European jurisdictions and in North America.[1] The United Kingdom's (UK) initial reluctance to adopt statutory environmental laws constituted an exception,[2] which, due to domestic reasons (e.g. the privatisation of the water industry necessitated the more legalistic approach) and external reasons (e.g. the increasing Europeanisation of British environmental policy), largely came to an end in the late 1980s.[3]

Gunningham, Grabosky and Sinclair's statement that 'in Europe, there remains heavy reliance on command-and-control as the basic instrument and very limited experimentation and policy mixes'[4] still remains broadly correct, although – for reasons which will be explained in this chapter – it applies more strongly to some European countries than others and is more relevant in some time periods than in others. In recent years most European countries have adopted an increasingly wider range of 'new' environmental policy instruments (NEPIs) – without, however, abandoning traditional regulatory tools. This has produced often complex mixes of 'new' and 'old' instruments that exist side by side, or that in some cases have mutated into hybrid instruments.[5] The need to better orchestrate such complex policy instrument mixes has been further exacerbated by the rise of international environmental agreements that have also stipulated new instruments or new variants of old

---

[1]   E.g. Faure, Vervaele & Weale (1994); Bemelmans-Videc, Rist & Vedung (1998); Gunningham, Grabosky & Sinclair (1998); Wurzel, Zito & Jordan (2013).

[2]   E.g. Vogel (1986).

[3]   E.g. Jordan (2002); see also Table 4.1, this volume.

[4]   Gunningham, Grabosky & Sinclair (1998), at 18.

[5]   E.g. Gunningham & Grabosky (1998); Jordan, Wurzel & Zito (2005); Wurzel, Zito & Jordan (2013).

69

instruments.[6] As 'old' and 'new' policy instruments have increasingly been combined either by design or default, questions about 'smart mixes'[7] and 'not-so-smart mixes'[8] have become more pertinent. Importantly, the question about instrument mixes 'has become a very central question' not only for academics but also for practitioners.[9]

The debate about 'optimal' policy instruments and/or instrument mixes was initially dominated by economists, who usually advocated the use of market-based instruments over traditional 'command-and-control' regulation on efficiency grounds.[10] Economists often either ignored or downplayed the significance of contextual factors (such as national regulatory traditions and party political preferences) for policy instrument selection, although there are important exceptions.[11] Legal scholars and political scientists, on the other hand, have typically emphasised the importance of legal and political context variables (such as their compatibility with particular legal systems and the politics of instrument selection) in addition to effectiveness and efficiency criteria. Majone[12] and Hood[13] put forward some of the first comprehensive policy instrument studies from a political science perspective, in which they argued that policy instrument choice is an inherently political process. As Hood[14] has argued: 'very commonly it is the instrument selected for reaching a policy aim that is far more contentious than the aim itself'. Seen from a more political perspective, policy instrument choice is not only about solving environmental problems efficiently and effectively but also about who gains and who loses from the adoption of which type of policy instrument.[15] It is therefore unsurprising that the process of selecting policy instruments and their exact design features – which may vary considerably from jurisdiction to jurisdiction – often attracts considerable lobbying activities from non-state actors who will be affected by these instruments.

In order to more precisely assess how different instruments interact to create mixes, it is important to assess the characteristics of the instrument types found in the policy sector. Consequently, this chapter focuses on the use of traditional 'command-and-control' regulation and three different types of NEPIs (informational instruments, voluntary agreements and market-based instruments) in five different jurisdictions – Austria, Germany, the Netherlands, the United Kingdom (UK) and the European Union (EU) – since the early 1970s. As informational instruments are

---

[6]   Abbott (2012).

[7]   Gunningham, Grabosky & Sinclair (1998).

[8]   E.g. Howlett & Rayner (2004).

[9]   Interview, German Environment Agency official, 2017

[10]   E.g. Siebert (1976); Baumol & Oates (1979); Burrows (1979).

[11]   E.g. OECD (1994).

[12]   Majone (1976).

[13]   Hood (1983), at 9.

[14]   Ibid., at 136.

[15]   E.g. Hood (1983); Lascoumes & Le Galès (2007); Wurzel, Zito & Jordan (2013).

widely perceived as relatively 'soft' instruments and traditional 'command-and-control' regulations as relatively 'hard' instruments, with voluntary agreements and market-based instruments falling somewhere in between, this chapter covers the full spectrum of instruments, ranging from 'soft' to 'hard'.

4.2 GOVERNMENT AND GOVERNANCE

Before assessing in more detail the adoption and usage patterns of 'new' and 'old' environmental policy instruments across jurisdictions and time, we locate the examination about 'smart' and 'not-so-smart' mixes of policy instruments in the wider debate about government and governance. Governance and government can be perceived as different forms of governing.[16] If the 'strong state' is characterised by the extreme form of top-down government in the era of 'big government',[17] then the equally extreme form of bottom-up governance is represented by self-organising societal systems, which are almost insusceptible to steering efforts by governments.[18] These two extreme positions can rarely be found in highly developed (European) liberal democracies, although they constitute useful heuristic analytical devices that can help us to better understand the central differences between traditional tools of government (i.e. top-down 'command-and-control' regulation) and new modes of governance (such as NEPIs and other new modes of governance) as well as the policy instrument selection process. In the international relations arena, somewhat similar extreme positions have been identified in the form of top-down multilateral agreements that stipulate legally binding targets and timetables, and bottom-up polycentric governance arrangements that rely on voluntary pledges and a high degree of self-coordination.

It is widely accepted that NEPIs can be used as yardsticks to assess empirically whether less hierarchical new modes of *governance* have supplanted top-down 'command-and-control' regulation, or merely supplemented it.[19] While much of the general *governance* literature has claimed that a decisive shift from traditional top-down state-led *government* towards bottom-up societal *governance* has taken place,[20] most of the more empirically informed NEPIs studies have argued that 'new' policy instruments have either supplemented (rather than supplanted) traditional regulation or have formed new hybrid policy instruments with features from both 'old' and 'new' instruments.[21]

---

[16] Finer (1970).

[17] E.g. Kitschelt et al. (2003).

[18] E.g. Luhmann (1984).

[19] E.g. Bemelmans-Videc, Rist & Vedung (1998); Jordan, Wurzel & Zito (2005); Holzinger, Knill & Schäfer (2003); Héritier & Lehmkuhl (2008); Wurzel, Zito & Jordan (2013).

[20] E.g. Rhodes (1996); Stoker (1998).

[21] E.g. Gunningham, Grabosky & Sinclair (1998); Wurzel, Zito & Jordan (2013).

4.3 'NEW' ENVIRONMENTAL POLICY INSTRUMENTS

Much of the policy instrument literature distinguishes between traditional 'command-and-control' regulatory instruments and 'new' modes of environmental governance or NEPIs.[22] However, the definition of what constitutes a 'new' environmental policy instrument needs to be established empirically. While certain NEPIs are new in some jurisdictions, they may have already been in use for some time in others. For example, Germany adopted the world's first eco-label scheme in 1978 while Austria, the Netherlands and the EU adopted somewhat similar schemes only in 1992.[23]

There is no universally accepted definition of which instrument types constitute NEPIs, although the following threefold typology is relatively widely accepted:[24] (1) informational instruments (which are sometimes also termed suasive instruments), (2) voluntary agreements, and (3) market-based instruments. Our relatively parsimonious threefold typology has the analytical advantage of making it easier to identify shifting patterns in instrument use (and thus also policy instrument mixes across jurisdictions and time) than would be possible with a more complex typology.

4.3.1 *Informational Instruments*

Eco-label schemes and environmental management schemes (EMSs) constitute important examples of informational policy instruments. They are both relatively 'soft' instruments, relying on voluntary participation while providing participants with a way to publicise their relatively good environmental performance.[25] To avoid a competitive disadvantage vis-à-vis competitors that already participate in such schemes, corporate actors that operate in markets with a high level of public environmental awareness and/or 'green consumerism' sometimes apply for eco-label schemes or EMSs as well. The levels of public recognition of eco-label schemes and EMSs, as well as the participation rates, are important factors for determining the success or failure of these two sub-types of informational instruments. Similar pressures apply to voluntary international certification and labelling schemes.

4.3.1.1 Eco-Labels

Informational policy instruments such as eco-label schemes are frequently also labelled 'moral suasion' instruments because they provide citizens and consumers

---

[22] Gunningham, Grabosky & Sinclair (1998); Holzinger & Knill (2003); Jordan, Wurzel & Zito (2005); Wurzel, Zito & Jordan (2013).

[23] Jordan et al. (2004); Wurzel, Zito & Jordan (2013).

[24] E.g. Bemelmans-Videc, Rist & Vedung (1998); Bähr (2010); Jordan, Wurzel & Zito (2005); Wurzel, Zito & Jordan (2013).

[25] E.g. Wurzel, Zito & Jordan (2013).

with standardised information about the environmental impact of certain products and services with the aim of encouraging them to adopt more sustainable purchasing decisions.[26] The OECD[27] differentiates between the following three sub-types of eco-label schemes: Type I – externally verified, multi-issue schemes; Type II – unverified self-declaratory schemes by manufacturers and/or retailers; and Type III – single-issue schemes based on quantified product information related to life-cycle impacts (e.g. the product profile of a particular product model). Type I eco-label schemes were first designed in Europe (i.e. in Germany) where they have become widely used, although their popularity varies considerably between different European jurisdictions. Type II schemes are widely used by corporate actors across the world, although they tend to have a poor reputation in terms of their environmental ambition and implementation.[28] Type III eco-label schemes are usually narrowly focused (e.g. on forestry products) and are not (yet) very widely used in Europe. For space constraints, and because they are most widely used in Europe, this chapter focuses only on Type I eco-label schemes.

4.3.1.2 Environmental Management Schemes

Environmental management systems such as the EU's Environmental Management and Audit Scheme (EMAS) and the International Standard Organisation's (ISO) 14001 standard, provide incentives for industry to adopt measures for achieving certain environmental objectives. Both EMAS and the ISO 14001 require environmental impact audits, the setting up of internal management systems with the aim of reducing the negative environmental impact and the publication of regular public statements. Companies (in the case of EMAS and ISO 14001) and public organisations (in the case of EMAS) that achieve the required standards are granted an official confirmation and permitted to use a logo for marketing purposes. Although the EU's EMAS and the ISO 14001 are voluntary instruments, public and/or market pressure may push companies to participate for fear of losing out to competitors. Several European governments (e.g. Austria and Germany) have linked EMAS participation to lighter regulatory regimes (e.g. fewer inspections), thus creating additional incentives for this scheme.

4.3.2 *Voluntary Agreements*

There is no universally accepted definition of voluntary agreements. While the EU Commission defines voluntary agreements as 'agreements between industry and

---

[26] E.g. Jordan et al. (2004); Wurzel, Zito & Jordan (2013).
[27] OECD (1991).
[28] Jordan et al. (2004); OECD (1997).

public authorities on the achievement of environmental objectives',[29] the European Environmental Agency (EEA) has adopted a narrower definition, which covers 'only those commitments undertaken by firms and sector associations, which are the result of negotiations with public authorities and/or explicitly recognised by the authorities'.[30]

Börkey and Lévêque[31] have differentiated further voluntary agreements into: (1) unilateral commitments, (2) negotiated agreements, and (3) public voluntary schemes. *Unilateral commitments* are self-declaratory improvement measures put forward by corporate actors. *Negotiated agreements* are more formal agreements between industry and public authorities. Many negotiated agreements in the Netherlands are 'covenants' which are legally binding, although in practice it can be difficult to enforce them through the courts. In Austria and Germany voluntary agreements cannot be legally binding for constitutional reasons. Therefore, Austria and Germany could not have accepted legally binding voluntary agreements, which the European Commission was considering as a possible policy instrument on the EU level during the early 2000s.[32] Negotiated agreements are arguably closer to the government end of the government–governance dimension than are unilateral commitments, which are closer to the governance end. This applies in particular to the Dutch covenants, which Gunningham et al.[33] have described as 'an innovative version of command and control'. *Public voluntary schemes*, which are not widely used in Europe, are established by public bodies. Individual companies are free to join such schemes, although public authorities define the membership criteria. Due to space constraints, and because they are the most widely used subtype voluntary agreements in Europe, this chapter focuses only on negotiated agreements.

4.3.3 *Market-Based Instruments*

Eco-taxes and emissions trading schemes (ETSs) are widely seen as the most important market-based policy instruments. The OECD defines market-based instruments as tools which affect 'estimates of costs of alternative actions open to economic agents'.[34] It distinguishes between the following four main types of market-based instruments: (1) eco-taxes (including charges and levies); (2) emissions trading; (3) subsidies; and (4) deposit-refund schemes. In Northern Europe, eco-taxes have long been widely used. Emissions trading schemes were pioneered in the United States of America (USA) on a regional basis (for sulphur dioxide emissions)

---

[29] CEC (1996), at 5.
[30] EEA (1997), at 11.
[31] Börkey & Lévêque (1998).
[32] CEC (2002); Wurzel, Zito & Jordan (2013).
[33] Gunningham, Grabosky & Sinclair (1998), at 40.
[34] OECD (1994), at 17.

in the 1980s, while European jurisdictions started to experiment with them (for carbon dioxide emissions) only in the early 2000s.[35] However, in 2003 the EU adopted the world's first transnational/supranational ETS, which became operational in 2005.

Subsidies (e.g. for renewable energy) and fiscal incentives (e.g. for less polluting cars) are relatively widely used in Europe. Although there are important exceptions,[36] most neoliberal economists have been critical of subsidies and fiscal incentives for less environmentally damaging activities because they consider them to constitute market-distorting instruments. Subsidies for environmentally damaging activities (e.g. energy production in subsidised coal-fired power stations) have long been criticized by economists and environmental groups, as well as by environment agency officials, although governments have found it difficult to abolish them. The empirical focus of this chapter will be on eco-taxes and emissions trading schemes, which are widely seen as the most important market-based environmental instruments in Europe.

4.4 ADOPTION PATTERNS OF NEPIS AND REGULATION IN DIFFERENT JURISDICTIONS OVER TIME

Before assessing in more detail the adoption and usage patterns of particular types of 'new' and 'old' policy instruments, we will first analyse the general overall instrument adoption patterns across jurisdictions and over time. This will allow for 1) the identification of leaders, followers and laggards for particular policy instruments, and 2) the detection of jurisdictional policy instrument mixes and how they have changed over time. As explained in Chapter 1, policy mixes are rarely designed as a totality; instead, a more gradual layering process occurs as policy-makers experiment with new ideas and instruments in their particular jurisdictional contexts.

Table 4.1 summarizes the adoption rates of these three types of NEPIs and environmental regulation across five European jurisdictions between the 1970s and 2010s. Table 4.1 uses simple weighting categories – low, medium and high – to allow for meaningful comparisons between four different instrument types across five jurisdictions over a period of almost fifty years. The data in Table 4.1 is largely based on a study by Wurzel, Zito and Jordan,[37] who assessed a wide range of primary documents and undertook more than fifty interviews in all four member states as well as on the EU level between 2002 and 2012. It was updated with additional primary research (including interviews) in 2017. Importantly, the descriptors used in Table 4.1 offer a relative comparison across jurisdictions and time, rather than an absolute baseline judgement.

---

[35] E.g. Ellerman, Buchner & Carraro (2007); Wurzel (2008).
[36] E.g. OECD (1999).
[37] Wurzel, Zito & Jordan (2013).

TABLE 4.1: *The Use of Different NEPIs and Environmental Regulations since the 1970s*

Instrument Type	Jurisdiction	1970s	1980s	1990s	2000s	2010s
Eco-Label Scheme	Austria (*1990)	—	—	Medium	Medium	Medium
	Germany (*1978)	Medium	High	High	High/Medium	Medium
	Netherlands (*1992)	–	—	Low	Low	Low
	UK	—	—	—	—	—
	EU Eco-Label (*1992)	—	—	Low	Low	Low
EU EMAS (*1993)	Austria	—	—	High	High/Medium	High/Medium
	Germany	—	—	High	High/Medium	High/Medium
	Netherlands	—	—	Low	Low	Low
	UK	—	—	Low	Low	Low
	EU Average	—	—	Low	Low	Low
Voluntary Agreements	Austria	Low	Low	Medium/ High	Medium/Low	Low
	Germany	Low	High	High	Medium/Low	Low
	Netherlands	Low	High	High	Medium	Low/ Medium
	UK	Low	Low/ Medium	Low/Medium	Low	Low
	EU-Wide Use	—	—	Low	Low	Low
Eco-Taxes	Austria	Low	Medium	Medium	Low	Low
	Germany	Low	Medium	High	Medium	Low
	Netherlands	Medium	High	High	Medium	Low
	UK	Low	Low	Medium	Medium	Low
	EU-wide use	—	—	—	—	—

ETS	Austrian ETS	—	—	—	Low. EU ETS in 2005	EU ETS
	German ETS	—	—	—	Low. EU ETS in 2005	EU ETS
	Dutch ETS	—	—	Low	Medium National Experiments. EU ETS in 2005	EU ETS
	UK ETS (*2002)	—	—	—	High/Medium. National ETS Replaced by EU ETS in 2005	EU ETS
	EU ETS (*2005)	—	—	—	High/Medium	High/ Medium
Environmental Regulation	Austria	Low	Medium	High	Medium	Medium
	Germany	High	High	High	High	Medium
	Netherlands	High	High	High/ Medium	Medium	Medium
	UK	Low	Medium	Medium	Medium	Low
	EU	High	High	High	Medium	Medium/Low

Note: (*) Refers to the year in which a particular instrument was first adopted.
Source: Adapted from Wurzel, Zito and Jordan (2013: 213–214) with updates based on interviews carried out in 2017.

The following four broad conclusions can be drawn from Table 4.1. First, the use of NEPIs has become more widespread overall since the 1970s. All five jurisdictions have adopted some NEPIs. This finding is in line with the existing policy instrument literature, which argues that there has been a significant rise in the adoption of NEPIs across a wide range of jurisdictions.[38] Gunningham et al.'s statement that 'environmental policy is in transition: from command and control towards a much more pluralistic conception of instrument design'[39] therefore still seems largely applicable, although there has been a decline in the use of certain NEPIs. For example, voluntary agreements have been less popular in the 2000s/2010s than they were in the 1980s/1990s.

Second, although there has been an overall increase in the use of NEPIs, it has taken place at various speeds across different jurisdictions. Considerable fluctuations have occurred in the adoption patterns of particular NEPIs (as well as traditional regulations) across different jurisdictions.[40] For example, as will be explained in more detail in the following text, the Netherlands and Germany have adopted a much larger number of voluntary agreements than Austria, the EU and UK. While the UK has remained the only jurisdiction that has failed to adopt its own eco-label scheme, it acted as an emissions trading leader when it set up a national pilot scheme before the EU ETS became operational in 2005.

Third, although NEPIs are now more widely used overall in all five jurisdictions than they were in the early 1970s, they can be more commonly found in some jurisdictions (e.g. the Netherlands and Germany) than in others (e.g. the EU). While some jurisdictions have been early adopters of particular types of NEPIs, they have also been ambivalent about – or even hostile towards – other types of NEPIs. For example, in the case of eco-labelling, Germany adopted its own national scheme in 1978, well ahead of Austria and the Netherlands, as well as of the EU's supranational scheme, which was not adopted until 1992. In the case of emissions trading schemes, on the other hand, while the Netherlands (and particularly the UK) experimented with national schemes in the early 2000s, Austria and Germany did not have national emissions trading schemes in place before the EU adopted the world's first transnational emissions trading scheme.[41]

Fourth, the adoption patterns of NEPIs (and traditional regulation) can vary not only across jurisdictions, but also across time. For example, the Netherlands and

---

[38] E.g. Bemelmans-Videc, Rist & Vedung (1998); De Bruijn & Hufen (1998); Holzinger & Knill (2003); Bähr (2010); Jordan, Wurzel & Zito (2005); Wurzel, Zito & Jordan (2013).
[39] Gunningham, Grabosky & Sinclair (1998), at 89.
[40] See also Wurzel, Zito & Jordan (2013).
[41] Wurzel, Zito & Jordan (2013).

Germany both adopted more voluntary agreements in the 1980s/1990s than in the 2000s/2010s. The adoption of the EU ETS constitutes one important factor for the decline in Dutch and German voluntary agreements, which were particularly widely used for climate change issues. In these cases, one type of NEPI (i.e. emissions trading) seems to have at least partly supplanted another type of NEPI (i.e. voluntary agreements). This empirical finding is in line with Hood's argument that a 're-tooling' process may take place from time to time.[42]

Salomon has pointed out that '[f]ar from simplifying the task of public policy solving, the proliferation of tools has importantly complicated it even while enlarging the range of options and the pool of resources potentially brought to bear'.[43] Questions have therefore been raised about what constitutes an 'optimal' or 'smart' mix of environmental policy instruments.[44]

As will be explained in more detail in the following text, what complicates any cross-jurisdictional comparisons is the fact that the same type of NEPI may be used in dissimilar ways in different jurisdictions.[45] When comparing the use of particular policy instruments and/or instrument mixes, it is therefore important to take into account exactly how a certain NEPI is used in a particular jurisdiction.

The chronological breakdown of NEPIs usage in Table 4.1 enables us to identify patterns of policy instrument sequencing. Much of the existing policy instrument literature advocates a sequencing that starts with softer (i.e. less coercive) instruments before the adoption of harder (i.e. more coercive) instruments.[46] While drawing on Doern and Wilson,[47] Vedung has argued that:

> the least coercive instruments are introduced first in order to gradually weaken the resistance of certain groups of individuals and adjust them to government intervention in the area. After some time, the authorities feel entitled to regulate the matters definitely by employing their most powerful instrument.[48]

Similarly, Gunningham et al. have argued for the sequencing of environmental policy instruments: it enables 'escalation from the preferred least interventionist option, if it fails, to increasingly more interventionist alternatives'.[49]

The gradual ratcheting upwards of policy instruments according to their degree of coerciveness – starting with soft policy instruments (i.e. horizontal self-governance) and ending with hard policy instruments (i.e. coercive top-down

---

[42] Hood (1983), at 126.
[43] Salomon (2002), at 18.
[44] E.g. Gunningham, Grabosky & Sinclair (1998); Jordan, Wurzel & Zito (2007).
[45] E.g. Jordan, Wurzel & Zito (2005); Wurzel, Zito & Jordan (2013).
[46] E.g. Gunningham, Grabosky & Sinclair (1998); Salamon (2002); Vedung (1997).
[47] Doern & Wilson (1974).
[48] Vedung (1997), at 40.
[49] Gunningham, Grabosky & Sinclair (1998), at 123.

government) – is, in principle, also in line with liberal/neoliberal economic thinking. However, the empirical evidence put forward in this chapter shows a different pattern.[50] Germany, the Netherlands and the EU all relied almost exclusively on traditional environmental regulations in the 1970s. Austrian and (in particular) UK environmental policy initially relied less heavily on environmental regulations. However, by the 1980s, environmental regulations were also the dominant policy instrument in Austria. During the 1970s and 1980s, EU environmental policy relied almost exclusively on traditional environmental regulation (in the form of directives, regulations and decisions). Importantly, in four out of five jurisdictions traditional environmental regulations were adopted first before being supplemented by softer environmental policy instruments. Germany's early and heavy reliance on traditional environmental regulation instead of NEPIs can be explained with reference to its dominant policy style and its long tradition of being a constitutional state *(Rechtsstaat)*, as well as the fact that voluntary agreements or other soft policy instruments were 'deemed inappropriate for the defence against dangers *(Gefahrenabwehr)* to the environment and human health. You need regulations for such a task'.[51] The UK was initially reluctant to adopt statutory environmental standards. However, from the mid-1970s onwards this gradually changed as it became necessary to implement a rapidly increasing number of EU environmental regulations. Around that time, therefore, EU environmental legislation became a major driver for UK environmental policy. Clearly, particular policy instruments are considered more appropriate and legitimate in certain jurisdictions regardless of their theoretical advantages, for example, in terms of efficiency.[52]

4.4.1 *Usage Patterns of Informational Instruments*

4.4.1.1 Eco-Labels

Germany adopted the world's first national eco-labelling scheme in 1978. Austria and the Netherlands followed with their own national eco-labelling schemes in 1991 and 1992 respectively. In 1992 the EU adopted an EU-wide eco-labelling scheme that supplemented (rather than supplanting) member states' national eco-labelling schemes. The UK is one of several EU member states which has relied solely on the EU eco-label. However, the EU eco-labelling scheme continues to suffer from relatively low participation rates; these are at least partly due to low recognition rates among the general public, who tend to know more about their national eco-label scheme where such a scheme exists. However, the popularity of

---

[50] See also Wurzel, Zito & Jordan (2013).
[51] Interview, German Environment Ministry official, 2001
[52] E.g. Howlett (1991); Salamon (2002), at 24.

national eco-labelling schemes has also undergone certain ups and downs over time. Table 4.1 illustrates that the German Blue Angel eco-labelling scheme was significantly more popular during the 1980s/1990s than it is in the 2010s.

During the early 1990s, a rapid global diffusion of eco-labels took place, although not all eco-labelling schemes became successful.[53] The Dutch and –to a lesser degree – the Austrian eco-labelling schemes still suffer from relatively low participation, while some countries (e.g. the UK) decided against the adoption of national eco-labelling schemes. The German Blue Angel scheme acted as a catalyst, but eco-labelling scheme followers did not simply copy it. For example, the Austrian, Dutch and EU eco-labelling schemes have put more emphasis on sophisticated life-cycle analysis, while the German eco-labelling scheme has relied more strongly on simplicity with the aim of communicating clear messages to consumers and participants. Importantly, the various national eco-labelling schemes reflect national preferences: the German scheme puts a lot of emphasis on products (and services); Austria pioneered a tourism eco-label; and the Dutch were first to include the food and flower sectors.

4.4.1.2 Environmental Management Schemes

In 1993 the EU adopted EMAS, which became applicable in 1995. The EU negotiated EMAS at about the same time as the ISO discussed the 14001 standard, which created a similar, although less demanding, voluntary EMS that has worldwide recognition. The UK, which already had national standards for domestic EMSs in place, initially supported both the EU's EMAS and the ISO 14001. However, while the ISO 14001, which closely resembled the British national EMS standards, has achieved a relatively high adoption rate in the UK, the EU's EMAS has not done so – partly because its requirements significantly exceeded the British Standard Institution/ISO standards. Austria and Germany were initially highly sceptical about EMAS for the following two main reasons: First, EMAS constitutes a procedural policy instrument which initially did not fit easily with the dominant Austrian and German environmental regulatory styles, which are both characterised by a high number of substantive policy measures derived from the best available technology (BAT) (*Stand der Technik*) principle. Second, Austrian and German environmental policy makers were initially weary that self-regulatory environmental management systems (in the form of EMAS and ISO 14001) might weaken substantive environmental standards. However, by the 2000s, about 70 percent of all registered EMAS sites were in Austria and Germany. One reason for the surprising popularity of EMAS in Germany is the fact that the German government has combined EMAS with traditional regulatory instruments (e.g. by reducing the environmental regulatory burden in the form of fewer inspections for companies that successfully participate in EMAS).

---

[53] Jordan et al. (2004).

4.4.1.3 Usage Patterns of Voluntary Agreements

Overall, the adoption of voluntary agreements has grown significantly since the 1970s despite concerns about their effectiveness.[54] The most popular type of voluntary agreement within the EU is the negotiated agreement. In terms of its design and usage, however, significant jurisdictional differences exist for this particular sub-type of voluntary agreement. Since the 1990s, most Dutch voluntary agreements have taken the form of legally binding covenants.[55] The Dutch covenants, which are negotiated between industry and government within a fairly formalised process, stipulate monitoring requirements and are – at least in theory – enforceable through the courts. All German voluntary agreements are non-binding, although they are often negotiated in the 'shadow of the law' and thus are often proposed by industry as a means of pre-empting regulation.[56] Compared to the Dutch covenants, Austrian and German voluntary agreements are developed in a much more informal manner and sometimes even fail to stipulate explicit monitoring requirements, although notable exceptions exist. In the UK, relatively few (negotiated) voluntary agreements exist, and most of the ones that do are non-binding and very flexible.

The adoption of voluntary agreements at EU level has also remained very low. By the early 2000s, despite considerable efforts by the EU Commission to increase their uptake, only twelve EU-wide environmental voluntary agreements had been adopted.[57] EU-wide voluntary agreements are normally negotiated between the Commission and companies. Once agreement has been reached between these two actors, the results are communicated to the Council and European Parliament (EP); the latter has traditionally been hostile towards the adoption of EU-wide voluntary agreements for legitimacy and transparency reasons, which helps to explain their low uptake. Since EU-wide voluntary agreements are not listed in the EU Treaties as possible EU policy instruments, they have to be adopted outside the formal EU decision-making procedures. The EP therefore lacks a clearly defined role in the adoption process of EU-wide voluntary agreements, which weakens the legitimacy of this type of policy instrument on the EU level. Moreover, as voluntary agreements cannot be enforced under the EU Treaties, there are serious concerns about free-riding, which is also a potential weakness of member state voluntary agreements. However, these concerns are magnified on the EU level because a much larger number of actors are usually involved, thus making monitoring more challenging while – in terms of competition – more is at stake in the much larger Single European Market (SEM) than in member states' national markets.

---

[54] OECD (2003).
[55] E.g. Mol, Lauber & Liefferink (2000).
[56] Wurzel, Zito & Jordan (2013).
[57] CEC (2002).

4.4.2 *Usage Patterns of Market-Based Instruments*

4.4.2.1 Eco-Taxes

Japan and Germany adopted some of the world's first eco-taxes in the early 1970s.[58] However, although Germany adopted a wastewater levy in 1974, it was actually not fully implemented until the early 1980s.[59] A more systematic use of eco-taxes occurred in Europe – first in the Netherlands and then in the Nordic countries – in the 1980s and 1990s.[60] The UK adopted major national eco-taxes only from the 1990s onwards.[61]

There is wide variation in the type of eco-taxes adopted in the five jurisdictions assessed in this chapter. Between the early 1990s and the early 2000s, the Netherlands adopted the most comprehensive range of eco-taxes. Germany introduced a major ecological tax reform in 1998 and a raft of additional eco-taxes in 2010. In other words, Germany developed this particular policy instrument further in the late 1990s. Austria and the UK have adopted more modest eco-taxes in a narrower range of sectors.

The Dutch eco-taxes evolved from environmental charges and levies, which were used to support specific regulatory efforts, to more encompassing general eco-taxes; German eco-taxes evolved more gradually, and very much in an ad hoc fashion, until the adoption of the ecological tax reform in 1998. Austria set up a Tax Reform Commission, which reported on a possible ecological tax reform; however, this was not adopted. Instead the Austrian government continued to fine-tune existing eco-taxes while incrementally adding new ones. The UK innovated with hypothecated eco-taxes and a wide range of eco-taxes in the 1990s, but has since fallen behind again. Germany, the Netherlands and the UK innovated with eco-taxes that were explicitly linked to a reduction in labour costs (e.g. in the form of reduced national insurance and/or pension contributions). Although they tend to emit large volumes of greenhouse gasses, high energy users that are in direct competition with companies abroad have been granted generous exemptions from eco-taxes in Austria, Germany, the Netherlands and the UK. Despite the European Commission' strong push for an EU-wide carbon dioxide/energy tax in the early 1990s, the EU has so far failed to adopt EU-wide eco-taxes. The main reason for this is the unanimity requirement for the adoption of all taxes on the supranational EU level, which the UK has opposed on sovereignty grounds.

4.4.2.2 Emissions Trading

At the insistence of the USA, the 1997 Kyoto Protocol stipulated emissions trading (and other flexible instruments such as Joint Implementation (JI) and the Clean

---

[58] Andersen (1996); Andersen & Sprenger (2000).
[59] E.g. Wurzel, Zito & Jordan (2013).
[60] Andersen (1996); Andersen & Sprenger (2000), at 27.
[61] OECD (1993).

Development Mechanism (CDM)) as a possible policy instrument for achieving the desired greenhouse gas reductions. The EU and most of its member states were initially strongly opposed to emissions trading, although they eventually gave in to American pressure.[62] Only the UK and Denmark, as well as – to a much lesser degree – the Netherlands experimented with national emissions schemes prior to the setting up of the EU ETS.

The EU ETS, which distinguished a first trading phase (2005–07) and a second trading phase (2008–12), was initially a highly decentralised ETS for CO_2 emissions from stationary sources. During the revision of the EU ETS, which centralised the rules of the scheme, a third trading phase (2013–20) was added. The Commission acted as an emissions trading policy instrument entrepreneur. Frustrated by the lack of progress of its carbon dioxide/energy tax proposal, which had been vetoed by the UK, and concerned about competing national ETSs, the Commission proposed an EU-wide emissions trading scheme in a Green Paper in 2000. It was strongly supported in its efforts by the majority of member states (including the UK and the Netherlands) and the EP. The German government remained opposed to emissions trading until well into the adoption phase of the EU ETS. Not until after the 2002 German national elections, in which the Greens gained a significant number of seats in parliament, did the German government start to support the adoption of the EU ETS.

Austria and Germany initially acted as emissions trading laggards. Gerhard Schröder, the German Chancellor at the time, almost torpedoed the adoption of the EU ETS. However, Austria and Germany (as well as the Netherlands and UK) have since supported the tightening and centralisation of the rules of the EU ETS. In the second and third trading phases Germany has made wide use of auctioning, while Austria recognised that the EU ETS had become an essential policy instrument for achieving its Kyoto Protocol greenhouse gasses reduction target.

4.4.3 *Environmental Regulations*

Gunningham et al. have argued that '[t]he dominant government response [to the arrival of environmental issues on the political agenda], ... has been the application of "direct" or "command and control" regulation designed to prohibit or restrict environmentally harmful activities'.[63] Regulation remains important even in jurisdictions where NEPI use is high, such as in the Netherlands and Germany. In short, there has not been a wholesale switch to NEPIs. Instead NEPIs seem to have *supplemented*, rather than *supplanted*, regulatory tools.[64] NEPIs often 'fill in the cracks' not covered by regulation or deal with emerging issues like climate change,

---

[62] Grubb, Vrolijk & Brack (1999); Wurzel (2008).

[63] Gunningham, Grabosky & Sinclair (1998), at 38.

[64] Interviews, Member State official, 2017; see also Mol, Lauber & Liefferink (2000); Holzinger & Knill (2003); Wurzel, Zito & Jordan (2013).

which are difficult to tackle with regulations. Moreover, many NEPIs also require regulations, for example, to set the rules for their operation. Gunningham et al. have therefore argued that hybrid instruments which are made up of both NEPIs and regulations have emerged.[65]

Importantly, regulation remains the mainstay of EU environmental policy despite the fact of the reduction in the adoption rate for regulations that has occurred in the 2010s. The EU has adopted a relatively large number of traditional environmental regulations (in the form of directives, regulations and decisions); however, it has struggled greatly to develop a popular eco-labelling scheme, it has been very slow in adopting voluntary agreements, and it has failed to clear the unanimity requirement (among member states) for EU-wide eco-taxes. The only strong evidence of entrepreneurial NEPI influence at the EU level is the existence of emissions trading. Relatively few EU level NEPIs exist, despite political commitments to adopt more of them (see the Commission's 2001 White Paper on *European Governance*.[66] One important reason for this is that, while member states tend to support the adoption of common policy objectives at the EU and/or international level, they are keen to reserve the right to determine the instruments of achieving these objectives.

Environmental regulation has taken on an important supporting role for the adoption of many NEPIs. This is the case in particular for market-based policy instruments such as emissions trading and eco-taxes. Gunningham et al. have pointed out that '[s]ome economic instruments, such as taxes and charges, are high on coercion and low on prescription'.[67] It was arguably only due to the 'shadow of the law'[68] that a large number of voluntary agreements were adopted in Germany and – to a somewhat lesser degree – in Austria in the 1990s. The Dutch *covenants* constitute quasi-legal agreements. Many NEPIs, therefore, rely on regulation.

A cross-jurisdictional trend towards greater flexibility in environmental regulations is discernible across all five jurisdictions where traditional environmental regulations have become less rigid and more flexible. Since the early 1990s, EU and member state environmental regulations have become more flexible and innovative. In other words, European environmental regulations have become 'smarter'.[69] While remaining dominant in most jurisdictions, regulation has evolved over time. It is often adopted in 'smarter' or more 'light-handed' forms when compared with the early 'command-and-control' regulations that prevailed in the 1970s.[70] For example, in the UK, the national integrated pollution control regime (i.e. regulation) shares many similarities with what continental states describe as negotiated agreements underpinned by the law (e.g. the Dutch covenants).

---

[65] Gunningham, Grabosky & Sinclair (1998).
[66] CEC (2001).
[67] Gunningham, Grabosky & Sinclair (1998), at 391.
[68] Héritier & Lehmkul (2008).
[69] Gunningham, Grabosky & Sinclair (1998).
[70] Ibid.; Rengeling & Hof (2001).

Gunningham et al.[71] use 'smart regulation' as an umbrella term that subsumes a wide range of NEPIs including eco-labels, environmental management systems, voluntary agreements, eco-taxes and ETSs. In this chapter we distinguished conceptually between environmental regulations and NEPIs, although we also paid attention to the hybridisation of 'old' instruments (i.e. regulations) and 'new' tools (i.e. NEPIs). This has helped us to identify the changing patterns of the use of traditional regulation and NEPIs, as well as the changing policy instrument mixes which they have produced, in four member states (Austria, Germany, the Netherlands and the UK) and on the EU level.

4.5 LEADERS, FOLLOWERS AND LAGGARDS

Table 4.2 summarises the adoption patterns for NEPIs in the five European jurisdictions assessed in this chapter. Importantly, the analytical descriptors – leader, follower and laggard – used in Table 4.2 (like in Table 4.1 earlier in this chapter) offer a relative assessment between the five jurisdictions listed, rather than an absolute ranking. NEPI leaders, as defined in Table 4.2, are jurisdictions which have first made significant use of a particular NEPI type. Followers eventually catch up with the leaders, while laggards either fail to adopt a particular type of NEPI or utilise it only in an insignificant manner.

Leaders may eventually be overtaken by followers in the use of a particular NEPI. In order to achieve a better domestic fit and/or to improve its effectiveness, followers often modify NEPIs which they have transferred voluntarily from another jurisdiction or which have been 'coercively' transferred to their jurisdiction. Germany acted as an early leader for eco-labelling schemes when it set up its domestic Blue Angel scheme in 1978. However, it was primarily the general idea behind this NEPI type – rather than the specific rules of the original German eco-labelling scheme – which was transferred to other jurisdictions (including the EU). Austria (1990), the Netherlands (1992) and the EU (1992) all followed the German example only after a considerable time lag, but without simply copying the German Blue Angel scheme. The followers all adopted domestic eco-labelling schemes that exhibited some novel features reflecting both domestic priorities and the desire to achieve a good domestic fit and increased effectiveness.

The UK set the pace for EMSs with its innovative BS 7750 voluntary standard, which also influenced the EU's EMAS. Germany and Austria initially opposed the EU's proposal for EMAS. However, once the EU had adopted EMAS, the followers Germany and Austria quickly became the highest EMAS users, leaving the UK well behind in EMAS registrations. Germany and Austria lobbied hard both to make EMAS more ambitious and to extend its scope from private companies to public institutions such as government ministries and agencies. The Netherlands never

---

[71] Gunningham, Grabosky & Sinclair (1998).

TABLE 4.2 *Leaders, Followers and Laggards*

Types of NEPIs		Leaders	Followers	Laggards
Informational Instruments	Eco-Label	Germany	Austria, EU Netherlands	UK
	EMS/EMAS	UK, EU	Germany, Austria	Netherlands
Voluntary Instruments	Voluntary Agreements	Germany, Netherlands	Austria, UK	EU
Market-Based Instruments	Eco-Taxes	Netherlands, Germany	Austria, UK	EU
	Emissions Trading	UK, EU, Netherlands	Austria, Germany	—

Source: Adapted from Wurzel, Zito and Jordan (2013) with updates.

developed a national EMS and has made very little use of EMAS. Dutch firms have instead preferred the ISO 14001 standard which, although it is environmentally less ambitious, nevertheless has a global reach.

The Netherlands and Germany have been early innovators and heavy users of voluntary agreements. However, the core features of Dutch and German voluntary agreements differ significantly.[72] The Netherlands makes wide use of sector-wide covenants, which are legally binding and monitored by independent institutions. German voluntary agreements are non-binding agreements (*Selbstverpflichtungen*, or self-binding agreements), which are usually put forward by companies with the aim of pre-empting government regulation. Only a small number of German voluntary agreements have been monitored by an independent verifier. Austria became a voluntary agreements follower in the 1980s when it partly emulated the non-binding German voluntary agreements (rather than the Dutch covenants). However, after extensive usage in the 1990s there was a steep decline in voluntary environmental agreements usage in Austria in the 2000s. Constitutional reasons have prevented Austria and Germany from adopting legally binding voluntary agreements such as the Dutch covenants. The UK has adopted only a moderate number of voluntary agreements. The lack of a clear Treaty base for voluntary agreements prevented EU Commission officials from trying to emulate the Dutch covenants.[73] Moreover, concerns about free riders are magnified at the EU level due to the significantly large number of actors involved and the large size of the SEM which makes arguments about fair competition more important.

Importantly, although the early leaders (Germany and the Netherlands) have remained the largest users of voluntary agreements within the EU, their popularity nevertheless declined – even in the two innovator states – in the 2000s, and it has

---

[72] E.g. Wurzel, Zito & Jordan (2013).
[73] Interviews, 2001 and 2004

decreased even more in the 2010s (see Table 4.2). One important explanation for the declining popularity in the use of voluntary agreements is that they have been supplanted by other NEPIs and/or traditional environmental regulations. For example, the EU ETS rendered obsolete some Dutch, German and Austrian voluntary agreements aimed at reducing greenhouse gas emissions, while legislation has overwritten the EU-wide voluntary agreement on the reduction of CO_2 emissions by the car industry.

The Netherlands was an early eco-taxes innovator, and has consistently made use of a wide range of eco-taxes. Germany was also an early eco-taxes pioneer when it adopted the 1976 waste water levy;[74] however, although Germany strongly supported the adoption of an EU-wide carbon dioxide/energy tax, its domestic use of eco-taxes remained moderate until the adoption of an ecological tax reform in 1998. Austria and the UK were followers that made use of eco-taxes only belatedly when compared with the Netherlands and Germany. However, the UK adopted some innovative eco-taxes (e.g. the fuel escalator) which influenced the eco-tax thinking of environmental policy makers in Austria, Germany and the Netherlands. The EU is an eco-taxes laggard because it failed to adopt a supranational eco-tax. One important reason for this is because the UK has consistently vetoed taxes at the EU level on sovereignty grounds. Whether the UK's exit from the EU may pave the way for the adoption of EU-wide eco-taxes, however, remains doubtful. Member states other than the UK (e.g. traditionally Spain and (more recently) the Visegrád countries) have traditionally also been opposed to EU-wide eco-taxes for fear that this may put them at an economic disadvantage. Because the EU failed to adopt a supranational carbon dioxide/energy tax, a group of like-minded European countries met between 1994 and 1998 to discuss the practical experience with – and possible transnational coordination of – national eco-taxes, although little progress was made.

The UK became an emissions trading innovator when it set up Europe's first domestic ETS for greenhouse gasses in 2002. Up to the early 2000s, the Netherlands had experimented only with rudimentary small-scale domestic ETSs (e.g. for manure). In 2003 the EU adopted the world's first supranational ETS for CO_2 emissions. Austria and Germany, which initially acted as reluctant ETS followers, eventually became supportive of a relatively ambitious EU ETS.

Importantly, emissions trading constitutes the only NEPI type assessed in this chapter for which the EU has acted as a genuine innovator. However, the EU was only a reluctant emissions trading pioneer.[75] It is unlikely that the EU ETS and UK ETS 'could have been adopted as quickly as they were without America setting a domestic example and insisting on emissions trading in the 1997 Kyoto Protocol'.[76]

---

[74] Andersen (1994).
[75] Wurzel (2008).
[76] Ibid., at 5.

Germany (eco-labels, voluntary agreements and – to a lesser degree – eco-taxes) and the Netherlands (voluntary agreements, eco-taxes and – to a lesser degree – emissions trading) have both acted as leaders for three types of NEPIs, albeit different ones. The UK has acted as leader for two types of NEPIs (EMS and emissions trading), and the EU has acted as leader for only one NEPI type (emissions trading). Austria is the only jurisdiction which has so far failed to become a leader for any of the five NEPIs assessed in this chapter. Nevertheless, Austria has always followed the NEPIs leader, thereby avoiding the status of NEPIs laggard. Similarly, Germany always acted as a follower in cases where it was not an NEPIs innovator. The UK and EU acted as laggards for one (eco-labels) and two (voluntary agreements and eco-taxes) NEPI types respectively. The EU had adopted only about a dozen EU-wide voluntary agreements by the 2010s; this amounts to an insignificant use of voluntary agreements when compared with the large number of voluntary agreements adopted by some individual member countries, especially the Netherlands and Germany. The Netherlands could be classified as an EMS laggard because it made only insignificant use of EMAS and failed to adopt a domestic EMS. However, Dutch firms have made relatively wide use of the ISO 14001.

The analytical classification of leader, follower and laggard can mask significant changes in the use of particular NEPIs that may take place in a certain jurisdiction over time. Table 4.2 illustrates the changing usage patterns for the different types of NEPIs in each of the five jurisdictions. It shows that once a certain type of NEPI has been adopted by a particular jurisdiction, its usage tends to change only gradually. The usage of voluntary agreements seems to have peaked in the Netherlands, Germany and Austria in the 1990s. Similarly, the drive for eco-taxes lost momentum in the 2000s. The EU ETS, which became operational in 2005, is the only NEPI whose usage expanded significantly in the 2000s (and early 2010s). These findings provide empirical evidence for the claim that 're-tooling' processes take place over time.[77]

This chapter has assessed the role of the EU in terms of the ability to innovate 'old' and 'new' policy instruments. However, there is a broader question about what impact the supranational EU has had on the member state policy instrument mixes. Because the EU has often been the NEPIs laggard, its role in transforming the policy mix of the member states has largely focused on climate change and particularly the ETS, which involved all member states having to reshape their climate governance approach to some extent. With the exception of the ETS and traditional regulation, the EU has not added particularly to the member states' policy instrument mixes. This is not to downplay the EU role, as since the 1970s the EU has issued a relatively large number of supranational environmental laws creating common standards and objectives that its member states have had to follow. Over time, those regulations have generated their own innovations. Accordingly, the Water Framework Directive was notable for creating certain processes that

---

[77] Hood (1983), at 126.

enhanced dialogue with society.[78] This harkens back to the notion of 'smart regulation', where the instrument contains several different mechanisms.[79] By contrast, the regulatory framework that protects the operation of the SEM means that the EU, in the form of state aid and competition policy concerns, has had a prohibitive – or at least limiting – impact on member states seeking to expand their mixes with, for example, subsidies and other incentives. For example, the German ecological tax reform could be adopted by the German parliament only after it had been altered to comply with instructions from the European Commission.[80]

4.6 CONCLUSION

This chapter assessed how policy instrument mixes appear in the real world as opposed to the theoretical world of (especially economic) textbooks. It has shown how the mixes have changed over a period of almost five decades by focusing on traditional regulation and three different types of NEPIs – informational instruments, voluntary agreements and market-based instruments (eco-taxes and emissions trading) in five jurisdictions, namely Austria, Germany, the Netherlands, the UK and the EU. The empirical focus on NEPIs and traditional regulation allowed us to analyse whether new modes of *governance* have supplanted traditional methods of *government* (i.e. command-and-control regulation) in different European jurisdictions, or merely supplemented them. Clearly there has been a significant uptake of NEPIs in all five jurisdictions. However, although important differences remain between the specific policy instrument mixes across different jurisdictions, they also remain within one and the same jurisdiction over time. A certain degree of hybridisation of 'old' and 'new' instruments exists towards what Gunningham et al.[81] have termed 'smart regulation'.

Seeking to specify what is an optimal mix a priori is an important and interesting theoretical topic, but it does not really explain why instruments are actually mixed in the real world of policy and politics. Taking a longitudinal empirical perspective has allowed us to identify leaders, followers and laggards for different types of NEPIs. It has revealed policy instrument diffusion patterns that are influenced by jurisdictional path-dependencies that can lead to very different sequencing of the adoption of certain types of NEPIs. For example, Germany adopted a national eco-label scheme in 1978, while Austria (1991), the Netherlands (1992) and EU (1992) followed the German example only after a considerable time lag. By the late 2010s the UK still had not adopted a national eco-labelling scheme; instead, the UK continues to rely on the EU eco-label, at least until it exits from the EU.

---

[78] Page & Kaika (2003).
[79] Gunningham, Grabosky & Sinclair (1998).
[80] Wurzel, Zito & Jordan (2013).
[81] Gunningham, Grabosky & Sinclair (1998).

In all five jurisdictions, complex policy instrument mixes have emerged that blend the 'old' and the 'new' in puzzlingly different ways. In the language of the (new modes of) governance literature, what we seem to be witnessing is arguably best described as 'governance-cum-government'. However, there is little empirical evidence for the widely held claim that non-coercive policy instruments (such as informational instruments and voluntary agreements) should be used before coercive instruments are adopted. This has created a complex mixture of (policy instrument) continuity and change in different jurisdictions, which may use one and the same type of policy instrument differently.

Classic administrative theory largely considered policy instrument selection as 'a neutral, even scientific exercise'.[82] However, Hood has flagged up that:

> Given that all feasible alternatives cannot be systematically approached, it follows that in many cases politics will play a large part in the selection of tools for the job ... Indeed, very commonly it is the instrument selected for reaching a policy aim that is far more contentious than the aim itself.[83]

The jurisdictional context variable (e.g. national or sectoral policy styles) plays an important role for both the selection and specific design of policy instruments.

Once established within a particular jurisdiction, policy instruments tend to retain their relative role in the policy instrument mix for a considerable period of time. However, as Smith and Ingram have pointed out, in the real world of politics and power, 'the tool box is not constant. It changes over time as a result of the invention of new tools, national and international developments, and the interactions of the supranational institutions'.[84] Bridging the theoretical accounts, which are often informed by economic cost-effectiveness considerations, with more political and institutional approaches of the evolution of policy instrument mixes in the real world should help us to better understand why it is often difficult to move from 'not-so-smart' policy instrument mixes towards smart policy instrument mixes.

REFERENCES

Abbott, K. 2012. 'The Transnational Regime Complex for Climate Change'. *Environment and Planning C: Government and Policy* 30(4), 571–590.

Andersen, M. S. 1994. *Governance by Green Taxes. Making Pollution Prevention Pay.* Manchester, Manchester University Press.

Andersen, M. S. & Sprenger, R. -U. (eds.). 2000. *Market-Based Instruments for Environmental Management: Politics and Institutions.* Cheltenham, Edward Elgar.

Bähr, H. 2010. *The Politics of Means and Ends.* Farnham, Ashgate.

Baumol, W. J. & Oates, W. E. 1979. *The Theory of Environmental Policy. Externalities, Public Outlays and Quality of Life.* Englewood Cliffs, NJ, Prentice-Hall.

---

[82] Bemelmans-Videc, Rist & Vedung (1998), at 268.
[83] Hood (1983), at 136.
[84] Smith & Ingram (2002), at 598.

Bemelmans-Videc, M. -L., Rist, R. C. & Vedung, E. 1998. *Carrots, Sticks and Sermons. Policy Instruments and their Evaluation*. New York, Transaction Publishers.

Börkey, P. & Lévêque, F. 1998. Voluntary Approaches for Environmental Protection in the European Union, report prepared for the OECD Environment Directorate, ENV/EPOC/ GEEI(98)29/final. Paris, Organisation for Economic Co-operation and Development.

Burrows, P. 1979. *The Economic Theory of Pollution*. Oxford, Martin Robertson.

CEC. 2002. Environmental Agreements at Community Level, COM (2002) 412 final. Brussels, Commission of the European Communities.

    2001. European Governance: A White Paper, COM(2001) 428 final. Brussels, Commission of the European Communities.

    1996. Communication from the Commission to the Council of the European Parliament on Environmental Agreements, COM(96) 561 final. Brussels, Commission of the European Communities.

De Bruijn, H. & Hufen, H. 1998. 'The Traditional Approach to Policy Instruments'. In Peters, B. G. & Van Nispen, F. (eds.), *Public Policy Instruments: Evaluating the Tools of Public Administration*. Cheltenham, Edward Elgar, 11–32.

Doern, G. & Wilson, V. (eds.). 1974. *Issues in Canadian Public Policy*. Toronto, Macmillan of Canada.

EEA. 1997. *Environmental Agreements*, Copenhagen, European Environmental Agency.

Ellerman, D., Buchner, B. & Carraro, C. (eds.). 2007. *Allocation in the European Emissions Trading Scheme*. Cambridge, Cambridge University Press.

Faure, M., Vervaele, J. & Weale, A. (eds.). 1994. *Environmental Standards in the European Union in an Interdisciplinary Framework*. Antwerp, Maklu.

Finer, S. 1970. *Comparative Government*. Harmondsworth, Penguin.

Gunningham, N., Grabosky, P. & Sinclair, D. 1998. *Smart Regulation: Designing Environmental Policy*. Oxford, Oxford University Press.

Grubb, M., Vrolijk, C. & Brack, D. 1999. *The Kyoto Protocol*. London, Earthscan.

Héritier, A. & Lehmkuhl, U. 2008. 'The Shadow of Hierarchy and New Modes of Governance: Sectoral Governance and Democratic Government'. *Journal of Public Policy* 28(1), 1–17.

Hood, C. 1983. *The Tools of Government*. Basingstoke, Macmillan.

Holzinger, K. & Knill, C. 2003. 'Marktorientierte Umweltpolitik – ökonomischer Anspruch und politische Wirklichkeit'. *Politische Vierteljahresschrift* 43, special issue, 232–255.

Holzinger, K., Knill, C. & Schäfer, A. 2003. 'Steuerungswandel in der europäischen Umweltpolitik'. In Holzinger, K., Knill, C. & Lehmkuhl, D. (eds.), *Politische Steuerung im Wandel: Der Einfluss von Ideen und Problemstrukturen*. Opladen, Leske + Budrich, 103–129.

Howlett, M. 1991. 'Policy Instruments, Policy Styles and Policy Implementation: National Approaches to Theories of Instrument Choice'. *Policy Studies Journal* 19(2), 1–21.

Howlett, M. & Rayner, J. 2004. '(Not so) "Smart Regulation"? Canadian Shellfish Aquaculture Policy and the Evolution of Instrument Choice for Industrial Development'. *Marine Policy* 28(2), 171–184.

Jordan, A. 2002. *The Europeanization of British Environmental Policy. A Departmental Perspective*. London, Palgrave.

Jordan, A., Wurzel, R. & Zito, A. 2007. 'New Modes of Environmental Governance. Are "New" Environmental Policy Instruments (NEPIs) Supplanting or Supplementing Traditional Tools of Government?'. *Politische Vierteljahresschrift* 39, special issue, 283–298.

Jordan, A., Wurzel, R. K. W. & Zito, A. R. 2005. 'The Rise of "New" Policy Instruments in Comparative Perspective: Has Governance Eclipsed Government?'. *Political Studies* 53 (3), 477–496.

Jordan, A., Wurzel. R. K. W., Zito, A. & Brückner, L. 2004. 'Consumer Responsibility-Taking and Eco-Labelling'. In Micheletti, M., Follesdal, A. & Stolle, D. (eds.), *Politics, Products and Markets*. New Brunswick, NJ, and London, Transaction Publishers, 161–180.

Kitschelt, H., Lange, P., Marks, G. & Stephens, J. (eds.). 2003. *Continuity and Change in Contemporary Capitalism*. Cambridge, Cambridge University Press.

Lascoumes, P. & Le Galès, P. 2007. 'Introduction: Understanding Public Policy Through Its Instruments'. *Governance* 20(1), 1–22.

Luhmann, N. 1984. *Soziale Systeme. Grundriß einer allgemeinen Theorie*. Frankfurt, Surhkamp Verlag.

Majone, G. 1976. 'Choice among Policy Instruments for Pollution Control'. *Policy Analysis* 2 (4), 589–613.

Mol, A., Lauber, V. & Liefferink, D. 2000. *The Voluntary Approach to Environmental Policy*. Oxford, Oxford University Press.

OECD. 2003. *Voluntary Approaches for Environmental Policy*. Paris, Organization of Economic Co-operation and Development.

   1999. *Economic Instruments for Pollution Control and Natural Resources Management in OECD Countries*. Paris, Organization of Economic Co-operation and Development.

   1997. *Eco-labelling: Actual Effects of Selected Programmes*. Paris: Organization of Economic Co-operation and Development.

   1994. *Managing the Environment: The Role of Economic Instruments*. Paris: Organization of Economic Co-operation and Development.

   1993. *Taxation and the Environment: Complementary Policies*. Paris, Organisation for Economic Co-operation and Development.

   1991. *Environmental Labelling in OECD Countries*. Paris: Organisation for Economic Co-operation and Development.

Page, B. & Kaika, M. 2003. 'The EU Water Framework Directive: Part 2. Policy Innovation and the Shifting Choreography of Governance'. *Environmental Policy and Governance* 13 (6), 328–343.

Rengeling, H. & Hof, H. (eds.). 2001. *Instrumente des Umweltschutzes im Wirkungsverband*. Baden-Baden, Nomos.

Rhodes, R. 1996. 'The New Governance: Governing without Governance'. *Political Studies* 44(4), 652–667.

Salamon, L. (ed.). 2002. *The Tools of Government*. Oxford, Oxford University Press.

Siebert, H. 1976. *Analyse der Instrumente der Umweltpolitik*. Göttingen, Vandenhoeck.

Smith, S. & Ingram, H. 2002. 'Policy Tools and Democracy'. In Salamon, L. (ed.), *The Tools of Government. A Guide to the New Governance*. Oxford, Oxford University Press, 565–584.

Stoker, G. 1998. 'Governance as Theory: Five Propositions'. *International Social Science Journal* 50(155), 17–28.

Vedung, E. 1997. 'Policy Instruments: Typologies and Theories'. In Bemelmans-Videc, M. L., Rist, R. C. & Vedung, E. (eds.). *Carrots, Sticks and Sermons: Policy Instruments and Their Evaluation*. New Brunswick/London, Transaction Publishers. 21–58.

Vogel, D. 1986. *National Styles of Regulation*. Ithaca, NY, Cornell University Press.

Wurzel, R. K. W. 2008. *The Politics of Emissions Trading in Britain and Germany*. London, Anglo-German Foundation.

Wurzel, R. K. W., Zito, A. & Jordan, A. 2013. *Environmental Governance in Europe: A Comparative Analysis of New Environmental Policy Instruments*. Cheltenham, Edward Elgar.

PART II

Fisheries and Forestry

5

The Pursuit of Good Regulatory Design Principles in International Fisheries Law

What Possibility of Smarter International Regulation?

Richard Barnes

5.1 INTRODUCTION

The regulation of international fisheries is in need of improvement. Many fisheries are marred by unsustainable levels or forms of exploitation. Given the limits of traditional regulatory approaches, the issue this chapter is concerned with is whether or not smart mix approaches offer hope of improvements. Since regulatory approaches are often quite context-specific, this chapter focuses on the potential for smart regulatory design principles to be used to shape fisheries regulation. The contribution made by this chapter to this debate is to map out some of the structural challenges presented by the application of smart mix approaches to an issue that is governed at root by agreements and practices at the international level.

International fisheries regulation seems to favour a toolkit of regulatory techniques, as articulated in the Food and Agriculture Organization of the United Nations (FAO) Code of Conduct for Responsible Fisheries.[1] Traditionally, fisheries have been the domain of State or public regulation, and international law has left this up to individual States to determine.[2] This has tended to favour command-and-control–type rules that determine what techniques or equipment can be used, as well as when and where, to take fish out of the sea. In recent decades, there has been increasing use of rights- or market-based measures, drawing the private sector into the mix, as well as a range of soft law or policy instruments. We are moving towards a more complex range of regulatory techniques to govern fisheries management. At one level this appears to be consistent with 'smart regulatory approaches', which

---

[1] The FAO Code of Conduct for Responsible Fisheries (available at www.fao.org/docrep/005/ v9878e/v9878e00.htm) and International Plans of Acts for Reducing Incidental Catch of Seabirds in Longline Fisheries, for Conservation and Management of Sharks, for the Management of Fishing Capacity, and to Prevent, Deter, and Eliminate Illegal, Unreported and Unregulated Fishing comprise a toolkit of approaches for fisheries management. See www .fao.org/fishery/code/ipoa/en.

[2] Barnes (2009).

favour instrument mixes. However, the extent to which such approaches are consistent with 'good' regulatory design principles, i.e. those that optimise or direct the mix to ensure that certain outcomes – such as the efficient or fair use of resources – are achieved is not clear. This same uncertainty applies to the question of the extent to which international fisheries law can and should embrace notions of good regulatory design. To date there has been little consideration of the 'smartness' of the regulatory mix as it applies to international fisheries;[3] instead, the focus has been on domestic fisheries.[4]

There is a substantial body of literature that explores the effectiveness of different regulatory approaches, tools and techniques.[5] Labelled variously as smart regulation, responsive regulation, risk-based regulation or instrument choice theory, the common focus of this literature is on articulating and evaluating the design of regulatory principles. Such principles include: compatible instrument mixes; low level intervention strategies; scaled interventions; empowering surrogate regulators; maximising win-win opportunities; sensitivity to regulatory attitudes, cultures and institutional conditions; conducting impact assessments; and engaging in performance-responsive governance.

Much of this literature appears to be focused at the level of local or national regulation, with Howlett et al. only recently observing the need to diagnose problems of complexity across and at different regulatory levels,[6] and Gulbrandsen examining the interaction of State and non-State actors in the context of certification schemes.[7] There has been no consideration of how such design principles could or should apply to international fisheries. This should not surprise us since smart regulation embraces innovative forms of social control beyond government, thereby engaging a range of non-state actors – something anathema to traditional international law approaches. And yet for matters of common concern, like fisheries, it appears that a mix of complementary regulatory techniques is desirable, especially if this can be adjusted to suit specific contexts.[8] As this chapter argues, fisheries are fundamentally international; we therefore need to understand how this international level of governance influences the regulatory smart mix. The challenge is great, because when we move to the global level, the aims, control of and implementation of regulatory strategies – for want of a better description – explode.[9] The complexity of objectives/outcomes of regulation is magnified as a multiplicity of States,

---

[3] Gulbrandsen (2012).

[4] See Howlett & Rayner (2004); Baldwin & Black (2008).

[5] See for example: Gunningham & Sinclair (1999); Gunningham & Grabosky (1998); Bothe & Sand (2003); Howlett & Rayner (2007); Stewart (2008); van Gossum, Arts & Verheyen (2010); Baldwin, Cave & Lodge (2012).

[6] See Howlett, Vince & del Rio (2017).

[7] Gulbrandsen (2012).

[8] See Chapter 1, this volume.

[9] As Wiener observes, 'The Olympics of instrument choice are underway, the contest joined at the international level'. Wiener (1999), at 680.

transnational actors, both governmental and civil society, become involved. This makes it difficult to plot and measure not just legal effectiveness, but also problem solving and behavioural effectiveness.[10] Also, the capacity of regulators (States) to reflect upon and adapt international governance mechanisms are hampered due to structural limitations in the way that international law operates. For example, it is quite rational for States to press for higher catch limits, despite knowledge of the harm this may do to stocks, unless they can be sure other States will abide by lower limits that will allow stocks to recover.[11] Despite these challenges, the interface between international, regional and national systems must be accounted for because the effectiveness of fisheries management involves factors that transcend each of these levels of governance. Unless we consider some of these structural issues, then we risk misusing or misapplying smart regulatory approaches.

Section 5.2 in this chapter provides an account of the nature and development of international fisheries law. The key structural features are extrapolated so as to provide a baseline for how more effective smart regulatory approaches can be applied at the level of international fisheries. This section addresses an important gap in the literature, with few scholars having assessed the link between international fisheries law and the requirements of smart regulation.[12] The present analysis shows how different structural and natural parameters influence and constrain regulatory design options at the level of international law. This analysis is continued in Section 5.3, where I examine how certain regulatory design principles measure against the main instruments and institutions of international fisheries law. Again, a range of limits are seen to flow from the structure of international law, although some important initiatives are underway that expand and strengthen the toolkit international fisheries law has to use. This includes a brief survey of the specific instruments used in fisheries management, some of which are increasingly used at the international level, and not merely within the domestic part of a fisheries regime. In the final section some tentative conclusions are offered. First, the legal framework is flexible and broad, so it is largely neutral in the way regulatory approaches are structured. Second, and critically, the sources of international law and limited range of actors constrains the use of certain regulatory techniques, and it is only if we view fisheries issues as both a matter straddling international and domestic frameworks that we can realise the potential for good regulatory mixes. Third, the political nature of the system may limit the use of incentives to improve

---

[10] See Chapter 1 (this volume).

[11] Serdy (2016), at 155.

[12] This is considered by Techera & Klein (2017), chapter 3. However, their focus is on a narrower range of principles (accountability, consistency, proportionality transparency, effectiveness, flexibility). This misses more reflective approaches, and also remains rooted to a general notion of effectiveness. More generally, see Abbot and Snidal, who state that 'New Governance cannot be uncritically transferred to the very different circumstances involved in the international system, where the role of the state is even more attenuated, but it does provide key insights for improving international regulation'. Abbott & Snidal (2009).

fisheries regulation. Finally, whilst international fisheries law has traditionally avoided reflective, adaptive governance, this is growing in practice. In the pursuit of solutions to structural limitations, new opportunities are emerging to engage more directly with non-State actors, and to revise existing management arrangements to accommodate different regulatory techniques, such as impact assessment and the use of market-based approaches.

5.2 THE STATE, STRUCTURE AND DEVELOPMENT OF INTERNATIONAL FISHERIES REGULATION

The most recent FAO report on the state of world commercial fisheries indicates that capture fisheries have been relatively static since the late 1980s, although many regional fisheries have declined.[13] According to the FAO report, 31.4 percent of fish stocks are fished at a biologically unsustainable level (i.e. fish mortality through catch and other factors is higher than new recruitment to the stock), fully fished stocks account for 58.1 percent and under-fished stocks a mere 10.5 percent.[14] The number of stocks fished at unsustainable levels increased dramatically by 16 percent between 1974 and 1989, and has increased steadily since then.[15] Demand for fisheries resources remains high with global population expected to exceed 9 billion people by 2050. However, wild capture supply remains limited and vulnerable. Demand is increasingly met through aquaculture, which overtook the supply of wild catch for human consumption in 2014.[16] Developing economies are increasingly engaged in fisheries capture and export, although China, the United States (US), Japan and the European Union (EU) remain the main actors due to the size of the markets for fish products. This basic data indicates the vulnerable state of fishing, increasing pressures and changing dynamics in the participants and systems for supply of fish products. These factors and dynamics need to be taken into account when designing regulatory systems.[17]

Fundamentally, regulation is required when other modes of organisation fail (e.g. open access, market failures), thereby generating conflict or ineffective or inefficient patterns of use, or when the conduct of activities needs to be brought into line with social priorities.[18] The effectiveness of any regulatory policy will be context-dependent, responding to the nature of the problem, the availability of instruments and institutions, regulatory culture and governance discourse.[19] Therefore, it is important to outline the context of fisheries regulation. Fisheries regulation is in

---

[13] FAO (2016), at 2. Available at www.fao.org/3/a-i5555e.pdf.
[14] Ibid., 5–6.
[15] Ibid., 6.
[16] Ibid., 2.
[17] See generally, Grafton et al. (2010).
[18] Baldwin, Cave & Lodge (2011), chapter 2.
[19] See generally, Gunningham & Grabosky (1998); van Gossum, Arts & Verheyen (2010), at 248.

part the product of the physical nature of the resource base and the ocean environment, in part the product of historical factors that have emerged from an iterative process of regulation and reaction (bricolage), and in part the product of unorchestrated or diffusely orchestrated social and political processes. Cumulatively, these have generated a complex picture of fisheries regulation.

Good regulation must respond to the physical nature and context of the regulatory subject matter. A law requiring the sun to stop shining or tides to stop rising and falling would be futile. Regulation must *fit* the regulatory subject matter. Thus, a first step is to assess the basic nature of marine fisheries. Marine fisheries form an intrinsic part of ocean ecosystems, conditioned by energy flows, temperature, salinity, chemical composition of waters, ocean currents, patterns of food supply and predation. Marine fisheries are a typical common pool resource.[20] They are a diffuse and fungible resource, frequently moving across legal boundaries. A common pool resource is defined by two attributes. Firstly, it is costly to exclude access, and, secondly, the benefits of consumption by one person subtract from the benefits available to others.[21] Historically, this resulted in most marine fisheries being treated as open access regimes. For much of history marine fisheries have been largely unregulated.[22] Open access may be unproblematic when the level of fishing activity is so low and so diffuse as to avoid conflict between fishermen and any threats to the sustainability of stocks. For most of human history, marine fishing has been open access. This presents a challenge for creating win-win scenarios because one person's (or a State's) gain is another's loss. In law, this has been perpetuated under the notion of the freedom of the high seas and reinforced through the perception that marine fisheries were boundless, or, at least, so plentiful as to be incapable of depletion. Open access has also prevailed because, historically, States have lacked the legal authority to regulate much of ocean spaces. Fisheries have been subject to limited protection, and such law as there was focused on managing specific local conflicts, rather than conservation and management in general.[23] Significantly, fish stocks were considered to be inexhaustible, even until the late nineteenth century.[24] And it was only once these assumptions were challenged that regulatory interventions became justified, and controls on fishing activity began to emerge.

Modern efforts to regulate fisheries are in part a response to growing awareness that we are depleting fish stocks through overfishing or destructive fishing practices. Initially, regulation was a response to increased conflicts between users, both within domestic fisheries and between local and distant water fishing fleets. A significant

---

[20] Gardiner, Ostrom & Walker (1990); Barnes (2009), at 2.
[21] Ostrom, Gardner & Walker (1994), at 6.
[22] This is not to ignore localised conflicts in respect of coastal fisheries, which have given rise to exceptional regulatory interventions.
[23] See further Bangert (1999).
[24] Smith (1994), chapter 2.

driver of change was the expansion of exclusive coastal State control over fisheries to meet domestic food, security and economic needs.[25]

At this point it may be observed that regulatory regimes were somewhat contingent on whether fisheries were domestic or international. Domestic fisheries comprise those fisheries that are exclusive to individual States (i.e. located within a narrow band of exclusive coastal waters), and which could be regulated independently by States. International fisheries are those which straddle different jurisdictions, or which migrate through different jurisdictions, or which are located upon the high seas are areas beyond national jurisdiction. A key feature of 'international' fisheries' is that the subject matter cannot be regulated by States in isolation. However, even this distinction is hard to sustain because most fisheries have some international dimension. First, domestic fisheries may be subject to some international regulation through the setting of standards or objectives for conservation and management. Second, the nature of the international legal system is such that international fisheries law is polycentric. Regulation must proceed upon the basis of cooperation and standards resulting from deliberations in multiple fora. As noted in Section 5.2, the fluid and complex interaction between components of marine ecosystems means that fisheries cannot be regulated in isolation from other matters which may also be subject to international regulation (e.g. pollution and protection of the marine environment). Critically, international law is fundamental to setting the limits of national jurisdiction, and hence the sphere of domestic fisheries regulation. Only once the exclusive spatial jurisdiction of States over coastal waters was recognised did the basis for domestic regulation become a realistic possibility.

Since fishing is carried out by private persons, immediate regulatory responsibilities have fallen upon States. Here it appears that fisheries regulation has followed the same trajectory as other subjects of domestic regulation.[26] So, historically, domestic fisheries have been dominated by command-and-control regulation, and in particular input controls, which sought to restrict the location, time, quantity, intensity and types of equipment to be used when fishing. This includes licensing, restrictions on the size and power of fishing boats, gear restrictions (e.g. net sizes, pot numbers), closed areas and limits on fishing time. Such measures are popular because they are easy to design and implement. However, as the collapse in marine fisheries indicates, these approaches have not been successful in securing sustainable fishing practices. Input controls are susceptible to regulatory slide, whereby fishermen channel their efforts into uncontrolled areas in order to circumvent limits on effort. Although input controls may be combined to address this, this still fails to incentivise 'good' behaviour. It requires increasing levels of intervention, which is costly and complex, and may in turn generate hostility and non-compliance. Also, given the nature of fishing, non-compliance is difficult to monitor, detect and

---

[25] See Attard (1987); Kwiatkowska (1989).
[26] See Scott (2008), chapter 4.

penalise. States may simply be unable to support effective enforcement action at sea. They may even be reluctant to take a hard line on compliance because this will generate resistance from domestic fishing interests and it may run counter to domestic socio-economic interests in securing food and employment. Recognition of the limitations of input controls gave rise to interest in output controls, and ultimately rights-based measures (RBM). These tools focus on how many fish can be taken. Usually this is done through the use of annual quotas allocated to fishermen or vessels, or landing limits. During the mid-twentieth century a number of economists began to subject fisheries to critical scrutiny.[27] They challenged the predominantly biological approach to fisheries management and focused instead on how alternative instruments such as RBM could address the tragedy of the commons, and, at the same time, improve the efficiency of fishing. The basic logic behind RBMs is that by limiting entry to a fishery and vesting those limited entrants with a secure, durable and transferable interest by way of a fixed quota, holders of the entitlement will be able to manage their fishing effort in the most efficient way and take steps to ensure the value of their capital interest in the resource. The stronger the entitlement, the stronger the interest holders have in policing and securing the value of the entitlement, and hence the sustainability of the fish stock. RBMs have gained considerable traction in domestic fisheries, with varying degrees of implementation in Australia, Canada, Europe, New Zealand, and the United States. Since private rights generally depend upon the State for their existence, their utility within international fisheries has been somewhat limited. However, there is growing interest in using RBMs, such as tradable quotas, within international fisheries.[28]

For the most part, international fisheries law has been silent on the detail of regulation, favouring the setting of broad objectives, and allowing states discretion as to how they regulate fishing. International fisheries (i.e. those controlled primarily through regional fisheries management organisations [RFMOs]) have tended to adopt the same broad regulatory approaches that feature within domestic regimes, that is to say a combination of input controls and selective output controls (e.g. licenses, gear restrictions, closed areas and quotas). However, this is changing as the limitations of existing regulatory approaches become apparent through continued declines in fish stocks.

At the global level, the main threats to fish stocks are considered to be overfishing, illegal, unreported and unregulated (IUU) fishing and high levels of by-catch and discards. Pollution, climate change (with adverse impacts on marine habitats and species distribution), growth in alien invasive species and increased competition of space are further threats to the sustainability of fisheries. There is a general

---

[27] Scott Gordon's analysis was seminal: See Gordon (1954), at 124. It was preceded by Innis, although this did not consider the open access nature of fish. Innis (1940). In later years: Christy & Scott (1965).

[28] See Barnes (2012). Copy on file with the author.

acceptance that there is excess capacity in fishing fleets. Excess capacity places pressure on regulators to accommodate high levels of fishing within the law (e.g. setting total allowable catch too high),[29] or it may incentivise fishing outside of regulatory frameworks (ie IUU fishing).[30] The overarching objective of global fisheries regulation is to ensure that all fisheries are sustainable, meaning that fishing levels are within biologically sustainable limits. This in turn requires more specific regulatory objectives targeted at unsustainable practices within different fisheries. As indicated in Section 5.3.1, a further concern is ensuring food security – the desire to ensure adequate supplies of food (including fish) to populations.[31] Given the drivers of fishing effort, and the fact that fisheries are impacted by a wider and more complex set of environmental factors, this means that we cannot view fisheries regulation in isolation from other factors.

In summary, mixed regulatory approaches have emerged as the product of a historical and organic process in response to demands to bring fishing within sustainable levels. It has been adaptive – although largely through trial and error – rather than emerging through reflective, planned regulatory change. Originally, regulation was a response to the simple absence of any instruments and institutions able to control fishing. More recently, regulation has tended to focus on refining the effectiveness of those instruments and institutions, or developing new techniques of regulation, in light of new knowledge and experience of success and failure in regulation. Key drivers of regulatory innovation have been advances in knowledge of stocks and the impact of fishing, the growth in exclusive coastal State authority, and the availability of knowledge, and the structure and political nature of inter-State transactions. Above all international regulation has been shaped by the physical nature and location of fish stocks.

5.3 INSTRUMENT CHOICE AND AVAILABILITY IN INTERNATIONAL FISHERIES LAW

International fisheries law can be seen to operate in a number of ways. First, there are international rules directly governing the basic framework, objectives and means of cooperation for fisheries. These have general application but, as noted in Section 5.2, they have particular significance for high seas fisheries because in areas beyond national jurisdiction individual States must cooperate to regulate fishing activity. Within coastal waters (territorial sea and exclusive economic zone [EEZ]), States' authority to determine the goals and manner of fisheries regulation remains fundamentally shaped by international law because the existence of sovereign rights

---

[29] Communication from the Commission to the Council – Fishing Opportunities for 2009: Policy Statement from the European Commission (COM/2008/0331 final), para. 4.1.

[30] Hatcher (2004).

[31] Since the mid-1990s food security has become an increasingly prominent feature of global fisheries policies. See Garcia & Rosenberg (2010); van der Burgt (2013), at 194ff.

beyond territorial limits is contingent upon the operation of international law. However, since States enjoy exclusive control over fisheries in their coastal waters, there is less immediate concern with inter-States issues. The main exception to this is in respect of straddling and highly migratory species. Here, the fact that stocks occur or move between different jurisdictions means that some coordination of fishing is required between concerned States.

International fisheries are governed by a range of treaty regimes, customary rules, and soft law instruments. It is important to stress that at the inter-State level, States are both the regulator and the regulatee. This represents a challenge for regulatory design because there is potentially a fundamental conflict of interest in the dual role of States, one that may make it difficult to engage in win-win outcomes and reflective regulatory development. States may be internally conflicted, seeking ambitious conservation targets, yet resisting accountability for meeting such targets. Compliance incentives may be created through the redistribution of resources and systems of reward, by enhancing the opportunities between States (e.g. loans or trading opportunities), through the creation of reputational benefits, and by impos-ing (or removing) sanctions.[32] The creation of incentives may require some States to act altruistically – in order to induce behavioural change in other States. Thus, the creation of incentives (and promotion of win-win scenarios) may be beyond the gift of individual States. Or compromises may be simply politically unacceptable to States.

Within this basic international framework, States regulate the fishing activities of private persons, i.e. individuals, vessels and companies. This private law framework must be consistent with international law. Thus, States cannot grant greater fishing rights than international law permits, nor can States operate fisheries contrary to their wider international law commitments. Domestic fisheries management may encompass a wide range of technical measures, rights-based instruments and market-based measures. These instruments may present challenges when applied to fishing activities with an international element (e.g. fishing of shared stocks or fishing conducted on the high seas). First, some instruments may depend upon or require transactions across different maritime zones, or beyond the exclusive jurisdiction of States. This may entail cross-jurisdictional arrangements. Second, there may be important structural obstacles to overcome before particular measures can be adopted for an international fishery because international law is not structurally geared to the direct regulation of private persons. For example, RBMs seem to be limited to domestic markets because transfers of permanent rights-based entitle-ments (i.e. property) might entail a limitation of States' sovereign rights or freedoms to fish. However, as indicated in Section 5.3.2, these are increasingly identified as options for international fisheries.

---

[32] On compliance with international law, see Guzman (2002), at 1823.

The precise impacts of different instrument combinations is highly contextual, but it is still possible (and important) to consider the frameworks and instruments available, and to assesses the extent to which they reflect the principles of good regulatory design.[33] Or the extent to which the structure of the system impedes effective regulatory outcomes.

5.3.1 *International Instruments*

5.3.1.1 The United Nations Convention on the Law of the Sea 1982 (UNCLOS)

UNCLOS provides a governance framework for all ocean spaces and activities. It has 176 States Parties, and its provisions on the conservation and management of living resources represent customary international law. This means that all States' domestic and international fisheries regulatory regimes should accord with the provisions of UNCLOS. If one examines the terms of UNCLOS, it is clear that it does little to drive the development of a range of specific regulatory techniques. Its provisions are mainly focused on the setting of goals, the allocation of competence to manage fisheries in respect of fisheries (e.g. rules on the attribution of EEZs in Part V). Thus, UNCLOS requires coastal States to determine the total allowable catch (TAC).[34] *Taking into account* the best scientific evidence available, the coastal State shall then ensure *through proper conservation and management measures* that the maintenance of the living resources in the exclusive economic zone is not endangered by over-exploitation.[35] Measures shall be adopted that 'restore or maintain fish stocks at the maximum sustainable yield, *as qualified by* relevant environmental and economic factors, including the economic needs of coastal fishing communities and the special requirements of developing States, and *taking into account* fishing patterns, the interdependence of stocks and *any generally recommended international minimum standards, whether sub-regional, regional or global*'[36]. The coastal State *shall take account* of species interdependence and exchange/share scientific information.[37] The objective of optimum utilization is to be found in Article 62, which further requires the coastal State to make surplus catch available to other states. UNCLOS does not dictate specific management measures, although it alludes to a range of regulatory techniques in Article 62, which third States

---

[33] These principles, as noted in Section 5.1, are: compatible instrument mixes; low level intervention strategies; scaled interventions; empowering surrogate regulators; maximising win-win opportunities; sensitivity to regulatory attitudes, cultures and institutional conditions; conducting impact assessments; and engaging in performance responsive governance.

[34] UNCLOS, Article 61(1).

[35] UNCLOS, Article 61(2).

[36] UNCLOS, Article 61(3).

[37] UNCLOS, Article 61(4).

can be required to comply with when fishing in the EEZ (e.g. licensing, gear restriction, closed areas and so on). On the high seas, UNCLOS restates similar conservation and management provisions, although it does so in less detail.[38] As the text in italics indicates, the direction of regulation option by UNCLOS is openly structured to ensure high levels of discretion, without mixes of policy instruments being mandated. More specific pathways to achieve conservation and management of fish stocks are left to the discretion of States. As such, UNCLOS contributes little to the structured development of instrument mixes, although it certainly does not impede this. If one views UNCLOS as one component in the system of international fisheries law, then it could be regarded as part of and contributing to the regulatory mix. However, beyond UNCLOS, the tools available under international fisheries law remain strongly dependent upon treaty mechanisms, with non-binding codes, such as the Code of Conduct (discussed in the following text), making small inroads into this mix.

Where UNCLOS does steer regulation is with the requirement for States to have due regard for the rights and duties of other States in the EEZ under Articles 56(2), 58, 60 and 66 and on the high seas under Article 87(2) – thereby driving compatibility of States' regulatory measures. This represents a binding legal commitment to be sensitive to other regulatory matters. It is important because it shows how law reflects the intrinsically fluid and connected nature of marine activities, which in turn generates a commitment to sustainable resource use and science-based (or at least evidence-based) decision-making.

Because UNCLOS provides States with wide discretion over fisheries management, UNCLOS could be regarded as non-interventionist, or favoring low levels of intervention in the sovereign affairs of States. However, this is more a reflection of UNCLOS' broad framework structure, rather than a deliberate regulatory strategy. UNCLOS does little to dictate fisheries management to States, and it has little to say about whether States adopt fisheries management systems with high or low degrees of intervention in the affairs of individuals. Experience indicates that only over time does international law tend to 'interfere' more and more actively to constrain States. An example of this is the more detailed rules on the management of shared fish stocks found in the United Nations Fish Stocks Agreement (UNFSA), which defines the modus operandi of some of the general commitments to cooperate found in UNCLOS.

Despite the possibility of more detailed regulation, there is limited scope within UNCLOS, or other fisheries agreements, 'to escalate up the chain of regulatory responses' in order to compel States by way of more stringent interventions. This is due to the horizontal structure of international law, whereby there is no higher law-making and enforcement body than the individual State. Whilst collective measures through the United Nations (UN) Security Council might be conceived of as a form

---

[38] UNCLOS, Article 119.

of escalation, one cannot realistically conceive of breaches of international fisheries law as meeting the threshold of 'threats to international peace and security' required to initiate responses by the UN.[39] UNCLOS relies on a high degree of voluntary compliance by States to adopt effective management measures, and when this fails there are limited sanctions or options open to other States to induce compliance. Although UNCLOS has compliance mechanisms, including compulsory third-party dispute settlement, fisheries matters are excluded from the scope of this.[40] In general, there is a high degree of cost and inconvenience associated with international litigation. States may resort to diplomatic protest, retorsion, countermeasures and remedies for breach of treaty as a means of securing compliance, but there is little evidence of the use or effectiveness of such measures in respect of fisheries.[41] These measures are contingent upon States' interests being sufficiently affected to motivate such a response, and for the responding State having sufficient influence to make the unilateral responses count against the offending State. These elements are often absent, showing the inherent limits of compliance mechanisms in a horizontal legal system. Despite the infrequency with which States resort to compliance mechanisms, it might be argued that the mere threat of such measures can motivate States to comply with the law.[42] However, this is impossible to prove empirically.

As an international agreement, UNCLOS has little concern for non-State actors.[43] Although some regulatory authority is delegated to the International Maritime Organization (IMO) in respect of shipping matters, no such empowerment of surrogate actors is envisaged for fisheries. This is not to say it cannot happen. For example the FAO, an intergovernmental organization, has been active in developing a range of binding and non-binding fisheries instruments, such the Code of Conduct for Responsible Fisheries 1995 and the Port State Measures Agreement 2009.[44] However, the FAO measures are contingent upon States to have effect, so it is not truly an independent regulatory body. In recent years some States have afforded a degree of regulatory responsibility to non-governmental organizations (NGOs) through, for example, the use of certification by the Marine Stewardship Council (MSC). MSC certification harnesses market incentives to drive regulatory standards by requiring certified products to come from fisheries with effective

---

[39] This has happened in the context of piracy, where there was a clear and direct concern with security issues. See UN Security Council Resolution 1976 (2011).

[40] UNCLOS, Article 297(3) excludes States from binding third-party dispute settlement in respect of disputes arising out of the exercise of sovereign rights with respect to the living resources of the EEZ.

[41] Churchill (2012), at 813.

[42] On compliance, see Guzman (2002).

[43] Papanicolopulu (2012), at 867.

[44] FAO Code of Conduct for Responsible Fisheries 1995, FAO Doc 95/20/Rev/1. Available at www.fao.org/docrep/005/v9878e/v9878e00.htm; Agreement on Port State Measures to Prevent, Deter, and Eliminate Illegal, Unreported and Unregulated Fishing 2009. Available at: www.fao.org/fileadmin/user_upload/legal/docs/037t-e.pdf.

management systems, and which comply with relevant local, national and international laws. More generally, NGOs have started to enhance the capacity of States' enforcement strategies by providing information about fishing activities and non-compliance. This includes initiatives like 'whofishesfar'[45] or 'Global Fishing Watch'.[46]

UNCLOS is a package deal, wherein the distribution of rights and benefits across a wide range of matters is intrinsically connected legally and politically. This means that rules on fisheries cannot be viewed apart from navigational or environmental matters. This means fisheries regulation must be sensitive to other concerns, such as navigation or scientific research. At the same time, it renders the adoption of regulatory changes more complex by requiring that account be taken of their systemic effect. The balance of regulatory rights and duties in different parts of UNCLOS is the product of a trade-off, one that States are careful not to upset. This connectivity of issues forms part of UNCLOS' dominant regulatory logic. Indeed, the notion of integration pervades much of the academic literature on the law of the sea.[47] It is also evident in inter-State transactions: thus negotiations on a new internationally legally binding instrument on the conservation and sustainable use of marine biological diversity beyond areas of national jurisdiction (ABNJ instrument) are committed to ensuring the consistency of any new agreement with earlier agreements.[48] Keeping faith with the UNCLOS approach to oceans issues, development of this new instrument has also proceeded on the basis of a package of issues.[49] Whilst most commentators consider this to be a desirable approach, especially since it corresponds to ecosystem-based approaches, in practice this 'packaging of issues' can impede the adoption of new agreements on fisheries management. Thus, UNCLOS took some nine years to negotiate and a further decade to enter into force after its adoption. The UNFSA took a mere two years to negotiate, but six years to enter into force. The ABNJ instrument has its roots in an ad hoc working group established in 2004, yet remains under development thirteen years later.[50] Thus, another important aspect of the 'regulatory logic' of the law of the sea is the slow progress and politically sensitive nature of legal developments.

The reflexive and adaptive nature of UNCLOS, as a single instrument, with no institutional process for fisheries governance, is somewhat limited. UNCLOS' accordance with this design principle ought to be evaluated in light of UNCLOS' position as part of system of fisheries instruments that have evolved from UNCLOS in response to its perceived shortcomings. UNCLOS is considered a living

---

[45] www.whofishesfar.org/.
[46] See http://globalfishingwatch.org/.
[47] See for example, Caminos & Molitor (1985); Tanaka (2008).
[48] UNGA Res. 69/292, UN Doc. A/Res/69.292, 6 July 2015, para. 3. See further Barnes (2016a).
[49] See Letter dated 13 February 2015 from the Co-Chairs of the Ad Hoc Open-ended Informal Working Group to the President of the General Assembly, UN Doc. A/69/780*.
[50] UN GA Res 59/24 of 17 November 2004, para. 73.

instrument. This means its language, structures and institutions can adapt to changed circumstances. In part this can be managed through the annual meeting of States Parties, which is concerned with mainly procedural issues.[51] It is also influenced by the annual resolutions of the General Assembly on oceans and the law of the sea.[52] Mostly, however, change occurs through diffuse and uncoordinated means, through the practice of States and the process of treaty interpretation. By way of dynamic interpretation, the meaning and application of UNCLOS can be kept contemporary with wider technical and legal developments.[53] For example, reference to 'best scientific evidence' in the context of fisheries management decisions means States should continuously reappraise the basis for their decisions in light of new science. However, such approaches are limited. Interpretations cannot cut across the grain of UNCLOS' text or stretch meaning beyond reasonable limits. Thus, the authority of States to take action against foreign fishing vessels on the high seas is fundamentally hampered by the principle of exclusive flag State jurisdiction, something no manner of textual hermeneutics can circumvent. Although UNCLOS contains mechanisms for amendment or modification, these impose procedural barriers so stringent that amendment or modification is all but impossible.[54] UNCLOS is in this respect institutionally rigid. These constraints are widely acknowledged by commentators and form part of the institutional context for fisheries management.[55] This has meant implementing agreements have been used to advance international fisheries law in response to UNCLOS institutional rigidity. Of these, the most important is the UNFSA.

5.3.1.2 United Nations Fish Stocks Agreement 1995 (UNFSA)

The UNFSA is a free-standing legal agreement, but it operates within the general framework of UNCLOS.[56] This is stated in Article 4, which requires the fisheries stock assessment (FSA) to be interpreted and applied consistently with UNCLOS, and Article 44, which preserves rights and obligations under compatible instruments. It is reinforced by Article 7, which requires the compatibility of conservation

---

[51] See: www.un.org/depts/los/meeting_states_parties/meeting_states_parties.htm.

[52] See: www.un.org/depts/los/general_assembly/general_assembly_resolutions.htm.

[53] See Barnes (2016b).

[54] Article 312 (amendment process) provides for proposed amendments to be considered by a conference of parties, which requires the agreement of no less than 50 per cent of States Parties to convene. At such a conference, amendments must be agreed by consensus and, that failing, by way of a vote (carried by a two-thirds majority of representatives present and voting). Article 313 (simplified amendment procedure) permits amendments to be proposed without a conference, but they will only be adopted if no State objects within twelve months, effectively giving all States a veto. No State has initiated either process to date.

[55] Boyle (2006).

[56] UNFSA, Article 4.

measures between high seas and coastal waters. This reiterates the concern for compatibility in fisheries regulation evident in UNCLOS.

As of December 2017, the UNFSA has been ratified by eighty-seven States.[57] It adds greatly to the detail of UNCLOS' general conservation and management obligations for straddling and highly migratory fish stocks,[58] although it is somewhat contingent on States adopting regional agreements and arrangements to give effect to its provisions. Like UNCLOS, it is facilitative, discretionary and open to regulatory approaches, rather than mandating specific regulatory tools or mixes. It is more detailed than UNCLOS, and to an extent represents a sequencing or refinement of the general provisions of UNCLOS. The UNFSA is based on twelve 'general principles' which are set out in Article 5. These include requiring coastal States and States fishing on the high seas to adopt measures to ensure the sustainability of fish stocks, to base those measures on the best available scientific evidence, to use the precautionary approach, to assess impacts on the wider ecosystem, to minimise by-catch and pollution, to protect biodiversity, to take measures to prevent overfishing, to take into account the interests of artisanal and subsistence fishers, to collect and share data, to promote scientific research and technologies, and to implement effective monitoring, control and surveillance. Of these principles, the precautionary approach is usually regarded as the most significant because it reinforces the importance of sustainability in decisions about exploitation and foregrounds the importance of impact assessment of activities.

Regulatory activity is channeled through inter-governmental organisations called regional fisheries management organisations (RFMOs). In this respect, regulatory authority is redirected from individual States to regional groups of State. This may be a form of surrogate regulatory authority. Although this does not introduce new regulators, since RFMOs are groups of States, it does alter the regulatory dynamic by expanding the potential options for regulation and relieving individual States of burdens through, for example, information sharing and cooperative compliance mechanisms.[59]

Arguably, the UNFSA introduces a significant incentive mechanism with the potential for win-win outcomes. Under UNCLOS, fishing was a general freedom, subject to due regard considerations.[60] By mandating RFMOs to govern regional fisheries, the UNFSA makes fishing on the high seas contingent upon either participation in an RFMO or compliance with its conservation and management measures.[61] Thus, participatory rights, which benefit individual States, are aligned with enhanced conservation outcomes through the inclusion of States within the

---

[57] UNDOALOS, *Table recapitulating the status of the Convention and of the Related Agreements, as at 10 October 2014.* Available at www.un.org/depts/los/reference_files/status2010.pdf.

[58] See UNCLOS, Articles 63–4 and 118–9.

[59] UNFSA, Articles 14 and 20.

[60] UNCLOS, Article 116.

[61] UNFSA, Articles 8(4) and 17.

RFMO. This step marks an important move forwards in changing incentive structures in international fisheries law.

The UNFSA is sensitive to the regulatory culture/institutional capacity in respect of developing States, with Article 25 requiring specific forms of cooperation with developing States. This includes assistance, training and capacity building measures at local and regional levels. This marks an important sensitivity to institutional capacity among participating States. It is important because there is little point in imposing demands upon States that lack infrastructure or expertise (*de facto* authority) to deliver on international legal commitments (*de jure* authority). Indeed, limitations in States' de facto regulatory capacity can effectively render a resource open access, and so unravel the whole edifice of fisheries management.[62]

As noted in Section 5.3.1.1, UNCLOS showed little concern with performance review and it is limited in terms of internal capacity to adapt to changed circumstances. In contrast, the UNFSA explicitly engages in this through Article 36, which establishes a mechanism to 'review and assess the adequacy of the provisions of this Agreement and, if necessary, propose means of strengthening the substance and methods of implementation of those provisions in order better to address any continuing problems in the conservation and management of straddling fish stocks and highly migratory fish stocks'. That said, the threshold for institutional change remains high, requiring not less than half of States Parties to agree to a review conference, and two-thirds of States Parties for any such negotiated amendments to enter into force.[63] Beyond the UNFSA, there has been some performance review of international fisheries at the level of individual RFMOs. To the extent that such reviews have been conducted, they have been useful in identifying good and bad regulatory practices in RFMOs. However, many reviews have been subject to criticism on the basis of their lack of independence from the RFMO, a lack of thoroughness, and a lack of follow-up action by the RFMO.[64] Some States have also expressed concern about the lack of engagement with performance review in some RFMOs and the lack of follow-up action to implement recommended changes.[65] Other concerns include a lack of common assessment frameworks, problems with compliance mechanisms and allocation of fishing rights, and more fundamentally, the need for some changes to the legal structure of RFMOs, including strengthening of their mandates.[66] This is particularly so for RFMOs that have not yet accommodated modern fisheries management principles. These initiatives, whilst not perfect,

---

[62] Barnes (2009), 2.

[63] UNFSA, Article 45.

[64] See Molenaar (2005), at 533; Hoel (2010), at 449.

[65] Report of the resumed Review Conference on the Agreement for the Implementation of the Provisions of the United Nations Convention on the Law of the Sea of 10 December 1982 relating to the Conservation and Management of Straddling Fish Stocks and Highly Migratory Fish Stocks, A/CONF.210/2016/5, 1 August 2016, paras. 27, 101–3.

[66] See the recommendations of Lodge (2007), at 117–128.

reflect a move towards more reflective and proactive regulation of international fisheries. Yet there is clearly scope to enhance the reflective and adaptive aspects of international fisheries law.

5.3.1.3 FAO Code of Conduct

By way of contrast to UNCLOS and the UNFSA, it is worth saying something about the FAO Code of Conduct for Responsible Fisheries. First, it is non-binding, and so represents a move in regulatory approaches in international fisheries away from treaties. This expands the range of international regulatory instruments in the mix. Second, its content goes beyond the framework provisions of UNCLOS and the UNFSA to more carefully delimit the kinds of instruments, or rather the qualities or objectives of such instruments, to be used in fisheries management. As a guidance instrument for the benefit of individual States and their fisheries administrations, it potentially bridges and reinforces the connection between international and domestic commitments in fisheries arrangements. The objectives of the Code are set out in Article 2, which sets out a lengthy list of aims – including to establish principles for responsible fisheries and the conservation of fish stocks – to serve as a reference point and to provide guidance in formulating relevant frameworks, measures, and agreements, and to promote the protection of the marine living resources and their environment. Whilst the Code is not binding, it notes that 'certain parts of it are based on relevant rules of international law'.[67] Thus it operates in a complementary manner to formal legal texts. Article 6 sets out the guiding principle that 'States and users of living aquatic resources should conserve aquatic ecosystems. The right to fish carries with it the obligation to do so in a responsible manner so as to ensure effective conservation and management of the living aquatic resources'.[68] This goes beyond UNCLOS and the UNFSA, and explicitly links beneficial rights to harvest with commitments to conservation. Sustainable fishing responsibilities are reiterated: Article 6 sets out general principles and practices, which include the application of the precautionary principle, to development and use of selective and environmentally safe fishing gear and practices, and the requirement to protect and rehabilitate 'critical fisheries habitats', especially nurseries and spawning grounds. This reinforces the connectivity of fisheries and environmental matters and advances potential win-win scenarios for different stakeholders in the marine sector.

The FAO Code is non-binding, and facilitative. Whilst this may weaken its legal force, it also means it is available as a regulatory template beyond State-to-State interactions. Indeed, the FAO Code of Conduct is directed at a range of actors, not

---

[67] FAO Code, Article 1.
[68] FAO Code, Article 6.1.

merely States.[69] It is a toolkit, guiding all fisheries actors as to how they can manage fisheries. Thus Article 7 sets out a range of approaches to and options for fisheries management, although it is short if being explicit on the use of specific regulatory tools. The range of factors to be taken into account in deciding management approaches is clearly broad enough to include command and control and market-based devices, and other innovative approaches. Thus, it drives the possibilities for different and complementary approaches within fisheries regulatory regimes.

One aspect of the FAO Code sets it apart from UNCLOS and the UNFSA. This is its approach to reflective, informed institutional adaption and impacts assessment. Its stated objective is to 'improve the legal and institutional framework required for the exercise of responsible fisheries and in the formulation and implementation of appropriate measures'.[70] It seeks to develop understanding of different management options.[71] And when change is made, this is to be notified and communicated.[72] It seeks to direct in much greater detail the regulatory agenda for fisheries – for example, urging the creation of agreements and regulatory instruments in the field of international fish trade.[73] It also is much stronger in facilitating assessment of fishing impacts. Thus, it requires the impact of fishing methods and changes thereto on fishing communities.[74] It also urges an evaluation of the cost effectiveness and social impact of conservation and management measures.[75] Article 7.6.8 of the FAO Code provides that: 'The efficacy of conservation and management measures and their possible interactions should be kept under continuous review'. Further, 'States, aid agencies, multilateral development banks and other relevant international organizations should ensure that their policies and practices related to the promotion of international fish trade and export production do not result in environmental degradation or adversely impact the nutritional rights and needs of people for whom fish is critical to their health and well-being and for whom other comparable sources of food are not readily available or affordable'.[76] Thus the FAO Code accommodates the good regulatory design principles concerned with reflective, adaptive approaches to regulation. Despite its soft-law nature, the Code has produced some legal effects. A FAO report reviewing implementation of the Code indicates that it has had some success in embedding improved design principles into fisheries laws worldwide.[77] However, the same report is also cautious about reading too much into its impact on fisheries management given the complexity of factors and variables at play in the real world. Acknowledging a lack of understanding of the impact of the

---

[69] FAO Code, Article 1.2.

[70] FAO Code, Article 2.c.

[71] FAO Code, Article para. 7.4.3.

[72] See for example, FAO Code, Article 11.3.8.

[73] FAO Code, Articles 11.2.10 and 11.2.13..

[74] FAO Code, Article 7.6.4.

[75] FAO Code, Article 7.6.7.

[76] FAO Code, para. 11.2.1.15. See also, paras. 12.10, 12.11.

[77] Hosch (2009).

code, in 2012 the FAO initiated a web-based reporting system to gather more data on this.[78] Data derived from this reporting system informs periodic reports on progress implementing the Code, with recent reports indicating high degrees of regulatory implementation.[79] Whether this is contributing to improved stock conditions is another matter, but it reinforces the reflexive approach to fisheries management.

5.3.2 *Specific Fisheries Management Tools and Techniques*

The success of specific tools and techniques within domestic fisheries has started to generate interest in how such measures could be used directly within international fisheries management regimes, such as RFMOs. Due to space constraints, this chapter will not go into a detailed account of the vast range of fisheries management tools available, but a broad typology and description of the principal tools and techniques is possible, with an indication as to how such instruments can cross over into the international aspects of fisheries management.[80]

Typically, fisheries management tools are categorised into three types: technical measures, input controls and output controls.[81] Technical measures comprise a range of controls on when and where fishers can fish. They include limits on the size of fish that can be caught, discard bans and landing requirements. Gear restrictions limit or control the different types of fishing gear that can be used (these include boat size, engine size, nets, mesh size, traps, lines, excluder devices, and so on) and where the gear can be placed. Area and time restrictions are used to prevent fishing in particular places or at certain times, usually to protect spawning stocks or to allow depleted stocks time to recover. Marine protected areas are a more sophisticated form of area restriction, where limits on fishing are often combined with other measures to protect the environment. Most fisheries management regimes are dominated by technical measures since they are necessary to deal with the particular conditions of a fishery. Input controls limit the amount of effort that can be 'put into' a fishery with a view to controlling the amount of fish that can be caught. Some limit on fishing capacity is generally seen as desirable. Input controls may overlap with the previously mentioned technical measures, but also include restrictions on fishermen through licences or permits, or effort quotas. Output controls impose direct limits on the amount of fish harvested. This includes setting a total allowable catch and fishing quotas, RBM such as individual transferable quotas (ITQs), community development quotas, catch shares and exclusive use rights such as territorial user rights in fisheries. Notably, these output controls form the basis of market-based regulation of fisheries, since they are essentially market-

---

[78] See www.fao.org/fishery/topic/166326/en.

[79] See COFI (2014). Available at www.fao.org/3/a-mko51e.pdf.

[80] On smart regulatory approaches generally, see further, Chapter 1, Section 1.2.

[81] FAO (1997), at 45–55.

based controls. These three types of controls can be imposed at different levels (local, national and international) through different management bodies (e.g. State, EU or RFMO). All of these instruments tend to be public law-type measures targeted at individual fishers/companies and implemented through legislation or international agreements. Even economic instruments like ITQs, which operate in and through markets, still depend upon a statutory basis.

The absence of legislative-type processes would appear to limit some regulatory options at the international level. However, quotas and certain types of technical measures frequently used at the national level (e.g. gear restriction or closed areas) can also operate at the international level. To date there is little use of other instruments such as right-based mechanisms because of the institutional limitations of high seas fisheries. This is because market-based instruments generally depend upon the creation of property rights. By establishing a form of property (usually an exclusive, durable and secure right to fish) fishers are incentivised to stop 'racing for fish'.[82] In particular, they won't have to fear that the fishery will close before they have caught their share. As a result, they are able to plan their activities in a way that allows them to catch fish most efficiently according to market conditions. In a market, the more efficient fishers should be able to profit and expand lower-cost fishing activities. In theory, this allows for the economic rent to be captured from the use of a resource.[83] Additionally, owners having a vested capital interest in the resource should have a rational interest in ensuring the resource is both sustainable and not harmed by destructive fishing practices. The difficulty of establishing and controlling property rights directly under international law may limit the availability of such regulatory approaches. So, although these measures have emerged in some domestic fisheries over the last few decades, they are not widely used in international fisheries. There are three possible options for harnessing such tools.[84] The first is to draw out the analytical incidents of RBM and to use these to help design the instruments used to regulate inter-State transactions.[85] Thus Serdy has suggested that some form of international quota trading could be developed for States' shares of catch allocations in RFMOs based upon RBMs.[86] The second is for international law to establish a clearing house for private transactions (e.g. trades in quota between private parties in different States). The third is to establish forms of RBM directly. One such example is the Vessel Day scheme operating in the Western and Central Pacific Ocean tuna fishery.[87] Interestingly, such approaches are increasingly driven

---

[82] See generally, Neher, Arnason & Mollett (1989); Leal (2005).

[83] Economic rent is defined as the payment to a factor of production in excess of that needed to keep it in its present use. Changes in rent may result from changes in demand. See Homans & Wilen (2005); Arnason (2011).

[84] See further Barnes (2009) and Serdy (2016).

[85] Barnes (2010).

[86] Serdy (2007).

[87] Havice (2013).

by some States, NGOs and industry groups, suggesting an alignment of interests/ attitudes and a blurring of the distinctions between international and domestic fisheries management.[88]

The next group of tools are fiscal or tax measures. These include both charges on fishers (e.g. landings tax) and subsidies (e.g. investment in new gear or fuel rebates). Charges operate on the same principle as the polluter pays principle, which can require the producers of externalities (e.g. harmful fishing such as bottom trawling) to internalise such costs into the economic cost of fishing. Subsidies can be given to fishers that generate positive externalities (e.g. removing predator species, thereby allowing prey stocks to flourish).[89] Fiscal or tax measures can be distinguished from purely market-based measures since they are not necessarily targeted at influencing market behaviour, but may instead, or additionally, provide some means of cost recovery.

Linked to this is the need to enhance sustainable investment in fisheries. Philanthropic, private, public and blended investments can be used to leverage change within institutions and fisheries practice – for example, by removing excess fishing capacity or improving harvesting and processing technology to increase the value of marketable products.[90] Investment other than by grant will require some form of return on the initial investment. This is usually linked some form of security, typically in the form of property or secure tenure rights, catch limits and robust monitoring and enforcement capacity. Such investments have emerged at local levels because there is a more secure legal regime and potential asset upon which to hang the investment (e.g. statutory entitlements). The lack of property rights in international fisheries and the riskier legal environment (e.g. politicised decision making) has generally limited investment directly into international fisheries. If fishing rights can be made more secure, then there is potential to secure investment directly in international fisheries.[91]

Finally, we should consider regulatory approaches designed to generate or facilitate the use of information, either by States, managers or fishers. Fisheries management is heavily contingent upon information, since it underpins decisions about what types and combinations of regulatory instrument are used. Thus, a smart mix should include regulation concerning the generation, coordination and sharing of information. This extends from local rules requiring the reporting of catch all the way up to international agreements establishing institutions mandated to develop science and advice for the exploitation of the oceans, such as the International Council for the

---

[88] See for example, ISSF, 2010; ISSF, 2011; WWF (2012). Available at: wwf.be/assets/RAPPORT-POLICY/OCEANS/UK/WWF-RightManagement-brochure-final.pdf.

[89] FAO (2003).

[90] See Encourage Capital, *Investing for Sustainable Global Fisheries* (2016). Available at: http://investinvibrantoceans.org/wp-content/uploads/documents/FULL-REPORT_FINAL_1-11-16.pdf

[91] See for example, the initiative between Althelia Ecosphere and the US Agency for International Development (USAID). They signed a risk sharing agreement under USAID's Development Credit Authority, which will assist the newly launched Althelia Sustainable Ocean Fund to provide impact financing to ocean projects in developing countries.

Exploration of the Sea. Informational requirements are commonplace in international agreements, at least at a general level, encouraging data sharing and an evidence-based decision-making process. Related to this are suasive instruments. These use information to influence individual behaviour, and include education, guidelines, codes of practices, and catch certification schemes. While common at the domestic level, these are increasingly being coordinated at the international level, given the global market for fish products. To an extent they can overlap with market-based mechanisms since they seek to influence consumer behaviour, and so straddle the line between private and public regulatory tools. Examples of this include the growing body of fishery product and process standards being developed by the International Organization for Standardization (ISO). This includes traceability standards for finfish and mollusk, environmental monitoring of fish farms and standards for eco-labelling. Of course, such standards are for use within domestic fisheries, and so do not directly concern inter-State conduct. The example of MSC certification noted in Section 5.3.1.1 is important and it is worth noting that there is increasing recognition of this within governance arrangements.[92] In 2009 the FAO adopted Guidelines for the Ecolabelling of Fish and Fishery Products from Marine Capture Fisheries.[93] Although these are a reference point, there are as yet no internationally agreed standards regulating this process. Developing this and other common international standards for market-related measures would seem to be a clear opportunity to enhance win-win opportunities in fisheries law.

5.4 CONCLUDING REMARKS

International fisheries present complex regulatory phenomena. This is in part due to the complexity of the physical resource system, its location and interaction with other ocean activities. Layered upon this is a complex regulatory framework encompassing international, regional and local levels of legal control, involving multiple and asymmetrical actors. These actors possess and seek to advance a range of regulatory agendas and interests, sometimes compatible, sometimes conflicting. A vast array of regulatory tools is available; these are not merely limited to catch activities, but encompass broader environmental standards, landing, trading and marketing controls. It is difficult to evaluate the precise effectiveness of these regulatory mixes. Instead we should focus on whether the legal systems are consistent with generally accepted principles of good regulatory design.

If fisheries regulation is in part rooted in international law, then it is useful to assess the extent to which key instruments facilitate or meet with the requirements of good regulatory design. Having done this, perhaps the most basic and sweeping conclusion is that international fisheries law has been principally concerned with

---

[92] Washington & Ababouch (2011).
[93] FAO (2009). Available at: www.fao.org/docrep/012/i1119t/i1119t00.htm.

setting of broad goals and regulatory objectives, as well as allocations of authority, and that it has remained insensitive to substantively advancing principles of good regulatory design. Moreover, the structure of international law as a horizontal system of law, principally conducted by and for States, constrains both its regulatory capacity and its capacity to harness good regulatory design principles. As Section 5.3 indicated, international fisheries laws have tended to develop organically, through an iterative approach, reinforcing the absence of deliberative, reflective regulatory approaches. However, this picture is changing, influenced especially by the success of different instrument mixes and approaches domestically, and there is growing evidence of principles of good regulatory design emerging within international fisheries law.[94]

Considering these principles in turn, the first observation is that the general and facilitative content of fisheries agreements leaves open the possibility of different combinations of regulation and different institutional approaches – at least under domestic law. The instruments discussed in Section 5.3 are part of an open toolkit of techniques for fisheries management. There is a growing and more sophisticated use of a range of regulatory techniques in domestic fisheries management, and some of these, in the form of market-based approaches, are beginning to spill over into international law. The key challenge is accommodating or integrating these tools into international fisheries management regimes. For example, is it conceivable that some form of tax could be implemented by an RFMO? Such a power would clearly challenge a traditional prerogative of States. Part of the constraint on international governmental organizations (IGOs) is that they depend upon States for authority and finance, and the possibility of their fiscal independence would challenge the sovereignty of States.

More generally, the range of legal sources under international law that can be used to manage fisheries are limited to treaty, customary international law and general principles, with case law and academic writing being subsidiary sources of law. Some of these sources are ill-equipped to support more detailed or complex regulatory measures. For example, custom is a diffuse, organic mode of law creation incapable of being directed prospectively.[95] It emerges organically from the practice of States. Custom is not the product of a deliberative process, so it cannot be conceived of as an 'orchestrated' process.[96] Some commentators take the exceptional view that customary law may be created deliberatively through international organizations.[97] However, decisions taken in international fora, such as the UN General Assembly, tend to direct law creation through treaties, or at best provide mere evidence of *opinio juris*, rather than be constitutive or directing of customary rules. General principles lack the capacity to advance more detailed technical content. This leaves us heavily dependent upon treaties to govern inter-State

---

[94] See *supra* note 5 of this chapter, and the related text.

[95] Trachtman (2016).

[96] The term is borrowed from Abbott & Snidal (2009).

[97] Alvarez (2006), at 591–595. Most writers take the view that decisions of international organizations do have such a 'legislative capacity'. See Johnson (1955–56); Lowe (2007), at 90.

conduct. To some extent treaties are deliberative, planned law creation processes. However, they usually occur on an ad hoc basis, negotiated by and for particular constituencies, with little capacity to account for system-wide factors. Treaties are drafted with little if any rigorous assessment of impacts in the form of cost-benefit analysis or a true understanding of behavioural consequences. Such concerns may emerge incidentally through the process of negotiation. However, this is not orchestrated by any supranational authority. International law lacks the governance structures to formalize this aspect of a legislative process. Treaties remain the product of political compromise that in turn may limit optimal regulatory choices. More recently soft law codes have added another dimension to fisheries management. Beyond the obvious addition to the mix, their informal status means they also complement treaties. These codes are more sensitive to proactive adaptive management – at least as a direction to individual States and other fisheries actors. As non-binding instruments, the regulatory agendas set in these codes are not constrained by traditionally formal structures of international law. Moving beyond the traditional structural limitations of formal international law, they represent an important change in direction of regulatory design, providing evidence of low-level regulatory intervention, drawing attention to win-win scenarios. They work with the logic of State-centered legal process, rather than trying to challenge the patterns of authority and competence in the international legal system. Of course, this is perhaps the crux of the issue. Can good regulatory design and more effective fisheries management occur within a horizontal system of law?

States are both law makers and subjects of the law. They lack autonomy of viewpoint to truly reflect on their position as 'regulatee'. Since international law begins and ends with States, the effect is to flatten the scale of regulatory responses. Most sanctions are contingent upon individual States having their interests harmed, and having both the will and capacity to take action against other States. There are few effective collective sanctions available against States to advance or encourage the satisfaction of regulatory goals. Although international law has some structures and processes that are capable of facilitating autonomous decisions within the system (e.g. independent third-party adjudication), these processes also remain at the mercy of States' self-interests. The absence of effective scaling of responses undermines international fisheries regulation and can result in regulatory impasse. For example, the long-standing failure to respond to the use of flags of convenience and their role in IUU fishing is a manifestation of the critical role that sovereignty plays in the system.

Related to this is the challenge of accommodating surrogate regulators. International law remains a State-centric discipline; States are the principal actors and the law is mainly contingent upon the consent of States.[98] There is limited participation of non-State actors in process of prescribing international fisheries law. For example, RFMOs are responsible for governing certain regional fisheries. However, these are composite

---

[98] Hollis (2005), at 137.

bodies of States, established under and limited by their constituent treaty, and so they are unable to exert regulatory influence apart from their composite State members. The EU has international legal personality and enjoys exclusive competence to manage living resources on behalf of its member States. However, it is also a composite State entity. NGOs play a minor role in fisheries regulation, often providing advice and data, or subjecting States to pressure through the media. NGOs lack any direct regulatory capacity and so are limited to using suasive or information instruments. Some of these instruments are quite important, such as product or traceability standards, and can complement international legal standards, as shown in Section 5.3.2. They also open the door to new incentive structures. MSC certification, which is open to States to use as clients, can generate incentives to improve fisheries management in order to secure stronger market positions for fish products. This tool seems well suited to strengthening compliance with international fisheries law. Of course, due to the fact that the locus of authority in international law remains with States, and that there is a large asymmetry in power between States and other actors, certification schemes appear to have a stronger role to play at national and local levels.

International law is also weak on reflexive governance mechanisms. Whilst international law is generally dynamic and continually evolving (as an adaptive institution), this does not mean that it is able to accommodate a wider range of interests (beyond States) or to be responsive to its regulatory deficiencies. This is reinforced by the lack of effective mechanisms for assessing the performance of international fisheries law beyond annual exhortations from the UN Secretary General and UN General Assembly, and from critical academic commentaries. Only recently have performance reviews of RFMOs become part of the toolkit of regulatory measures, and there is much scope for these to improve in terms of rigor, scope and impact on future regulatory performance of RFMOs. Most international regulatory approaches accept, but defer to, the attitudes of States. This is both a strength and a weakness. Since international law must respond to States' interests, this is an institutional necessity; a conditioning part of the regulatory domain. However, it may limit regulatory possibilities by reducing everything down to what is acceptable to States. As a result, international fisheries law continues to develop (slowly) as a product of political interactions, rather than in response to a formal assessment of institutional effectiveness and sense checking of regulatory options. Finally, at an operational level, States do not generally engage in any form of impact assessment when designing or adopting fisheries management mechanisms. Treaty law is purposeful, but only to the extent to which it meets the political ends of States. There is no capacity, and possibly no political will, to engage in meaningful impact assessment of prospective fisheries laws.[99]

---

[99] Occasionally domestic legislature engages in impact assessments of treaties. However, this usually concerns domestic implications of an existing agreement, rather than prospective impacts of an agreement at the international level. See for example, House of Lords European Union Committee, *The Treaty of Lisbon: An Impact Assessment*, HL Paper 62-I, 2008.

Viewed as a discreet regulatory regime, international fisheries law meets or facilitates some of the principles of good regulatory design. The principal limits of international law are the range of actors (mainly States and IGOs) and tools employed (mainly treaty and soft law instruments), as well as the absence of strong reflective and adaptive governance structures. The absence of these two aspects of good regulatory design present more significant concerns for effectiveness because their absence limits the possibilities for change and improvement.

Of course, this question of compliance with principles of regulatory design is very much one of perspective or categorization, and the strong critique of international law proceeds on the assumption that it is only appropriate and possible to assess international management apart from the wider range of fisheries management efforts that are also given effect through domestic legal regimes. Arguably, we can only understand international fisheries law as part of a complete system. If we look 'inside' international fisheries law to those domestic aspects, it is possible to evaluate it more favorably because a wider range of tools and approaches are part of the smart mix. Critically, greater scope exists for adaptive and reflective processes within domestic legal institutions. This is important, because if we continue to travel in the direct that smart regulation points, with instrument mixes, compatibility, sequenced interventions, empowered participants, and sensitive, reflective and adaptive performance, then we are likely to see further integration and growing mutual dependence of international and domestic aspects of fisheries regulation.

REFERENCES

Abbott, K. & Snidal, D. 2009. 'Strengthening International Regulation through Transnational New Governance: Overcoming the Orchestration Deficit'. *Vanderbilt Journal of Transnational Law* 42, 501–578.

Alvarez, J. 2006. *International Organizations as Law-Makers*. Oxford, Oxford University Press.

Arnason, R. 2011. 'Loss of Economic Rents in the Global Fishery'. *Journal of Bioeconomics* 13, 213–232.

Attard, D. 1987. *The Exclusive Economic Zone in International Law*. Oxford, Oxford University Press.

Baldwin, R. & Black, J. 2008. 'Really Responsive Regulation'. *Modern Law Review*, 71(1), 59–94.

Baldwin, R., Cave, M. & Lodge, M. 2011. *Understanding Regulation. Theory, Strategy and Practice*. Oxford, Oxford University Press.

Bangert, K. 1999. 'The Effective Enforcement of High Seas Fishing Regimes: The Case of the Convention for the Regulation of the Policing of the North Sea Fisheries of 6 May 1882'. In Goodwin-Gill, G. S. & Talmon, S. (eds.), *The Reality of International Law: Essays in Honour of Ian Brownlie*. Oxford, Oxford University Press, 1–20.

Barnes, R. 2009. *Property Rights and Natural Resources*. Oxford, Hart.

2010. 'Entitlement to Marine Living Resources in Areas Beyond National Jurisdiction'. In Oude Elferink, A. G. & Molenaar, E. J. (eds.), *The International Legal Regime of Areas beyond National Jurisdiction: Current and Future Developments*. Leiden, Martinus Nijhoff, 83–141.

2012. *Pathways to strengthen rights based management programs with a "high seas" component in the context of internationally managed tuna stocks*, Report commissioned by the WWF.

2016a. 'The Proposed LOSC Implementation Agreement on Areas beyond National Jurisdiction and Its Impact on International Fisheries Law'. *International Journal of Marine and Coastal Law* 31(4), 583–619.

2016b. 'The Continuing Vitality of UNCLOS'. In Barrett, J. & Barnes, R. (eds.), *The United Nations Convention on the Law of the Sea: A Living Instrument*. London, BIICL, 459–489.

Bothe, M. & Sand, P. H. (eds.). 2003. *Environmental Policy: From Regulation to Economic Instruments*. Leiden/Boston, Brill/Nijhoff.

Boyle, A., 2006. 'Further Development of the 1982 Convention on the Law of the Sea'. In Freestone, D., Barnes, R. & Ong, D. (eds.), *Law of the Sea: Progress and Prospects*. Oxford, Oxford University Press, 40–62.

Caminos, H. & Molitor, M. 1985. 'Progressive Development of International Law and the Package Deal'. *American Journal of International Law* 79(4), 871–890.

Christy, F. T. Jr. & Scott, A. D. 1965. *The Common Wealth of Ocean Fisheries*. Baltimore, MD, Johns Hopkins Press.

Churchill, R. R. 2012. 'The Persisting Problem of Non-Compliance with the Law of the Sea Convention: Disorder in the Oceans'. *International Journal of Marine and Coastal Law* 27(4), 813–820.

COFI. 2014. *Report on Progress in the Implementation of the Code of Conduct for Responsible Fisheries and Related Instruments*, COFI/2014/INnf.15/Rev 1. Available at: www.fao.org/3/a-mko51e.pdf.

FAO. 2016. *The State of World Fisheries and Aquaculture*. Rome, FAO.

2009. *Guidelines for the Eco-Labelling of Fish and Fishery Products from Marine Capture Fisheries*. Revision 1. Rome, FAO.

2003. *Report of the Expert Consultation on Identifying, Assessing and Reporting on Subsidies in the Fishing Industry - Rome, 3–6 December 2002*. FAO Fisheries Report No. 698. Rome, FAO.

1997. *Technical Guidelines for Responsible Fisheries*, No. 4. Rome, FAO.

Garcia, S. M. & Rosenberg, A. A. 2010. 'Food Security and Marine Capture Fisheries: Characteristics, Trends, Drivers and Future Perspectives'. *Philosophical Transactions of the Royal Society B* 365, 2869–2880.

Gardiner, R., Ostrom, E. & Walker, J. W. 1990. 'The Nature of Common-Pool Resource Problems', *Rationality and Society* 2(3), 335–358.

Gordon, H. S. 1954. 'The Economic Theory of a Common Property Resources: The Fishery'. *Journal of Political Economy* 62(2), 124–142.

Grafton, Q. R. et al. (eds.). 2010. *Handbook of Marine Fisheries Conservation and Management*. New York, Oxford University Press.

Gulbrandsen, L. 2012. 'Dynamic Governance Interactions: Evolutionary Effects of State Responses to Non-State Certification Programs'. *Regulation & Governance* 8(1), 74–92.

Gunningham, N. & Grabosky, P. (eds.). 1998. *Smart Regulation: Designing Environmental Policy*. Oxford, Clarendon Press.

Gunningham, N. & Sinclair, D. 1999. 'Regulatory Pluralism: Designing Policy Mixes for Environmental Protection'. *Law & Policy* 21(1),49–76.

Guzman, A. T. 2002. 'A Compliance-Based Theory of International Law'. *California Law Review* 90(6), 1823–1887.

Hatcher, A. 2004. 'Incentives for Investment in IUU Fishing Capacity'. In Gray, K., Legg, F. & Andrews-Chouicha, E. (eds.), *Fish Piracy: Combatting Illegal, Unreported and Unregulated Fishing*. Paris, OECD, 239–254.

Havice, E. 2013. 'Rights-Based Management in the Western and Central Pacific Ocean Tuna Fishery: Economic and Environmental Change under the Vessels Day Scheme'. *Marine Policy* 42, 259–267.

Hoel, A. H. 2010. 'Performance Reviews of Regional Fisheries Management Organizations'. In Russel, D. A. & VanderZwaag, D. L. (eds.), *Recasting Transboundary Fisheries Management Arrangements in Light of Sustainability Principles*. Leiden, Martinus Nijhoff, 449–472.

Hollis, D. 2005. 'Why State Consent Still Matters – Non-State Actors, Treaties, and the Changing Sources of International Law'. *Berkeley Journal of International Law* 23(1), 137–174.

Homans, F. & Wilen, J. 2005. 'Markets and Rent Dissipation in Regulated Open Access Fisheries'. *Journal of Environmental Economics and Management* 49(2), 381–404.

Hosch, G. 2009. *Analysis of the Implementation and Impact of the FAO Code of Conduct for Responsible Fisheries since 1995*. FAO Fisheries and Aquaculture Circular, No. 1038. Rome, FAO, 75–79.

Howlett, M. & Rayner, J. 2004. '(Not so) "Smart Regulation?" Canadian Shellfish Aquaculture Policy and the Evolution of Instrument Choice for Industrial Development'. *Marine Policy* 28(2), 171–184.

  2007. 'Designing Principles for Policy Mixes: Cohesion and Coherence in "New Governance Arrangements"'. *Policy and Society* 26(4), 1–18.

Howlett, M., Vince, J. & del Rio, P. 2017. 'Policy Integration and Multi-Level Governance: Dealing with the Vertical Dimension of Policy Mix Designs'. *Politics and Governance* 5 (2), 69–78.

Innis, H. A. 1940. *The Cod Fisheries: The History of an International Economy*. New Haven, CT, Yale University Press.

ISSF. 2011. *The Cordoba Conference on the Allocation of Property Rights in Global Tuna Fisheries*. Washington, DC, International Seafood Sustainability Foundation.

  2010. *Bellagio Framework for Sustainable Tuna Fisheries: Capacity controls, rights-based management, and effective MCS*. Washington, DC, International Seafood Sustainability Foundation.

Johnson, D. 1955–56. 'The Effect of Resolutions of the General Assembly of the United Nations'. *British Yearbook of International Law*. 32, 97–122.

Kwiatkowska, B. 1989. *The 200 Mile Exclusive Economic Zone in the New Law of the Sea*. Dordrecht, Nijhoff.

Leal, D. R. (ed.). 2005. *Evolving Property Rights in Marine Fisheries*. Lanham, MD, Rowman and Littlefield.

Lodge, M.W. et al. 2007. *Recommended Best Practices for Regional Fisheries Management Organizations*. London, Chatham House.

Lowe, V. 2007. *International Law*. Oxford, Oxford University Press.

Molenaar, E. J. 2005. 'Addressing Regulatory Gaps in High Seas Fisheries'. *International Journal of Marine and Coastal Law* 20(3), 533–570.

Neher, P.A., Arnason, R. & Mollett, N. 1989. *Rights Based Fishing*. Dordrecht, Kluwer.

Ostrom, E., Gardner, R. & Walker, J. 1994. *Rules, Games and Common-Pool Resources*. Ann Arbor, University of Michigan Press.

Papanicolopulu, I. 2012. 'The Law of the Sea Convention: No Place for Persons?'. *International Journal of Marine and Coastal Law* 27(4), 867–874.

Scott, A. 2008. *The Evolution of Resource Property Rights*. Oxford, Oxford University Press.

Serdy, A. 2016. *The New Entrants Problem in International Fisheries Law*. Cambridge, Cambridge University Press.

  2007. 'Trading of Fishery Commission Quota under International Law'. *Ocean Yearbook*, 21(1), 265–288.

Smith, T. 1994. *Scaling Fisheries: The Science of Measuring the Effects*. Cambridge, Cambridge University Press.

Stewart, R. B. 2008. 'Instrument Choice'. In Bodansky, D., Brunnée, J. & Hey, E. (eds.), *Oxford Handbook of International Environmental Law*. Oxford, Oxford University Press, 147–181.

Tanaka, Y. 2008. *A Dual Approach to Oceans Governance: The Cases of Zonal and Integrated Management in International Law of the Sea*. Abingdon, Routledge.

Techera, E.J. & Klein, N. 2017. *International Law of Sharks: Obstacles, Options and Opportunities*. Leiden, Brill.

Trachtman, J. 2016. 'The Growing Obsolescence of Customary International Law'. In Bradley, C. (ed.), *Custom's Future: International Law in a Changing World*. Cambridge, Cambridge University Press, 172–204.

van der Burgt, N. 2013. *The Contribution of International Fisheries Law to Human Development*. Leiden, Martinus Nijhoff.

van Gossum, P., Arts, B. & Verheyen, K. 2010. 'From "Smart Regulation" to "Regulatory Arrangements"'. *Policy and Science* 43(3), 245–261.

Washington, S. & Ababouch, L. 2011. Private Standards and Certification in Fisheries and Aquaculture: Current Practice and Emerging Issues. FAO Fisheries and Aquaculture Technical Paper. No. 553, Rome, FAO,

Wiener, J.B. 1999. 'Global Environmental Regulation: Instrument Choice in Legal Context'. *Yale Law Journal* 108, 677–800.

WWF. 2012. *Rights-Based Management: Conserving Fisheries, Protecting Economies*. Washington, DC, WWF, Available at: wwf.be/assets/RAPPORT-POLICY/OCEANS/UK/WWF-RightManagement-brochure-final.pdf.

6

Mixing Regional Fisheries Management and Private Certification

Markos Karavias

6.1 INTRODUCTION

Over the last four decades, the health of the world's fish stocks has reached a critical point. According to the FAO's latest data, the share of fish stocks within biologically sustainable levels decreased from 90 percent in 1974 to 68.6 percent in 2013.[1] At that time, it was estimated that 31.4 percent of fish stocks were being fished at biologically unsustainable levels, and therefore overfished.[2] According to the same data, the share of stocks that were being fished at biologically unsustainable levels increased from 10 percent in 1974 to 26 percent in 1989, and it continued to increase throughout the 1990s.[3] At the same time, world per capita fish consumption has trebled, fast outpacing population growth. Despite scientific concerns, the size of global marine catches has increased steadily since the 1970s, hitting the mark of 81.5 million tonnes in 2014.[4]

These alarming trends throw the reluctance of states to actively and effectively regulate fishing both within and beyond areas of national jurisdiction into stark relief. The intensification of international cooperation in fisheries management, evidenced by the proliferation of regional fisheries management organizations (RFMOs) from the 1990s onwards, has yielded less than encouraging results. Recent empirical studies suggest that 67 percent of stocks under RFMO management are either depleted or overfished, leading to the conclusion that RFMOs have not secured conservation of fisheries on the high seas.[5] RFMOs are plagued by the lack of credible sanctions, of means to avert "free-riding," and ultimately of the tendency of States to pursue short-term economic gain over long-term sustainable fishing.[6]

---

[1] FAO (2016), at 5.
[2] Ibid., at 5–6.
[3] Ibid., at 6.
[4] Ibid., at 10.
[5] Cullis-Suzuki & Pauly (2010).
[6] See Gjerde et al. (2013); Brooks et al. (2014), at 297–302.

It should come as no surprise that a paradigm shift in fisheries governance is unfolding. Since the mid-1990s, a number of private actors have employed the supply chain as a means of delivering sustainable fisheries management worldwide.[7] Fisheries management is thus no longer a domain of governments and international organizations, a realm of command-and-control public regulation. On the contrary, international organizations, regional management bodies and governmental authorities coexist and interact with private sustainability and certification initiatives, as do international and national legal rules with private standards. In other words, the taking of policy initiatives and the sharing of competences in relation to fisheries management are rather fluid and fractured, conjuring conceptions of "multilevel governance."[8]

One then is not faced with a governance "void" understood as a lack of rules or actors. On the contrary, there has been an exponential increase in novel forms of fisheries governance. Still, the existence of this governance kaleidoscope does not in and of itself guarantee the conservation of fish. The first challenge, then, is to understand the manners in which traditional and novel forms of fisheries governance interact with each other, and to map such interactions. This chapter will focus on the most visible of private fisheries certification initiatives, i.e. the Marine Stewardship Council, and the manner in which it interacts with intergovernmental bodies, namely the Commission for the Conservation of Antarctic Marine Living Resources (CCAMLR) and the Indian Ocean Tuna Commission (IOTC), within whose jurisdictions the MSC has proceeded to certify fisheries..

The second challenge is to gauge the effects of such interactions. In order to do so, this chapter makes use of the analytical tool of "smart mixes"[9] seeking to understand the conditions under which the operation of certification initiatives such as the Marine Stewardship Council may contribute to the effectiveness of regional organizations. To the extent that the mapping out of interactions may be understood as a descriptive exercise, the analysis of such interactions through the lens of "smart mixes" aims at yielding insights and suggestions towards maximizing the effectiveness of governance arrangements through the combination of various forms of regulation.

6.2 THE RISE OF PRIVATE CERTIFICATION IN FISHERIES: THE MARINE STEWARDSHIP COUNCIL

Over the last two decades, environmental law and policy have both grappled with a "turn to the market." This turn has manifested itself in two interrelated developments: (a) the increasing use of market-oriented policy instruments in domestic and

---

[7] See generally Pattberg (2004), at 52–53.

[8] Jordan (1999), at 13.

[9] See Chapter 1, this volume.

international law;[10] and (b) the emergence of a constellation of private actors, with a transnational reach, that operate in the global market and seek to partake in environmental governance through the adoption and implementation of regulatory standards.[11] At the heart of both these developments lies a belief in the market as a cost-effective mechanism of environmental protection capable of providing solutions, which traditional state-led command-and-control approaches have failed to deliver.[12] The constitution of private authority on the international plane suggests that sovereign states – willingly or unwillingly – yield their monopoly on regulatory power to private actors, whose authority is granted through the global supply chain,[13] thus defying territorial constituencies.[14] The proliferation of private regulatory initiatives in the field of environmental protection heralds the advent of what have been termed "new governance" arrangements, at the core of which stand private actors operating on the basis of regulatory standards, which are nonlegal from a formalistic point of view.[15]

The "turn to the market" in fisheries governance was set in motion in 1996 by the decision of the World Wide Fund for Nature (WWF) to team up with Unilever to establish the MSC. The MSC established itself in 1997 as an independent charitable organization. The shift in governance structure was perceived as an essential step toward building the credibility of the MSC as a neutral body in a multi-stakeholder industry.[16] The MSC is essentially a standard-setter that offers a set of principles and criteria[17] against which fisheries[18] are evaluated by an independently accredited third-party body (called a Conformity Assessment Body or CAB). The certification of the fishery as MSC-compliant is the initial, crucial step, which is then followed by a third-party supply chain certification takes place,[19] that ensures the traceability of labelled certified seafood from fishery to final consumer. While the MSC is not the sole private fisheries certification initiative, it has paved the way for a proliferation of similar initiatives and it is widely regarded as the leader in the field. The MSC has certified 283 fisheries in 30 states, and 88 more are in the process of certification; this

---

[10] For a taxonomy of policy instruments used with respect to environment protection, see Sand (2003), at 3–36; Stewart (2007), at 147–81.

[11] On the distinction between regulatory and nonregulatory private standards, see Morrison & Roht-Arriaza (2007), at 498–499.

[12] On the correlation between dissatisfaction with command-and-control approaches and the turn to supplementary market-based mechanisms, see Beyerlin & Marauhn (2011), at 303.

[13] See Cashore et al. (2007), at 161–162.

[14] Benvenisti (2014), at 36.

[15] See Abbott & Snidal (2009), at 503–512.

[16] Gulbrandsen (2009), at 654.

[17] Ward & Phillips (2008), at 16.

[18] The fishery or Unit of Assessment (UoA) is defined as a target species captured using a specific fishing method in a geographical area.

[19] The supply chain certification, which follows up on the fishery certification, is governed by the MSC Chain of Custody Standard: Default Version, Version 4.0, 20 February 2015, available at: www.msc.org/documents/scheme-documents/msc-standards/msc-default-coc-standard-v4/.

translates to the MSC label being carried by close to 10 percent of the annual global harvest of wild capture fisheries.[20]

The success of certification hinges on the consumer. Indeed, without consumer demand there would be no market for certified products.[21] Yet, certification goes beyond consumer demand to reflect corporate social responsibility (CSR) considerations of fish producers and retailers. Businesses may have economic reasons to supply certified products, even when consumers do not have a preexisting demand for them.[22] In this sense, certification creates a potential marketing advantage. Certification can provide evaluation and endorsement of a product, but it can also act as a consumer information and protection tool,[23] while having the potential to promote the objective of ecologically sustainable harvesting and management fisheries practice.[24]

Sutton lucidly encapsulates the rationale behind the establishment of the MSC as follows: "Governments, laws and treaties aside, the market itself will begin to determine the means of fish production."[25] The key assumption here is that market forces can be harnessed for positive change in the management of fisheries. According to the MSC, its purpose is to "create market initiatives to reward sustainable practices."[26] By setting up a market for sustainable fish, the MSC conditions purchasing preferences and thus increases global demand and market access for certified seafood. In this manner, it provides "the critical initiatives needed for fisheries to undergo the ... assessments required in the MSC program."[27] Fisheries operating below the MSC standard, which nonetheless want to become certified in order to reap market rewards, are then required to improve their practices in order to become eligible for participation.

Of course, certification as a governance instrument has been susceptible to critique. Shapiro, in an early contribution, highlighted the fact that while in some cases certification may permit the perfect information allocation, in other cases it may lead to higher prices due to the excessive investment that it induces.[28] Still, in the context of fisheries certification, the issue of the price premium and the whether it transmits from the retail level to the production level to compensate those who are engaged in fishing activities remains yet to be determined.[29]

---

[20] For current data, see: www.msc.org/global-impacts/key-facts-about-msc.
[21] Roheim (2008a), at 39.
[22] See Martin (2014), at 10098.
[23] On the MSC specifically, see Gutierrez et al. (2012).
[24] Roheim & Sutinen (2006), at 18, available at: www.vliz.be/imisdocs/publications/113442.pdf.
[25] Sutton (1998), at 133.
[26] MSC (2011), at 1, available at: www.msc.org/documents/msc-brochures/msc-theory-of-change. See Constance & Bonnano (2000), at 130.
[27] MSC (2011), at 1.
[28] Shapiro (1986), at 844.
[29] See Roheim et al. (2011), at 655–668.

6.3 MAPPING THE INTERACTIONS BETWEEN THE MSC AND INTERGOVERNMENTAL BODIES

The concept of "interaction" conjures images of reciprocal action, or the action of entities on each other.[30] It is in this sense that international lawyers have adopted the concept to describe the relationship between international law and domestic law. Thus, Arangio-Ruiz has defined "interaction" as a "simultaneous impact of international and national rights and obligations."[31] A similar understanding of the concept of "interaction" also holds sway in regulation theory. Perhaps the most compelling framework in this respect is the one set forth by the "transnational business governance interactions" (TBGI) group.[32] At the heart of this framework lies the need to analyze how the myriad of nonstate/private regulatory initiatives with a transnational reach interact with one another, as well as with public forms regulation.[33] In the TBGI context, "*interactions* are mutual actions and responses of individuals, groups or institutions, structures or systems."[34] These "[interactions] can be intentional or accidental, symmetrical or asymmetrical ... they take many forms (e.g. competition, cooperation, meta-regulation), exploit different causal pathways (e.g. modelling, reciprocal adjustment, conditional rule referencing), produce different patterns ... and have direct effects."[35] In other words, the TBGI framework draws attention to the reasons, mechanics and effects of interactions.

The interaction between private certification initiatives and intergovernmental bodies tasked with the conservation and management of fish has gone largely unexplored in literature. What is more, such a proposition goes against the grain of existing scholarship. International law has traditionally been perceived as binding entities bestowed with a measure of international legal personality, and as operating within a realm of publicness. Private certification initiatives are devoid of such legal personality, whereas private certification standards promulgated by these initiatives do not form part of the traditional sources of international law. It appears, then, that private initiatives such as the MSC operate in a private realm, in parallel with the realm of public international law.

An inverse divide lies at the heart of international relations or regulation and governance theory. Private governance instruments or "non-state market driven"[36] instruments are seen as forming part of a wave of "new governance," which operates outside the confines of traditional public regulation. In this vein, international treaties and the international governmental bodies they establish are seen as

---

[30] See the definition of "interaction" in the *Oxford English Dictionary*, available at: www.oed .com.

[31] Arangio-Ruiz (2007), at 16.

[32] Eberlein et al. (2014), at 1–21.

[33] Ibid., at 2.

[34] Wood et al. (2015), at 339.

[35] Ibid.

[36] On "non-state market driven" systems, see Cashore (2002).

cumbersome relics of the regulatory past, unable to furnish solutions for the increasingly complex environmental challenges of the present. The key for regulation theorists, then, is to examine in detail the impact that "new" or "non-state market driven" instruments have on the relevant actors on the ground, bypassing the influence that international legal rules may have.

The idea that international law and private governance interact has gone largely unexamined in scholarship.[37] However, such a state of affairs allows for simplistic readings either of international environmental agreements lacking impact on the market or of private governance instruments lacking impact on an international plane. Such readings stand in stark contrast with the actual operation of private certification initiatives, and more specifically the Marine Stewardship Council. This chapter aims to identify the interactions between the MSC and intergovernmental treaty-based bodies in the area of fisheries conservation or management. In order to do so, this chapter will assess (a) the role of intergovernmental bodies in the operation of the Marine Stewardship Council and, conversely (b) the role of the Marine Stewardship Council in the operation of intergovernmental bodies.

6.3.1 *The Role of Intergovernmental Bodies in the Operation of the MSC*

At the heart of the MSC's operation lies the Marine Stewardship Council Fisheries Standard and Guidance (MSC FSG).[38] The MSC FSG represents the end product of an intensive consultation process involving more than 300 organizations and individuals, and is the benchmark against which the independent bodies assess applications for MSC certification. It comprises three core principles, namely, (1) Sustainable target fish stocks, (2) Environmental impact of fishing, and (3) Effective management. Each principle is further broken down into a series of performance indicators (PIs), which represent the sustainability of the fishery. PIs, in turn, are based on three Scoring Guideposts (SGs). A score of 80–100 makes for an unconditional pass; a score of 60–75 marks a conditional pass (scoring can be assigned at 5-unit intervals), while a fishery scoring under 60 fails the certification process. In order for a fishery to become certified, the weighed average of the performance indicators must be 80 or more for each of the three core principles.

Core Principle 3 reads as follows: "The fishery is subject to an effective management system that respects local, national and international laws and standards and incorporates institutional and operational frameworks that require use of the resource to be responsible and sustainable."[39] This principle does not espouse the language of an international law rule; it includes a *renvoi* to "international laws,"

---

[37] See the incisive criticisms offered by Affolder (2010), at 163–167.
[38] MSC FSG, Version 2.0, 1 October 2014, available at: www.msc.org/documents/scheme-docu ments/fisheries-certification-scheme-documents/fisheries-standard-version-2.0.
[39] Ibid., at 6.

suggesting that the latter should be used as a benchmark against which to measure whether the fishery should be certified. In order to come to a decision on certification, the independent assessment body is called upon to gauge whether the "management system exists within an appropriate and effective legal and/or customary framework which ensures that it: (a) is capable of delivering sustainability in the [applicant fishery], (b) observes the legal rights created explicitly or established by custom of people dependent on fishing for food, and (c) incorporates an appropriate resolution framework."[40] International law rules play a fundamental role in this assessment. Thus, a fishery subject to international cooperation for the management of the stock (i.e. shared, straddling, highly migratory, high seas non-highly migratory fish stock) may only be certified if there exists at a minimum a level of cooperation among the various authorities and actors involved in the management based on bilateral and/or multilateral arrangements, which meet the requirements of cooperation as stipulated under Articles 63(2), 64, 118, 119 UNCLOS and Article 8 FSA.[41] What is more, such cooperation "shall at least deliver the intent of [Article 10 FSA] . . . relating to: (a) the collection and sharing of scientific data, (b) the scientific assessment of stock status, and (c) development of scientific advice."[42] For a fishery to qualify as "nearly perfect" and be certified unconditionally, the requirements are even more stringent. There has to be "binding legislation governing comprehensive international cooperation under the obligations of UNCLOS Articles 63(2), 64, 118, 119 and FSA Articles 8 and 10," while cooperation in the management of the fishery has to take place under the auspices of an RFMO/arrangement, the actions of which "demonstrably and effectively deliver FSA Article 10."[43]

The role of intergovernmental bodies and their decisions in the process of MSC certification can be exemplified through the certification of the South Georgian Patagonian Toothfish Longline Fishery, held by the Government of South Georgia and the South Sandwich Islands (SGSSI), a British overseas territory in the Southern Atlantic Ocean. Patagonian toothfish is a long-lived, slow growing fish, which is typically found within at depths of 200–2000m on continental shelf slopes around South America and the sub-Antarctic islands. Exploitation of Patagonian toothfish around South Georgia began in the 1970s as by-catch from a bottom trawl fishery. Longline fisheries were introduced to the South Georgia area in the 1980s, and allowed for exploitation of older fish in areas where trawls could not be used. The rapid growth of the toothfish market and the value placed upon the fish led to the development of illegal, unreported, and unregulated fishing for Patagonian toothfish in sub-Antarctic waters during the 1990s, reaching an estimated four times the regulated catch in 1997, risking the stock's collapse.[44]

---

[40] Ibid., at 66.
[41] Ibid., at 67.
[42] Ibid., at 68.
[43] Ibid., at 69.
[44] On Patagonian Toothfish and the challenges posed by overfishing, see Österblom & Rashid Sumaila (2011).

The Patagonian Toothfish fisheries in sub-Antarctic waters, and especially around SGSSI, fall within the scope of the Commission of the Conservation of Antarctic Marine Living Resources (CCAMLR), established under the 1980 Convention on the Conservation of Antarctic Marine Living Resources.[45] More specifically, SGSSI Patagonian Toothfish fisheries are situated within CCAMLR Sub-Areas 48.2, 48.3 and 48.4. These sub-areas form part of the SGSSI Exclusive Economic Zone. Whereas waters that fall within CCAMLR scope of application can be exempt from the conservation measures adopted by the CCAMLR Commission if they also happen to be waters over which a State exercises sovereignty or sovereign rights, the Government of SGSSI has chosen to allow for the application of such measures.

Since the end of the 1980s, and in response to the rise of IUU fishing of Patagonian Toothfish, the CCALMR has developed a set of innovative regulatory responses.[46] Specifically, the CCAMLR – following rigorous collection of biological and scientific data – introduced annual total allowable catch limits, which establish the maximum amount of fish that can be taken in any given area within the year. Furthermore, the CCAMLR has decided to establish a Catch Documentation Scheme (CDS) for fishing in the Southern Ocean.[47] Flag states sponsoring vessels fishing for Toothfish are required to provide each vessel with an individually numbered, non-transferable Patagonian Toothfish catch document, on which vessel captains record catch information and mark transshipments. The catch document accompanies all landings and imports of toothfish, creating a "paper trail" from harvest to importation and serving as a guarantee that the specific catch has been taken in a manner consistent with the CCAMLR conservation measures. Furthermore, the CCAMLR requires all vessels licensed to fish in the Commission area to carry a vessel monitoring system to allow monitoring and tracking by flag States. Finally, the CCAMRL operates an IUU "blacklist."

Faced with the enforcement difficulties arising due to the vast size and inhospitable conditions prevalent in the Southern Ocean, as well as NGO-led calls for a total ban on Patagonian Toothfish harvesting, the SGSSI Government decided to apply for MSC certification with respect to Sub Area 48.3, which encompasses the waters around South Georgia, and which is fully commercial. In 2004, the SGSSI Government was awarded MSC Fisheries Certification and in 2005, the MSC Chain of Custody Certification. It has been recertified ever since.

One of the key aspects of the certification process is the assessment of the effectiveness of the existent management system, including its legal basis. Indeed, the Certification Assessment Body (CAB) in the context of the SGSSI certification application examined in detail the complex web of domestic and international norms applicable in the fishery. The CAB in its initial 2004 report noted:

---

[45] The CCAMLR cannot properly be called a Regional Fisheries Management Organization, as its mandate extends beyond fish.

[46] See Agnew (2000); Miller, Saboureknov & Ramm (2004); Sovacool & Siman-Sovacool (2008) ; Constable (2011).

[47] CCAMLR, Conservation Measure 170/XVII.

"Fundamental to the MSC Standard is the compliance of the fishery with international and national legislation relevant to fishery management. It was therefore key to the assessment fishery that the fishery be in full compliance with any conservation measures, resolutions or decisions of the Convention for the Conservation of Antarctic Marine Living Resources (CCAMLR) relevant to its functioning."[48] The fishery was assessed as CCALMR-compliant and was given a near-perfect score at the first certification round.

Reviewing the management in place during the latest certification round, the CAB notes that the SGSSI Government "has established a management regime for the fishery that is compatible with CCAMLR and gives effect to the Convention. There is clear evidence that SGSSI participates fully with CCAMLR requirements to monitor and report both fishing activity, stock status and environmental impacts associated with the fishery."[49] Furthermore, the CAB takes stock of the varying nature of the measures in place, noting: "The statutory system in force requires, inter alia, that the fishing vessel reports daily fishing activity (location and catch weight) to the GSGSSI; monitoring of landings that are reconciled with daily catch reports; surveillance of the fishing vessel using two VMS systems (one for GSGSSI and another for CCAMLR); direct observation of fishing trips, monitoring of fishing practices and sampling of catches by on-board observers; inspection of vessels by GSGSSI staff at King Edward Point; and surveillance of fishing activity at sea by the patrol vessel Pharos SG ... This comprehensive system is capable of detecting breaches of management measures, strategies and rules. The level of compliance is excellent demonstrating the ability of the system to enforce these measures, strategies and rules."[50]

Perusal of the certification reports of the SGSSI Patagonian Toothfish Longline Fishery yields significant insights into the operation of private certification and the role that the decisions of international bodies play therein. Principle 3 acts as a bridge between private certification and international law. In assessing whether there exists an effective management system in relation to fisheries, which fall within the scope of international bodies such as the CCAMLR, the CAB is called upon to examine in detail the measures adopted by such organizations, the extent to which they apply to the fishery in question, and the extent to which such measures guarantee compliance. In sum, for a fishery to become certified, it has to be compliant with both national and international law and, more specifically, potentially compliant with decisions of international bodies. MSC certification then serves as a message to the consumer that the certified product can be traced back (through an MSC-certified supply chain) to a fishery where international law is upheld.

---

[48] SG Toothfish Scoring Table v2 The Licensed Patagonian Toothfish Longline Fishery of South Georgia 66.

[49] Intertek Fisheries Certification, South Georgia Patagonian Toothfish Longline Fishery Public Certification Report (September 2014) 107.

[50] Ibid., at 118.

6.3.2 *The Role of the MSC in the Operation of Intergovernmental Bodies*

The previous chapter has served to show that the existence of international rules and compliance therewith may serve as a prior requirement to certification. Nonetheless, the relationship between the MSC and the operation of international organizations may not be as linear as it first appears. Indeed, the operation of MSC certification may conversely result in the creation of international legal rules. The potential impact of the MSC on international rule making is clearly exhibited in the recent adoption by the inter-governmental Indian Ocean Tuna Commission (IOTC) of Resolution 16/02 "On harvest control rules for skipjack tuna in the IOTC area of competence."[51]

The roots of Resolution 16/02 can be traced back to the effort on behalf of the Maldives Seafood Processors and Exporters Association (MSPEA) to have the Maldives pole-and-line skipjack tuna fishery certified by the MSC. The independent CAB that was set up to review the MSPEA application held the fishery to be MSC-compliant, yet stipulated a number of conditions that responded *inter alia* to the lack of "explicit regional harvest control rules" on skipjack tuna. More specifically, the CAB proposed that "formal harvest control rules (HCRs) . . . be defined and put in place for skipjack fishery."[52]

The CAB Report goes on to lay out a detailed plan of action to be followed by the applicant client MSPEA, as well as by the relevant Maldives state authorities. As is made clear in the Report, the client has consulted "both the Ministry of Fisheries and Agriculture and the [Marine Research Centre] and [has] secured commitment to meeting the . . . plan."[53] By the first annual audit, the client has to provide the Assessment Body with "documented evidence that the client and representatives of the Government of the Maldives at the IOTC have promoted the need to define formal HCRs with IOTC related to the definition of interim biological reference points for skipjack tuna."[54] In order to meet this milestone, "Maldives shall: consult with IOTC secretariat . . . with the objective of establishing an agreed position on the adoption of harvest control rules . . . There will be a strong liaison between Maldives, other [Parties] who support the initiative to develop Harvest Control Rules and Strategies and the IOTC Secretariat."[55]

The fourth milestone to be reached is when the certification client is able to provide documented evidence that harvest control rules have been implemented

---

51 IOTC Resolution 16/02, "On Harvest Control Rules for Skipjack Tuna in the IOTC Area of Competence," 26 November 2016, available at: www.iotc.org/cmm/resolution-1602-harvest-control-rules-skipjack-tuna-iotc-area-competence.

52 Intertek/Moody International, Pole and Line Skipjack Fishery in the Maldives, Public Certification Report, Version 5, November 2012, p. 118, available at: https://fisheries.msc.org/en/fisheries/maldives-pole-line-tuna/@@assessments.

53 Ibid.

54 Ibid.

55 Ibid., at 118–119.

and there is a clear basis for considering that *they will be successful in achieving the desired outcome and that they have taken into account the main uncertainties.*"[56] To this end, the Maldives authorities are called upon to "contribute as required to IOTC meetings and workshops to promote the development of appropriate harvest control rules," "lobby IOTC secretariat and members" with a view to providing "implementation, through IOTC, of tested [harvest control rules]" designed to meet the minimum requirements of a SG80 score, in accordance with the MSC FSG.

In sum, the 2012 Final Report of the CAB urged the Maldives to lead an effort in the context of the IOTC towards the adoption of harvest control rules for skipjack tuna, designed to meet the MSC requirements. The adoption of the aforementioned Resolution 16/02 is the culmination of such a Maldives-led effort since 2013.

The adoption of harvest control rules for skipjack tuna by the Member States of the IOTC can be seen as forming part of the IOTC mandate, since the IOTC is tasked to "adopt ... conservation and management measures, to ensure the conservation of the stocks... and to promote the objective of their optimum utilization."[57] Yet, one cannot downplay the role of MSC certification in the adoption of Resolution 16/02. It was acknowledged early on that the Maldives' effort towards the adoption of harvest control rules aimed at maintaining MSC certification.[58] The Management Strategy Evaluation for the skipjack tuna fishery that was submitted to the IOTC Scientific Committee was drawn up by the Maldives in association with a number of nonstate partners, including the "Maldives MSC client, MSPEA."[59] Finally, the actual harvest control rules adopted in relation to the Maldives skipjack tuna fishery build on the default reference points set out by the MSC Fisheries Standard and Guidance. According to Articles 3 and 4 of Resolution 16/02, the biomass limit reference point is 20 percent of the unfished spawning biomass, while the biomass target reference point is 40 percent of the unfished spawning biomass. Thus, in adopting Resolution 16/02 the IOTC clothed the reference points adopted by the MSC[60] in an international law mantle, albeit implicitly. The role of the MSC in the process of the adoption of Resolution 16/02 has been affirmed by the MSC itself, which recognizes in its most recent "Global Impact Report" that the agreed "harvest control rules ... were catalysed by the MSC assessment process."[61]

---

[56] Ibid., at 119 [emphasis added].

[57] Article V 2(c) of the Agreement for the Establishment of the Indian Ocean Tuna Commission, Rome (Italy) 25 November 1993, in force 27 March 1996, available at: http://www.fao.org/fileadmin/user_upload/legal/docs/013s-e.pdf.

[58] See Adam, Sharma & Bentley (2013).

[59] Report of the 18th Session of the IOTC Scientific Committee, Doc. IOTC–2015–SC18–R[E] (2015), at 28.

[60] Cf. the default reference points, MSC FSG, Version 2.0, 1 October 2014, at 153–156.

[61] MSC, "Global Impacts Report 2017," at 23, available at: www.msc.org/documents/environmental-benefits/global-impacts/msc-global-impacts-report-2017.

6.4 TOWARDS SMART MIXES IN FISHERIES MANAGEMENT?

The analysis in Section 6.3 appears to challenge assumptions of a neat divide between the realms of public regulation and private initiatives. If anything, it suggests that the interactions developed may be dense and complex. It is hard to escape the conclusion that environmental governance nowadays proceeds on the basis of an infinite variety of combinations of public and private regulation and governance, command-and-control and market instruments and so on. In this sense, the employment of the term "mix" is an accurate description of the novel solutions that both States and nonstate actors have devised to counter transnational environmental problems of increasing complexity. Still, the term "mix" might also carry normative overtones. Gunningham and Sinclair have argued that due to the complexity previously alluded to, "it would be futile to attempt to construct a single optimal regulatory solution that would be applicable to a wide variety of circumstances."[62] In this sense, not only do public forms of regulation – such as international law – and private initiatives overlap, interact, or simply mix, but also such mixing may be a positive development that should perhaps be studied and engineered further.

Therefore, one may argue, in the vein of Gunningham and Sinclair, that a key challenge of contemporary environmental governance is the identification of mixes, which may serve the purpose of environmental protection, or in the context of the present article the conservation of the world's fisheries. In this vein, the mixes can be designated as "smart." Of course, a discussion of smart mixes generates two key questions, namely (a) what are the elements of the mix, and (b) what is the measure of the mix's "smartness"? As to the first question, there are a variety of elements, or phenomena or things that mix. A significant part of the literature has geared research towards smart "policy mixes."[63] The present analysis focuses on mixes of forms of regulation, namely public regulation and private regulation, in the form of the Marine Stewardship Council. The second –and more complicated– question is the yardstick by which the "smartness" or effectiveness of a mix can be measured. It is argued that in order to undertake such an endeavour, we must first unpack the various conceptions of effectiveness that have been developed in the context of environmental law and policy.

As stated in the Introduction to this volume (see Chapter 1), the concept of effectiveness may be read in a number of ways, such as *legal* effectiveness, which focuses on the issue of compliance with the law, *behavioral* effectiveness, which measures the role of rules in causing states and individuals to modify their behavior, and *problem-solving* effectiveness, which measures the degree to which rules or regimes solve the environmental problems they are designed to address.[64]

---

[62] Gunningham & Sinclair (2006), at 195.
[63] See Gunningham, Grabosky & Sinclair (1998).
[64] See Chapter 1. More generally, see Bodansky (2010), at 253.

Needless to say, the three conceptions are not identical, nor do they make use of a common methodological framework. International environmental lawyers tend to focus on the question of *legal effectiveness*, which speaks to compliance. Nonetheless, compliance may not always amount to an accurate tool with which to capture the realities on the ground, since compliance with a given international legal rule is in practice "often inadvertent, or an artefact of the legal rule or standard chosen ... the sheer fact of compliance with a given commitment tells us little about the utility and impact of that commitment."[65] Still, the drawing of distinctions among the various conceptions of effectiveness does not necessarily lead to the conclusion that they are mutually exclusive. On the contrary, there may be causal links between the various conceptions of effectiveness. Thus, while compliance does not always lead to problem-solving, it may constitute the first, albeit crucial, step. One would have to concede, then, that an examination of the effectiveness of international fisheries law viewed through the lens of compliance may fall short of providing insight into whether international fisheries law has achieved the full gamut of its objectives. However, from a legal point of view, compliance with an international legal rule, and how this compliance is achieved empirically, still matters as a starting point in the quest for assessing (a) the impact of international law, and (b) how such impact is magnified due to the operation of a mix between an intergovernmental organization and a private certification initiative, such as the Marine Stewardship Council. In this vein, one may argue that the efforts of States to comply with their international fisheries law obligations may contribute to the reversal of the depletion of fish stocks, and to the extent that the operation of the Marine Stewardship Council promotes compliance with international fisheries law, then the apparent mix may be seen as smart in terms of legal effectiveness.

6.4.1 *The MSC as a Means of Securing Compliance with International Fisheries Law*

As argued in Section 6.3.1, the key to understanding the relationship between compliance with international fisheries law and the operation of the Marine Stewardship Council is Principle 3 of the MSC Fisheries Standard and Guidance. According to this principle, in order for a fishery to become certified, it has to be subject to an effective management system that is compliant with national and international law. Compliance with international and national law, then, is a prerequisite of certification. Inversely, this means that well-managed, and thus potentially less threatened, fish is easier to certify. From the perspective of the certifying body, this choice appears logical, in the sense that the certification of an unsustainable or a critically threatened fishery lacking in management would harm the credibility of the certifier.

---

[65] Raustiala (2000), at 392.

These suggestions are reflected in the empirics on the operation of the MSC. The clear majority of the fisheries certified by the MSC are located within the jurisdiction of developed States. Only 8 percent of the total of MSC-certified fisheries, and 11 percent of the fisheries in full assessment, are developing world fisheries.[66] This discrepancy in the uptake of the MSC has been a cause for concern,[67] as it stands in stark contrast with the realities of global fishing, and more specifically with the fact that a significant number of the top ten global fish producers are developing countries.[68] Turning to the specific characteristics of the MSC-certified fisheries themselves, one of the first studies conducted noted that those fisheries were highly selective of their target species, had stocks that occurred within known areas for which there were exclusive national access rights, were well regulated and enforced, and often comanaged by governments, fishers, and scientists.[69] This conclusion seems to be vindicated in the case of SGSSI Patagonian Toothfish Longline certification. The Certification Assessment Body placed significant emphasis on the existence of a multitude of techniques employed by the CCAMLR to achieve the conservation of the fish stocks in question.

The analysis seems to suggest on its face that the operation of the MSC does not secure compliance with existing international fisheries law, but follows from it. Still, the MSC could potentially have a considerably greater impact on the compliance of States with international law. On the one hand, the MSC may act as an incentive for States to keep complying with international law, so that MSC-certified fisheries within their waters retain their certification; in sum, the operation of the MSC may *lock in* compliance with international fisheries law, including rules promulgated by regional fisheries organizations and bodies. On the other hand, the MSC may prompt States to lobby regional intergovernmental bodies towards the adoption of measures, with a view to establishing an effective management system, which would allow their fishing industry to seek certification. In this sense, MSC certification incentivizes States to meet their international law obligations, as was exemplified by the analysis of IOTC Resolution 16/02 in Section 6.3.

Nonetheless, MSC certification is not a silver bullet delivering compliance with international law, as it does not and objectively cannot remedy one of the key challenges faced by States in relation to meeting their obligations, namely the lack of capacity to enforce international law.[70] In sum, while MSC certification may serve as an incentive for State compliance, such an incentive finds its limit in the capacity of States. Factors such as the cost associated with establishing and implementing an effective management system, as well as the complexity of small-scale

---

[66] See   https://www.msc.org/about-the-msc/for-business/fisheries/developing-world-and-small-scale-fisheries/global-accessibility-program.

[67] Kalfagianni & Pattberg (2013), at 130.

[68] FAO (2016), at 11.

[69] Kaiser & Edward-Jones (2006).

[70] See Sand (1996), at 775.

fishing in the developing South, may defeat efforts towards MSC certification.[71] Overall, the higher the stringency of the standard in terms of requirements, the more limited the uptake in the developing world may be.[72] Thus, the lack of capacity of developing States to comply with international fisheries law may perpetuate the discrepancy in the uptake of MSC certification, lending credence to the criticism that MSC certification may restrict market access of products from developing countries that do not carry the MSC label, with dire consequences for producers.[73]

6.4.2 *The MSC as a Means of Securing the Sustainability of Fish Stocks*

As explained in Section 6.4, compliance is only one part of the smartness equation. What is perhaps more important is the extent to which the interactions developed between MSC certification and the operation of regional intergovernmental bodies contribute to the conservation and effective management of the respective fish stocks.

In this respect, the SGSSI Patagonian Toothfish Longline fishery stands out as a case where, for a period of more than a decade, the CCAMLR and the MSC have been operating in tandem. In a report drawn up in 2006,[74] it was noted that the precertification assessment process had identified and requested action on a key issue, namely the discarding by fishing vessels of fishhooks in fish heads. Certain types of albatrosses had been observed to take offal from around toothfish vessels, whereas, specifically, wandering albatrosses are large enough to ingest whole toothfish heads, which might contain hooks. Once the issue was raised, the SGSSI government immediately acted upon it, while the UK raised awareness and petitioned the CCAMLR respectively, leading to the adoption of a CCAMLR Conservation Measure urging for the removal of hooks from offal.[75] The report concluded that there was a correlation between the reduction in hook discards and the reduction in seabird mortality, thus there was an environmental gain, which could – at least partially – be attributed to the operational action resulting from the MSC certification.[76] Considering that the SGSSI fishery "operates among huge colonies of breeding albatrosses and petrels which are highly vulnerable to bycatch, ... the reduction of seabird mortality to low levels has been a major achievement."[77]

---

[71] Kalfagianni & Pattberg (2013a), at 131.

[72] Ibid.

[73] Gulbrandsen (2009), at 658.

[74] Agnew et al. (2006), at 99 available at: http://denglobalehandelsplads.dk/media/classroom/materials/MSC_Environmental_Benefits_Report_Phase1_FINAL_4May2006.pdf.

[75] Ibid., at 104.

[76] Ibid., at 106, cf. also Gulbrandsen (2009), at 658.

[77] COLTO, Patagonian Toothfish: Fighting Illegal Fishing and Protecting Albatross, (2015). Available at: www.colto.org/2015/05/18/patagonian-toothfish-fighting-illegal-fishing-and-protecting-albatross/.

As to the actual health of the Patagonian Toothfish stock, reflected in the population biomass, it has been noted that the stock "declined in biomass since certification, but was still 22 percent larger than the BMSY [namely, the biomass that would provide the maximum sustainable yield] of a fish stock in 2009."[78] Thus, the decrease was interpreted as fluctuating around the target under natural variability, and therefore the stock was seen as remaining above reference limits, which justified it being certified as well managed.[79]

Finally, literature has identified a synergistic effect between the MSC certification, held by the SGSSI Government, and the implementation of CCAMLR Conservation Measures. Thus, Agnew has stressed that MSC Certification "was also seen as a means of reviewing and strengthening the management systems of CCAMLR. Certification of the South Georgia toothfish fishery has considerably strengthened the catch documentation scheme."[80] Similarly, interviews with representatives of the fishing industry participating in the scheme suggest that the MSC certification "certainly complements" the measures adopted by CCAMLR.[81]

Based on these empirical findings, one may argue that the SGSSI Patagonian Toothfish Longline certification has led to some environmental gains, and thus could be considered a "smart mix." Still, the certification has not been free from controversy due to the persistence of IUU fishing for toothfish in the area. Thus, a DNA analysis of thirty-six MSC-certified toothfish samples shows that some did not originate from the certified population of South Georgia, while three of the samples were not Patagonian toothfish at all.[82] This study cast a shadow on toothfish management, as well as the credibility of MSC's traceability standard.[83]

While one cannot make an overall assessment of the "smartness," in terms of problem-solving effectiveness, of the mix between regional fisheries management and MSC certification, it is possible to assess this on a case-by-case basis, to the extent that reliable data exists.[84] Besides, even the identification of the smartness of one particular mix is not a sufficient yardstick with which to gauge the effectiveness of MSC certification as whole, which remains a matter of debate in literature.[85]

---

[78] Gutierrez et al. (2012).

[79] Martin et al. (2012), at 65.

[80] Agnew (2008), at 257.

[81] Roheim (2008b).

[82] Marko et al. (2011), at 621–622.

[83] See Christian et al. (2013), at 15.

[84] See also Kalfagianni & Pattberg (2013b), at 189.

[85] Consider positive views on the effectiveness of the MSC in Gutierrez et al. (2012); Bellchambers, Phillips & Perez-Ramirez (2016), with the critical views espoused by Froese & Proelss (2012), at 1288; Jacquet et al. (2010).

Still, the analysis in Section 6.4.1 seems to point towards the assumption that bigger environmental gains necessitate stricter certification processes. Of course, this would present the MSC with a Catch-22 scenario, since raising the bar of the standards would make certification even less of an attractive option, especially for "difficult fisheries."[86]

6.5 CONCLUSIONS

The rise of private fisheries certification, as exemplified by the growing numbers of applications to the Marine Stewardship Council, invites us to rethink certain key assumptions about environmental governance. First and foremost, simplistic dichotomies relying on the public/private divide do not seem to hold water. There exist complex and dense interactions between the realms of international law-making in the context of regional intergovernmental organizations and the MSC certification process. On the one hand, the assessment bodies closely examine and assess the measures adopted and implemented by organizations such as RFMOs in deciding whether to award certification. On the other hand, the assessment process may result in interested States petitioning RFMOs to adopt measures necessary for securing certification. In sum, private certification presupposes that the fisheries in question are compliant with international law. Nonetheless, private certification may operate inversely, creating incentives for States to comply with international law. Viewed through the lens of compliance, there appears potential for the identification of a "smart mix." One should not, however, lose sight of the fact that the potential of a market incentive towards compliance might be undercut by States' lack of capacity to implement international rules. Thus, fisheries that exist in the waters of developing States may be barred from gaining certification, thereby cutting them off from the lucrative market for sustainable fish that exists in the developed world.

The question of whether MSC certification contributes toward the sustainability of fish managed by an intergovernmental body is more convoluted. The Patagonian Toothfish Longline certification has yielded insights into the potential environmental benefits that the MSC may carry. However, these benefits were not wholly associated with the health of the fish stock per se. The truth of the matter is that any inquiry into the "smartness" of mixes of international fisheries law and private certification may be hindered by the lack of credible data. Having said this, it is hard to escape the conclusion that, today, the law and private certification standards mix. Therefore, it is imperative that these mixes be further examined, with a view to identifying the conditions under which such mixes can contribute to the conservation of the world's fish stocks.

---

[86] Ponte (2012), at 309.

REFERENCES

Abbott, K. & Snidal, D. 2009. 'Strengthening International Regulation Through Transnational New Governance: Overcoming the Orchestration Deficit'. *Vanderbilt Journal of Transnational Law* 42, 501–578.

Adam, M. S., Sharma, R. & Bentley, N. 2013. 'Progress and Arrangements for Management Strategy Evaluation of Indian Ocean Skipjack Tuna'. Paper submitted to the 15th Working Party of Tropical Tuna, 23–28 October. IOTC-2013-WPTT15–33.

Affolder, N. 2010. 'The Market for Treaties'. *Chicago Journal of International Law* 11(1), 159–196.

Agnew, D. J. 2000. 'The Illegal and Unregulated Fishery for Toothfish in the Southern Ocean, and the CCAMLR Catch Documentation Scheme'. *Marine Policy* 4, 361–374.

Agnew, D. 2008. 'Case Study 1: Toothfish – A MSC-Certified Fishery'. In Ward, T. & Phillips, B. (eds.), *Seafood Ecolabelling: Principles and Practice*. Oxford, Wiley-Blackwell, 247–258.

   et al. 2006. 'Environmental Benefits resulting from Certification against MSC's Principles & Criteria for Sustainable Fishing', available at: www.powerfulinformation.org/objects/mp/MSC_Environmental_Benefits_Report_Phase1_FINAL_4May2006.pdf.

Arangio-Ruiz, G. 2007. 'International and Interindividual Law'. In Nollkaemper, A. & Nijmann, J. (eds.), *New Perspectives on the Divide between National and International Law*. Oxford, Oxford University Press, 15–51.

Bellchambers, L., Phillips, B. & Perez-Ramirez, M. 2016. 'From Certification to Recertification the Benefits and Challenges of the Marine Stewardship Council (MSC): A Case Study Using Lobsters'. *Fisheries Research* 182, 88–97.

Benvenisti, E. 2014. *The Law of Global Governance*. The Hague, Hague Academy of International Law.

Beyerlin, U. & Marauhn, T. 2011. *International Environmental Law*. Oxford, Hart.

Bodansky, D. 2010. *The Art and Craft of International Environmental Law*. Cambridge, Harvard University Press.

Brooks, C. et al. 2014. 'Challenging the "Right to Fish" in a Fast-Changing Ocean'. *Stanford Environmental Law Journal* 33(3), 289–324.

Cashore, B. 2002. 'Legitimacy and the Privatization of Environmental Governance: How Non-State Market-Driven (NSMD) Governance Systems Gain Rule-Making Authority'. *Governance* 15(4), 503–529.

   et al. 2007. 'Can Non-State Governance "Ratchet Up" Global Environmental Standards? Lessons from the Forest Sector'. *Review of European Community and International Environmental Law* 16(2), 158–172.

Christian, C. et al. 2013. 'A Review of Formal Objections to Marine Stewardship Council Fisheries Certification'. *Biological Conservation* 161, 10–17.

Constance, D. & Bonnano, A. 2000. 'Regulating the Global Fisheries: The World Wide Fund, Unilever, and the Marine Stewardship Council'. *Agriculture and Human Values* 17(2), 125–139.

Constable, A. 2011, 15 March. 'Lessons from the CCAMLR on the Implementation of the Ecosystem Approach to Managing Fisheries'. *Fish and Fisheries* 12(2), 138–151. doi:10.1111/j.1467-2979.2011.00410.x

Cullis-Suzuki, S. & Pauly, D. 2010. 'Failing the High Seas: A Global Evaluation of Regional Fisheries Management Organizations'. *Marine Policy* 34(5), 1036–1042.

Eberlein, B. et al. 2014. 'Transnational Business Governance Interactions: Conceptualization and Framework for Analysis'. *Regulation and Governance* 8(1), 1–21.

FAO. 2016. *The State of the World's Fisheries and Aquaculture*, Rome, FAO.

Froese, R. & Proelss, A. 2012. 'Evaluation and Legal Assessment of Certified Seafood'. *Marine Policy* 36(6), 1284–1289.

Gjerde, K. et al. 2013. 'Ocean in Peril: Reforming the Management of Global Ocean Living Resources in Areas beyond National Jurisdiction'. *Marine Pollution Bulletin* 74(2), 540–551.

Gulbrandsen, L. 2009. 'The Emergence and Effectiveness of the Marine Stewardship Council'. *Marine Policy* 33(4), 654–660.

Gunningham, N. & Sinclair, D. 2006. 'Design Principles for Smart Regulations'. In Ramesh, M. & Howlett, M. (eds.), *Deregulation and its Discontents: Rewriting the Rules in Asia.* Cheltenham, Edward Elgar, 195–211.

Gunningham, N., Grabosky, P., & Sinclair, D. 1998. *Smart Regulation: Designing Environmental Policy.* Oxford, Oxford University Press.

Gutierrez, N. L. et al. 2012. 'Eco-Label Conveys Reliable Information on Fish Stock Health to Seafood Consumers'. PLoS ONE: *e43765* 7(8), https://doi.org/10.1371/journal .pone.0043765.

Jacquet, J. et al. 2010. 'Seafood Stewardship in Crisis'. *Nature,* 467, 28–29.

Jordan, A. 1999. 'The Construction of a Multilevel Environmental Governance System'. *Environment & Planning C: Politics and Space* 17(1), 1–17.

Kaiser, M. & Edward-Jones, G. 2006. 'The Role of Ecolabeling in Fisheries Management and Conservation'. *Conservation Biology* 20, 392–398.

Kalfagianni, A. & Pattberg, P. 2013a. 'Fishing in Muddy Waters: Exploring the Conditions for Effective Governance of Fisheries and Aquaculture'. *Marine Policy,* 38, 124–132.

  2013b. 'Global Fisheries Governance beyond the State: Unraveling the Effectiveness of the Marine Stewardship Council'. *Journal of Environmental Studies and Sciences* 3(2), 184–193.

Marko, P. B. et al. 2011. 'Genetic Detection of Mislabeled Fish from a Certified Sustainable Fishery'. *Current Biology* 21(16), 621–622.

Martin, W. 2014. 'Marine Stewardship Council: A Case-Study in Private Environmental Standard-Setting'. *Environmental Law Reporter News & Analysis* 44, 10097–10101.

Martin, S. et al. 2012. 'An Evaluation of Environmental Changes Within Fisheries in the Marine Stewarship Council Certification Scheme'. *Reviews in Fisheries Science* 20(2), 61–69.

Miller, D., Saboureknov, E. & Ramm, D. 2004. 'Managing Antarctic Marine Living Resources: The CCAMLR Approach'. *International Journal of Marine and Coastal Law* 19(3), 317–363.

MSC. 2011. 'Harnessing Market Forces for Positive Environmental Change: The MSC Theory of Change'. Available at: www.msc.org/documents/msc-brochures/msc-theory-of-change.

Morrison, J. & Roht-Arriaza, N. 2007. 'Private and Quasi-Private Standard Setting'. In Bodansky, D., Brunnée, J. & Hey, E. (eds.), *The Oxford Handbook of International Environmental Law.* Oxford, Oxford University Press, 498–527.

Österblom, H. & Rashid Sumaila, U. 2011. 'Toothfish Crises, Actor Diversity and the Emergence of Compliance Mechanisms in the Southern Ocean'. *Global Environmental Change* 21(3), 972–982. doi: 10.1016/j.gloenvcha.2011.4.2013.

Pattberg, P. 2004. 'Private Environmental Governance and the Sustainability Transition: Functions and Impacts of NGO-Business Partnerships'. In Jacob, K., Binder, M. & Wieczorek, A. (eds.), *Governance for Industrial Transformation.* Berlin, Environmental Policy Research Centre, 52–66.

Ponte, S. 2012. 'The Marine Stewardship Council (MSC) and the Making of a Market for "Sustainable Fish"'. *Journal of Agrarian Change* 12(2–3), 300–315.

Raustiala, K. 2000. 'Compliance & Effectiveness in International Regulatory Cooperations'. *Case Western Reserve Journal of International Law* 32, 388–440.

Roheim, C. 2008a. 'The Economics of Ecolabelling'. In Ward, T. & Phillips, B. (eds.), *Seafood Ecolabelling: Principles and Practice.* Oxford, Wiley-Blackwell, 38–57.

2008b. 'Seafood Supply Chain Management: Methods to Prevent Illegally-Caught Product Entry into the Marketplace'. Available at: http://cmsdata.iucn.org/downloads/supply_chain_management_roheim.pdf.

Roheim, C. & Sutinen, J. 2006. 'Trade and Marketplace Measures to Promote Sustainable Fishing Practices', International Trade and Sustainable Development Series, Issue Paper No. 3, Available at: www.vliz.be/imisdocs/publications/113442.pdf.

Roheim, C. A. et al. 2011. 'The Elusive Price Premium for Ecolabelled Products: Evidence from Seafood in the UK Market'. *Journal of Agricultural Economics* 62(3), 655–668.

Sand, P. H. 1996. 'Institution Building to Assist Compliance with International Environmental Law: Perspectives'. *Zeitschrift für ausländisches öffentliches Recht und Völkerrecht* 56 774–795.

Sand, P. 2003. 'Sticks, Carrots, and Games' In Bothe, M. & Sand, P. (eds.), *Environmental Policy: From Regulation to Economic Instruments.* The Hague, Nijhoff, 3–36.

Shapiro, C. 1986. 'Investment, Moral Hazard and Occupational Licensing'. *The Review of Economic Studies* 53(5), 843–862.

Sovacool, B. & Siman-Sovacool, K. 2008. 'Creating Legal Fish for Toothfish: Using the Market to Protect Fish Stocks in Antarctica'. *Journal of Environmental Law* 20(1), 15–33.

Stewart, R. 2007. 'Instrument Choice'. In Bodansky, D., Brunnée, J. & Hey, E. (eds.), *The Oxford Handbook of International Environmental Law.* Oxford, Oxford University Press, 147–481.

Sutton, M. 1998. 'Harnessing Market Forces and Consumer Power in Favour of Sustainable Fisheries'. In Pitcher, T., Hart, P. & Pauly, D. (eds.), *Reinventing Fisheries Management.* Dordrecht, Kluwer Academic Publishers, 125–136.

Ward, T. & Phillips, B. 2008. 'Ecolabelling of Seafood: Basic Concepts'. In Ward, T. & Phillips, B. (eds.), *Seafood Ecolabelling: Principles and Practice.* Wiley-Blackwell, Oxford, 1–37.

Wood, S. et al. 2015. 'The Interactive Dynamics of Transnational Business Governance: A Challenge for Transnational Legal Theory'. *Transnational Legal Theory* 6(2), 333–369.

7

RFMO-MSC Smart Regulatory Mixes for Transboundary Tuna Fisheries

Agnes Yeeting and Simon R. Bush

7.1 INTRODUCTION

Overfishing, overfished stocks and overcapitalization of fishing fleets are functions of the success or failure of inadequate institutions.[1] Rules for the exploitation of transboundary fish stocks, like tuna, are set by Regional Fisheries Management Organisations (RFMOs), with authority mandated through the 1982 United Nations Convention on the Law of the Sea (UNCLOS). RFMOs require coastal and fishing states to cooperate through international agreements to ensure the conservation and promotion of optimum utilisation of highly migratory species within and beyond the exclusive economic zones. More specifically, UNCLOS and the United Nations Food and Agricultural Organisation's (FAO) Code of Conduct for Responsible Fisheries require RFMOs to establish *precautionary* management measures, including a harvest strategy[2] with so-called harvest control rules (HCRs), which are a set of pre-agreed rules or actions agreed through an evidence-based framework for determining a management action in response to changes in indicators of stock status with respect to reference points.

To date, however, RFMOs have had limited success in setting and/or enforcing HCRs for tuna fisheries.[3] As outlined by Aranda, Murua and de Bruyn,[4] the adoption (or not) of HCRs by RFMOs is largely attributed to the (mis)alignment of political and economic goals of participating states, given their vested interests in different species of tuna, the fishing gears used to catch them and the protection of often long-standing access arrangements.[5] Promoting a precautionary approach for tuna fisheries has therefore led to the need for the better alignment of the diverging

---

[1] Squires et al. (2016).
[2] Hilborn, Orensanz & Parma (2005).
[3] Grafton et al. (2006); Squires et al. (2016).
[4] Aranda, Murua & de Bruyn (2012).
[5] See also Bailey, Sumaila & Martell (2013); Cullis-Suzuki & Pauly (2010).

interests of RFMO member states.[6] Such alignment is increasingly recognised as requiring 'smart mixes' of private rule-making institutions that bring independent (societal) norms and oversight, as well as market logics and incentives, to influence and coordinate the achievement of public regulatory goals.[7]

One such 'smart mix' includes certification by third parties such as the Marine Stewardship Council (MSC), which sets requirements that are rewarded by apparent market demand for credible managed fisheries and a transparent Chain of Custody. The MSC uses a voluntary private standard that sets requirements for sustainable fishing outside the remit of the state with a focus on the maintenance of healthy stocks, limited impacts on the ecosystem, and an effective management system.[8] By setting reference points and HCRs as major conditions for the fisheries it certifies, RFMOs with member countries that have an MSC-certified fishery can influence the adoption of a precautionary approach to tuna management in RFMOs.

This chapter explores how the MSC, as a private institution, influences and facilitates public regulation and coordination in three tuna RFMOs in the Indian, Pacific and Atlantic Oceans. We argue that in each case three different 'pathways' of RFMO-MSC (public-private) interaction are observed, each with different outcomes in terms of achieving the application of precautionary approach to management through the definition and adoption of HCRs. By identifying and characterising these pathways, we analyse how the process of MSC certification represents a smart mix of public and private regulation characterised by sequentially flexible and dynamic processes of learning and institutional change. By understanding and characterising these pathways, we argue, smart regulatory mixes can offer more effective long-term improvement in complex governance settings like RFMOs than public regulation alone.

The remainder of this chapter provides an overview of smart regulatory mixes and their relevance to the management of transboundary fisheries. We then present the results of three RFMO case studies: the Western and Central Pacific Fisheries Commission (WCPFC); the Indian Ocean Tuna Commission (IOTC); and the International Commission for the Conservation of Atlantic Tuna (ICCAT). The analysis is based on data collected through an extensive review of the certification documents available through the MSC and each of the RFMOs. These documents were supplemented by sixteen semi-structured interviews from October 2016 to April 2017 with representatives from the MSC and RFMOs, as well as various other actors involved at different stages of the (pre)certification process. The final section of this chapter discusses these cases with a view to characterising different improvement

---

[6] E.g. Havice (2010); Yeeting et al. (2017).

[7] Gunningham, Grabosky & Sinclair (1998).

[8] For details see Gulbrandsen (2010).

pathways emerging from these cases before drawing conclusions on the wider implications and application of our findings.

7.2 PATHWAYS OF SMART REGULATION

Smart regulation refers to a pluralist approach of mixing public and private forms of norms, standards, rules and laws to foster 'flexible, imaginative and innovative forms of social control'.[9] It has emerged largely in response to the prevailing global neoliberal order, under which state authority has been gradually supplemented with the private (moral) authority of NGOs and the (pragmatic) authority of industry self-regulation.[10] Reflecting a wider shift from 'government to governance', smart mixes are thought to improve the effectiveness and efficiency of more conventional forms of centralist modes of direct state regulation by combining multiple actors and modes of societal steering. As outlined by Gunningham and Sinclair, smart mixes of regulation envisage 'in the majority of circumstances, the use of multiple rather than single policy instruments, and a broader range of regulatory actors, will produce better regulation'.[11]

In the context of transboundary resources like fisheries, smart regulatory mixes engage a plurality of state interests (representing multiple publics and jurisdictions) in addition to multiple private interests (including environmental NGOs and industry actors). Public-private cooperation to reach 'collective goals' of transboundary fisheries management, as set out under UNCLOS and the United Nations Fish Stocks Agreement (UNFSA), is therefore of central importance. But cooperation is also undermined by three factors. First, the design and enforcement of any regulatory measure (such as HCRs) requires consensus, which tends to favour the 'law of least ambitious program'.[12] Second, the ambitions of coastal states in RFMOs are often linked to the economic and political influence enjoyed by distant-water fishing nations (and the fleets they represent), in the form of development aid, fees for fishing licenses and, ultimately, access to consumer markets.[13] Third, RFMO decision making is multi-level in that decisions are subject to national, subregional, and regional influence.[14] To be effective in the context of RFMOs, smart mixes therefore need to align the interests of public and private actors involved to cooperate across the multiple dimensions of borders, sectors and levels.

Cooperation across borders, sectors and levels also requires attention to the dynamic nature of introducing smart regulatory mixes over time. Sequencing smart mixes has been discussed in the context of the order of different instruments (legal,

---

[9] Gunningham & Sinclair (2017), at 133.
[10] Cashore (2002).
[11] Gunningham & Sinclair (2017), at 133.
[12] Underdal (1980), cited in Pentz & Klenk (2017).
[13] See for example Miller, Bush & van Zwieten (2014); Yeeting et al. (2018).
[14] E.g. Miller, Bush & van Zwieten (2014); Rochette et al. (2015); Brown (2016).

economic etc.) to coordinate social change[15] – for instance, holding certain instruments 'in reserve' when other instruments are not delivering their expected impact, or when an additional regulation is needed to invoke wider compliance.[16] However, sequencing can also refer to the order of strategies and decisions involved in applying a private instrument like MSC certification to the context of cooperative international decision making such as that done by RFMOs. An understanding of such sequencing also demonstrates the flexible and dynamic nature of learning and institutional change implied by a 'smart' regulatory mix. To increase the likelihood of achieving any predefined outcome, a smart mix should therefore be able to adapt to who is involved, how they are involved, and the effect of their involvement.

By characterising the sequence of institutional interaction, we open up the possibility of identifying different 'pathways' of smart regulatory mixing that affect international cooperation in the definition of transboundary rules. Pathways, in this sense, do not only mean structured dependency on a set of actors, interests and instruments,[17] but also a series of decisions that can be made by fishery managers in the development of an overall strategy for change. As Overdevest and Zeitlin argue, such pathways are emergent because rather than being a unified set of rules and procedures set out by a multilateral regime, they are processes of experimental rule setting within multilateral organisations like RFMOs.[18] In these pathways, private instruments like certification can be seen as a set of recurrent causal processes that can explain (but can also be used to shape) strategies for change in complex institutional settings. Seeing these instruments in this light enables us to address how and why particular mixes emerge, as well as the conditions and criteria that render the application of MSC in RFMOs 'smart'.

The following analysis of the three cases mentioned in Section 7.1 compares the sequence of interactions between the MSC and three RFMOs. Specifically, we analyse how these MSC-RFMO interactions have, to varying degrees, resulted in sufficient cooperation between member states to allow them to make progress towards the adoption of HCRs. We do this in three steps. First, we examine the challenges each RFMO has faced in applying the precautionary approach to transboundary tuna management prior to its application for MSC certification. In doing so, we identify the main actors involved in the RFMOs and their position in promoting or hindering the development of HCRs. Second, we outline the implementation of MSC certification in each RFMO. Here we focus on the justification for MSC certification by the clients of the certification (i.e. those controlling the unit of certification), and the degree of cooperation being shown by the success or failure of the certification process. Third, we characterise the improvement

---

[15] Gunningham, Grabosky & Sinclair (1998).
[16] Gunningham & Sinclair (1999, 2017).
[17] Peters, Pierre & King (2005).
[18] Overdevest & Zeitlin (2014).

pathways in each of the three RFMOs by considering the multiple dimensions, actors and sequencing outlined previously.

7.3 COMPARATIVE ANALYSIS: MSC-RFMOS INTERACTION

7.3.1 *Western and Central Pacific Fisheries Commission*

7.3.1.1 Setting HCRs and Reference Points

The Western and Central Pacific Fisheries Commission (WCPFC), established in 2004, is responsible for managing straddling and highly migratory fish stocks in the Western and Central Pacific Ocean (WCPO);[19] a region that accounts for approximately 70 per cent of annual global tuna production.[20] The WCPO is home to the remaining healthy tuna stock that provides about two thirds of the world's tuna supply every year.[21] Of this catch, 70 per cent of the total landings of the region are from the exclusive economic zones of the eight member states of the Parties to the Nauru Agreement (PNA) (established in 1982), including the Federated States of Micronesia, Kiribati, Marshall Islands, Nauru, Palau, Papua New Guinea, Solomon Islands and Tuvalu, between 2010–2014.[22] This catch, in turn, represents 49 per cent of global skipjack catch.

Tuna management in the WCPO is divided between three regional bodies – the WCPFC, PNA and the Pacific Islands Fisheries Forum Agency (established 1979). The Fisheries Forum Agency represents all twenty-two coastal states of the Pacific Ocean, including the PNA, and provides information and training to improve the general management of all tuna species and fisheries in the region. In contrast, the PNA focuses on the management of purse seine fishing targeting skipjack tuna.[23] But both the Fisheries Forum Agency and the PNA share common interest in ensuring: 1) effective management of tuna stocks and their ecosystem and 2) better governance for equitable distribution of wealth from tuna fishing in their exclusive economic zones. When the WCPFC was established in 2004, the Fisheries Forum Agency and the PNA became subregional blocs under its jurisdiction.

The WCPFC meets annually to determine total allowable catch/effort limits for highly migratory stocks, and to adopt standards for the collection and timely exchange of data on fisheries in the convention area.[24] Decision making depends on the advice and recommendation of the Scientific Committee, but instruments

---

[19] Tsamenyi, Rajkumar & Manarangi-Trott (2004).

[20] Harley et al. (2014a).

[21] Havice (2010); Havice (2013).

[22] This 70 per cent represents the purse seine catch in the PNA waters from 2010 to 2014 when the high seas were closed or limited.

[23] Barclay & Cartwright (2007).

[24] Havice (2010, 2013).

are adopted by consensus as binding conservation and management measures. Furthermore, the Scientific Committee rely on the work and advice of the Secretariat of the Pacific Communities, with strong political supports of the Fisheries Forum Agency and the PNA.

Since the 1990s, the advice of the Scientific Committee has been to reduce effort and catch to maximum sustainable yield. Achieving this requires a harvest strategy with clear and defined target and limit reference points for all tuna species to be able to specify measurable benchmarks.[25] Developing these reference points, however, has proven politically sensitive due to potential trade-offs in benefits and costs between the purse seine and longline fisheries in the region, which are exploited by different member states.[26] Specifically, the deployment of Fish Aggregating Devices (FADs) by purse seine fisheries targeting skipjack tuna has led to overfishing of both yellowfin and bigeye tuna, which are exploited by longline and handline fisheries.[27] The challenge in reaching agreement on setting precautionary limits to yellowfin and bigeye exploitation at maximum sustainable yield levels is to balance the interests of coastal and distant-water fishing states taking into account the wealth generated by maintaining growth of the skipjack purse seine fishery.

7.3.1.2 MSC Process

In response to the failure of the WCPFC to put harvest control rules in place, the PNA has worked at the subregional level to set measures for sustainable fishing in line with their wider objective of generating greater wealth from their tuna resources. At the first presidential summit in 2010, the leaders of the eight PNA countries signed the Koror declaration, which established the intention to have skipjack and yellowfin certified against the MSC standards, and in doing so, to expand their management control of tuna resources in the supply chain.[28]

In preparation for signing the Koror declaration the PNA underwent MSC pre-assessment in 2009 of their skipjack and yellowfin tuna purse seine fishery setting on 'free school' (or non-FAD) tuna, naturally occurring 'FADs' such as floating logs, and associated fisheries using anchored and drifting-FADs.[29] The result determined that only the skipjack targeted purse seine fishery setting on free school tuna would pass full assessment, whereas yellowfin was considered ineligible (or risky) because of concerns related to the data supporting the health of the stock in other areas of the Western Pacific Ocean, especially in Philippine and Indonesian waters. The PNA free school purse seine fishery entered full assessment in 2010 and was certified in

---

[25] Hampton et al. (2012); WCPFC (2015b).
[26] Barclay & Cartwright (2007); Havice & Campling (2010); Parris (2010); Bailey, Sumaila & Martell (2013).
[27] Harley et al. (2014a).
[28] Aqorau (2015).
[29] Banks (2009).

2011 with six improvement conditions and seven recommendations for improvement within the certification period to 2016.[30]

A major condition for recertification was the adoption of appropriate limit and target reference points and the development of more effective harvest control rules by the PNA and WCPFC. These conditions were agreed to by the PNA, and an action plan was developed for monitoring through an annual surveillance audit process within their five-year certification period.[31] The agreed action plan also included explicit milestones for the WCPFC to initiate the identification and development of appropriate reference points in 2012 (year 1), and their adoption in 2013 (year 2). In 2016, before the start of the recertification procedure for the PNA tuna fisheries, the PNA and WCPFC also committed to a well-defined HCR consistent with the limit reference point.[32]

The PNA argued that the 'third implementing measures' developed in 2009 and the Vessel Day Scheme established in 2007 constitute their harvest strategy for the skipjack purse seine fishery.[33] In particular, they refer to the vessel day scheme, and what are referred to as the 'third implementing arrangements', including catch retention of all tuna, seasonal FAD closure, partial high seas closures, full observer coverage and the use of an electronic vessel monitoring system. However, even if these implementing measures qualified the PNA for closing out the MSC condition for setting HCRs, their challenge was to ensure that they have the HCRs adopted as binding conservation and management measures for the region as a whole. For this reason, the adoption of HCRs and reference points at the WCPFC is a key for maintaining the MSC certification.

The PNA has demonstrated progressive improvement towards meeting the conditions and milestones required for MSC certification in 2016. New conservation and management measures have been adopted by the WCPFC either as a result of or in parallel with the MSC conditions for the PNA.[34] In 2012, in preparation for the Commission's meeting, the eighth annual meeting of the Scientific Committee held a series of workshops and sessions, during which the topics of the introduction of HCRs and limit reference points were presented and discussed.[35] These discussions led to the establishment of a harvest strategy for key fisheries and stocks, with a work plan to be achieved by 2017, and the establishment of interim target reference points for skipjack tuna at 50 per cent of biomass in the absence of fishing.[36] As a result, five of the six initial conditions from the initial assessment have been met,

---

[30] Banks et al. (2011); Banks et al. (2012)
[31] Banks et al. (2011); Bellchambers et al. (2016).
[32] Daume & Morison (2016).
[33] Banks et al. (2012); Aqorau (2015).
[34] Daume & Morison (2014, 2016).
[35] Hampton et al. (2012).
[36] WCPFC (2015a).

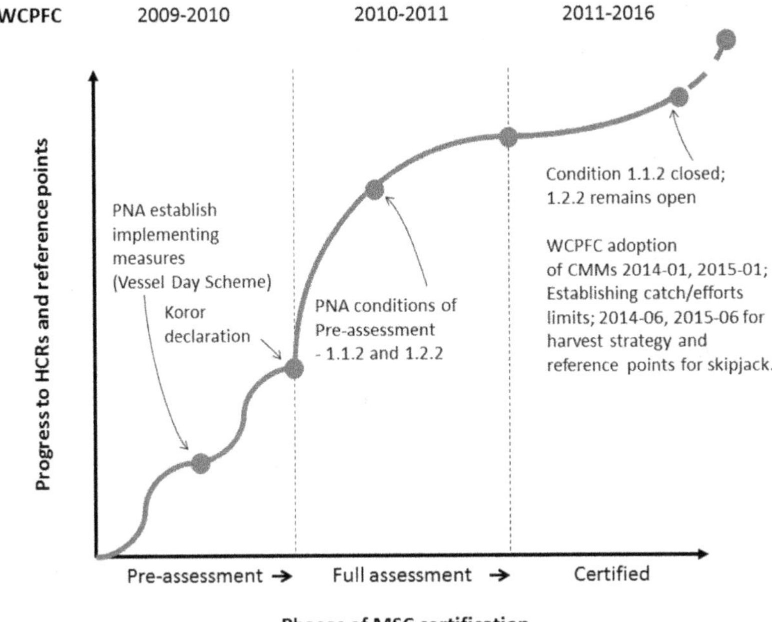

FIGURE 7.1 WCPFC-MSC improvement pathway

including those for setting limit reference points for skipjack.[37] The condition for setting a harvest strategy remains in progress.

7.3.1.3 Pathway

The PNA purse seine fishery has moved from pre-assessment to certification and is pending recertification (Figure 7.1). In the process, key conditions for improvement have been observed, related directly to the definition and implementation of HCRs and limit and target reference points. The PNA has responded by working to close out these conditions in accordance with the surveillance audits. What this case emphasises, however, is the role of the PNA as a subregional level in using the MSC standards to assess, set and measure progress to establishing a harvest control strategy and improvements in the fishery driven by the PNA in the absence of progress at the RFMO level. Reflecting the notion of dynamic experimentation in environmental regimes,[38] the PNA states have played an active role through the MSC process to initiate dialogue and negotiation for the establishment of measures that will promote maximum sustainable yield based on precautionary approaches in line with UNCLOS.

---

[37] Daume & Morison (2014).
[38] Overdevest & Zeitlin (2014).

Questions have been raised, however, as to whether the improvements seen to date at both the PNA and RFMO levels are attributable to MSC certification. Many of the elements associated with the harvest control strategy were established at the time or even before the MSC pre-assessment in 2009, perhaps indicating that the PNA would have made progress towards establishing a strategy for skipjack tuna anyway. In that sense, the MSC certification only verifies the actions the PNA was already undertaking. Though this may be true, the MSC nevertheless appears to have played a role in creating adequate incentives for the PNA to go beyond an internal management strategy. Without the MSC, the PNA might have been less likely to adopt and implement 1) the Vessel Day Scheme, 2) HCRs and reference points at the WCPFC, and 3) an action plan with well-defined milestones for improvement. Perhaps most importantly, it appears there would have been less ambition to contribute to the reduction of negative impacts from purse seine fisheries beyond PNA waters on the overall tuna stocks.

Finally, this case represents a strategy of disincentivising the use of FADs through certification. These are implemented with FAD penalties, FAD closures and catch retention rules (these are all referred to as 'compartmentalisation'), reflecting validity and supporting MSC's underpinning of the theory of change. At the time of this writing, this theory of change has been the subject of intense opposition by some industry groups not aligned to free school fishing.

7.3.2 *Indian Ocean Tuna Commission*

7.3.2.1 Setting HCRs and Reference Points

The Agreement for the Establishment of the Indian Ocean Tuna Commission (IOTC) was ratified in 1993 and made operational in 1996 under Article XIV of the FAO Constitution related to the management of food 'of particular interest to Member Nations of geographical areas'.[39] The IOTC area of competence is the Indian Ocean, defined as FAO statistical areas 51 and 57, including 'adjacent seas', expanding to north of the Antarctic Convergence.[40] Membership in IOTC is open to Indian Ocean coastal states and to countries or regional economic integration organisations that are members of the United Nations (e.g. the Bay of Bengal Programme), or one of its specialised agencies and are fishing for tuna in the Indian Ocean.[41] The adoption and design of economic institutions or policy instruments depend on the advice of the Scientific Committee recommended by its subsidiary bodies (working parties established for species and processes).

---

[39] See FAO (1945).
[40] IOTC-PRIOTCO2 (2016).
[41] IOTC (1993).

Historically, pole and line and longline were the primary tuna fisheries in the Indian Ocean, with a combined reported annual total catch of 400,000–600,000 metric tonnes (mt).[42] With the introduction of the purse seine fishery in mid-1980s, catches increased steadily to peak at 625,074 mt in 2006.[43] In 2006, the member countries of the IOTC agreed to reduce bigeye catches to the 2004 level and yellowfin to below the 2000 level.[44] Subsequently, catches declined to 455,999 MT and 428,719 MT in 2009 and 2010 respectively.[45] Although the IOTC was successful in reducing catch levels, it remained unclear whether this reduction would achieve maximum sustainable yield for bigeye and yellowfin in the long run without the adoption of targets and limits and HRCs. The reason for this is related to concerns over the increasing purse seine impacts on yellowfin and bigeye stocks in the long run, coupled with the lack of both political support and the Scientific Committee's capacity and capability to push for the adoption of precautionary approaches that would promote exploitation at maximum sustainable yield due to short-term interests of some Parties.[46]

Unlike the WCPFC, the IOTC Agreement does not refer to the application of a precautionary approach, given that it was established under the FAO and prior to UNCLOS and the UNFSA.[47] Instead, it seeks solutions for 'optimal utilization' throughout the convention area and for ensuring effective information is collected on the fisheries for scientific evaluation and advice.[48] For that reason, the IOTC Agreement lacks the definition of important elements such as fishing, fishing operations and fishing vessels, which are significant impediments to the efficient implementation of the agreement.[49] The first round of performance reviews of the IOTC conducted in 2009 recommended that this oversight be reformed and the ecosystem and precautionary approaches be adopted – which would in turn provide a basis to establish HCRs and reference points.

7.3.2.2 MSC Process

The introduction of the precautionary approach to the IOTC began through the engagement of the MSC certification of the Maldives pole and line skipjack fishery in 2012. At the start of the assessment process, the Maldives submitted a formal application for full membership of the IOTC in response to the MSC requirement to contribute to the management of the skipjack stock within the IOTC framework.

---

[42] IOTC (2009); IOTC (2012).
[43] IOTC (2009).
[44] Anonymous (2009).
[45] IOTC (2012).
[46] Huntington et al. (2009); IOTC (2009).
[47] IOTC (1993); IOTC (2009).
[48] IOTC (1993).
[49] IOTC (2009).

This brought the fishery, which accounts for 21 per cent of the global pole and line tuna catches (and 80 per cent of reported skipjack tuna catches) in the Indian Ocean, under the authority of the RFMO.[50] Handline targeting yellowfin tuna was initially part of the Maldives MSC application, but was suspended due to the poor condition of the stock.

The rationale for the Maldives to apply for MSC certification began in 2008,[51] as part of a wider plan supported by the World Bank to manage the bait fishery on which the pole and line industry depends.[52] Within this action plan the government outlined the need for a precautionary approach to realise the sustainable development of fisheries for economic diversification and growth.[53] The development of a precautionary approach drove the Maldives to enter the MSC process, due to a clear indication of a healthy skipjack stock and minimum pole and line negative impacts on the fishery ecosystem, and also seeing the potential role of the MSC program in helping to develop and achieve their strategy for implementing the precautionary approach. The pole and line fishery within the Maldives exclusive economic zone went through MSC pre-assessment from 2009 to 2010 and successfully completed full assessment in November 2012, with eight conditions for improvement that needed to be satisfied during the first five-year certification period, which ended in 2016.[54]

In response to the MSC conditions, the Maldives government and the IOTC developed an action plan to reach certification within the five years.[55] The plan included a clear timeframe for the development of a fisheries management plan for skipjack by 2012, a review of catch, effort and size frequency data by the end of 2013, the implementation of the management plan by the third quarter of 2014 and a recovery plan for yellowfin tuna. However, two of the eight MSC conditions were the development of HCRs and establishment of target and limit reference points for skipjack tuna, which required implementation by the IOTC.[56] After becoming members of the IOTC, the Maldives conducted a number of sessions with other coastal states on the relevance and importance of establishing HCRs and reference points with the support of key international and private organizations, including the International Pole and Line Foundation, International Seafood Sustainability Foundation, World Wildlife Fund, the World Bank and the MSC itself.

The end result was the adoption of Resolution 12–01 on the implementation of the precautionary approach, which marks the introduction of the precautionary

---

[50] IOTC (2015).
[51] IOTC (2009).
[52] Huntington et al. (2009).
[53] Ibid.
[54] Ibid.
[55] Scott & Stokes (2014).
[56] Huntington et al. (2009).

approach and harvest strategies in the IOTC.[57] The Maldives subsequently submitted a proposal to the IOTC in 2013 for the implementation of an interim HCRs rule for skipjack tuna. The proposed HCR formulated a procedure for making harvest policy decisions to achieve a desired state. As part of the 2013 surveillance audit, there was also an attempt to assess the eligibility of yellowfin stocks against MSC Principle 1 for healthy stock, which started in 2013 (and is still in progress to date).[58] In 2016, Resolutions 16–01 and 16–02 were adopted, which set a stock rebuilding plan for yellowfin. They do this through reductions by gear type to reduce the 2016 catch level by 20 per cent relative to 2014.[59] Resolution 16–02 was adopted for the establishment of HCRs for skipjack tuna in the Indian Ocean – the first harvest control rule to be developed.[60] Specifically, all fisheries are included in the proposal to adopt a binding resolution within the remit of IOTC by virtue of Maldives taking the initiative to submit the proposal.

The experiences of the Maldives contrasts markedly with the experiences of the second fishery in the IOTC to apply for MSC certification. In 2013 Echebastar, a family company from the EU and Seychelles that had been fishing in the region since 1981, applied for MSC certification for their free school purse seine fishery targeting bigeye, yellowfin and skipjack.[61] The assessment was finalized in November 2015, showing that all three units of certification scored above the 80-point threshold for certification. However, the final approval for certification was delayed because of the absence of HCRs and reference points at the IOTC (which subsequently came in 2016).[62] Six of the ten conditions identified by the assessors related to the development and adoption of HCRs and reference points for all three species (yellowfin, bigeye and skipjack) by 2019. But in contrast to the Maldives, which faced similar conditions, Echebastar's application for MSC certification has not been approved. At the time of this writing the Echebastar application is still undergoing assessment, and yet no evidence of progress with which to improve its prospects for approval has been seen at IOTC.

7.3.2.3 Pathways

We observe two slightly different improvement pathways of MSC in the Indian Ocean (see Figure 7.2). On the one hand, the Maldives pole and line case shows a pathway based directly on 'certify first, improve second', which implies a certified fishery with the necessary political engagement with improvement conditions at the IOTC level. On the other hand, a fishery controlled by a single company, and

---

[57] IOTC (2016).
[58] Scott & Stokes (2014).
[59] IOTC (2016).
[60] Ibid.
[61] Nicks et al. (2015).
[62] Ibid.; WWF (2016).

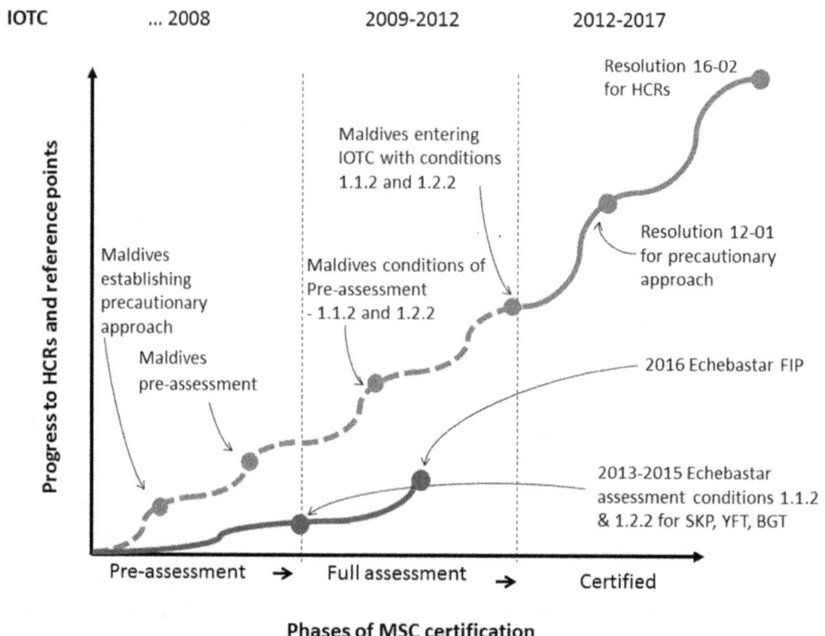

FIGURE 7.2 IOTC-MSC improvement pathway

forming the single unit of certification, remains in a prolonged assessment process despite having similar requirements for improvement set at the IOTC level.

In comparing the two pathways, we find that public-private engagement is more successful in driving change at tuna RFMOs than the fishing industry's (private) acting alone without state support. The difference between these fisheries is the Maldives' capacity as a member state to influence decision making at IOTC. This implies that a pathway of 'certify first and improve second' appears more attainable to a client-state than to a private unit of certification like Echebastar, which has less potential to influence decisions at the IOTC level. The Echebastar case in turn raises questions about the credibility of the MSC process in supporting fisheries that lack the ability to influence what are essentially political decisions at the IOTC level.[63] Having HCRs and reference points for skipjack adopted at IOTC in 2016 does open up the potential for Echebastar to obtain certification for skipjack fishing (though this may be contingent on their suspending their catching of yellowfin and bigeye). This example therefore shows that although private intervention is key for improvements and changes at tuna RFMOs, our IOTC cases demonstrate how the use of private institutions by states, with their ability to garner political support, can be shown to have a greater chance of success in speeding up developments at the RFMO level.

[63] WWF (2016).

7.3.3 *International Commission for the Conservation of Atlantic Tuna*

7.3.3.1 Setting HCRs and Reference Points

ICCAT was established in 1970 in response to the widespread concerns over the overexploitation of stocks of tuna and tuna-like fishes in the Atlantic Ocean and its adjacent seas.[64] Historically, overexploitation of Atlantic tuna was driven by the expansion of the large-scale longline sector and purse seine fishery in the late 1950s and mid-1960s respectively.[65] The ICCAT convention area covers more than thirty species, of which Atlantic bluefin tuna is the most economically and politically significant, followed by bigeye, albacore and yellowfin.[66] The stated objective of ICCAT is to 'cooperate in maintaining the populations of these fishes at levels which will permit the maximum sustainable catch for food and other purposes'.[67]

The Commission currently includes fifty-one contracting member states and is organised into a number of constituent bodies responsible for data collection, the definition of conservation and management measures and recommendations for monitoring, control and surveillance. The specific functions and tasks of ICCAT are separated based on the subregional division of fish stocks. These subregional divisions are particularly important because the stocks represent the political interests and capacity of the actors and participants of the fishery. For example, albacore tuna is treated as three separate stocks: the Northern Atlantic Ocean, Southern Atlantic Ocean and the Mediterranean Sea. Bluefin is divided into the Eastern and Western Mediterranean stocks, and Swordfish into the Northern, Southern and Mediterranean stocks. Bigeye and yellowfin, in contrast, are treated as a single stock.

ICCAT catches increased from 1950 to a peak of nearly 600,000 mt in 1994, from which time they have declined.[68] After sixty years of stock decline and growing pressure on the Commission from nongovernment organisations and key member states, ICCAT adopted a number of long-term recovery plans for albacore (since the 1990s) and bluefin (in the late 2000s).[69] These plans focused on setting catch limits, size limits, seasonal closures based on both areas and times of fishing and the closure of the Gulf of Mexico spawning area for the bluefin stock.[70] ICCAT has adopted a number of recommendations for northern albacore since 1998, including limits on fishing capacity, which were later updated for North Atlantic albacore catch limits for the period 2008–2009, as well as subsequent recommendations for the establishment and implementation of a rebuilding program.

---

[64] ICCAT (2008); Allen (2010).

[65] Saunders & Haward (2016).

[66] Ibid.

[67] Cited in ICCAT (2008), at 1.

[68] ICCAT (2015).

[69] ICCAT (2008); Saunders and Haward (2016).

[70] Allen (2010); Carleton et al. (2010).

Although plans are in place to recover overexploited stocks, the recovery of stocks has been slow, with claims that ICCAT has failed to reach its own object-ives.[71] Critiques of the recovery plan have focused on the failure of the Commission to adopt appropriate recommendations of the Standing Committee on Research and Statistics (SCRS) due to the political pressure from member states and eco-nomic pressure from the industry.[72] Others have noted the failure of some contract-ing parties to comply with their obligation to provide timely and accurate data, as well as the presence of illegal, unregulated and unreported (IUU) fishing efforts.[73] Some respondents also note that until 2008, ICCAT measures were based on the recovery plan with the aim of increasing biomass instead of developing a clear management plan, which has undermined the long-term sustainability of the stocks in question.

The outcome of the 2008 independent review of ICCAT recommended the adoption of the precautionary and ecosystem approach to fisheries management in line with the 1995 FAO CCRF and UNFSA.[74] However, ICCAT countered this review by stating that the precautionary and ecosystem approaches had already been introduced in 1997 when ICCAT established an ad hoc working group and com-mittee. ICAAT also stated that they progress had been made since 1998 ICCAT in developing a stock recovery plan. But as respondents noted, progress was slow and substantive change lacking because of the persistent dynamic political issue of conflicting conservation and economic interests.[75] Since 2008, more practical and ongoing work has been done within the separate committees and panels in broadening their efforts to incorporate precautionary and ecosystem approaches, including the continued dialogue on the definition and evaluation of precautionary reference points and HCRs through the implementation of a management strategy evaluation approach.[76] Such dialogue continues until today, and there has not been any decision on the adoption of such measures supporting a precautionary approach to management.

7.3.3.2 MSC Process

Many claim that the MSC has played a significant role in influencing improvement towards the adoption of precautionary approaches in ICCAT.[77] However, MSC's presence in ICCAT is both narrow and recent. This is mainly due to the poor status of most tuna stocks in ICCAT and the ongoing recovery work plan for all tuna

---

[71] ICCAT (2008); Merino et al. (2016).
[72] ICCAT (2008).
[73] Saunders & Haward (2016).
[74] ICCAT (2008).
[75] Also see ICCAT (2008).
[76] Saunders & Haward (2016).
[77] ICCAT (2015); Silva et al. (2016).

stocks.[78] This chapter identifies three types of MSC engagement in the Atlantic Ocean: a failed assessment, an ongoing fisheries improvement program, and one certified fishery. In doing so, this chapter examines how this multiple engagement has influenced the development of HCRs and reference points by the Commission.

The St. Helena pole and line fisheries for albacore, bigeye, yellowfin and skipjack tuna went through MSC assessment from February 2009 until July 2010.[79] This small isolated island, located inside the southern tropics of the Atlantic Ocean, has a capacity of 500 mt of fish, which provides income to 40 fishermen on the island. The fishery failed the assessment because of the status of bigeye and yellowfin for the entire Atlantic, skipjack in the eastern half of the Atlantic, and albacore in the Southern Atlantic.[80] After the St. Helena assessment there was no further involvement of MSC in ICCAT-managed fisheries until 2016, because it was not until then that the health of stocks were at a requisite level to be certified.[81]

In response to the ICCAT's slow progress in setting requirements for implementing a stock rebuilding plan, in 2016 the Organisation of Associated Producers of Large Tuna Freezer Vessels (OPAGAC) and WWF developed and agreed to a global tuna fisheries improvement plan.[82] In line with ICCAT's rebuilding plan for bigeye, yellowfin and skipjack tuna, OPAGAC agreed to prioritise activities to meet MSC requirements for having healthy stock status, noting key ICCAT recommendations for tropical tunas. These recommendations include the establishment of a multiannual management plan, updating catch limits on bigeye and yellowfin to the 2010 levels, and a framework for developing a harvest strategy for each stock.[83] Though none of the ICCAT-OPAGAC fisheries are MSC certified or in assessment, there is apparent interest in the fishery in having tropical tunas enter MSC once stocks are fully recovered and MSC standard for healthy stock is satisfied.

Also in 2016, the North Atlantic albacore (Spanish) fishery was awarded MSC certification. The clients are the Spanish Inshore Producers Organisations from Guipuzcoa (OPEGUI) and Biscay (OPESCAYA), and the San Martin de Laredo Fishermen's Guild. The fishery covers 129 vessels, of which 87 are troll vessels that capture albacore from June to October, and 42 pole and line vessels that fish from July to November, within the Bay of Biscay and adjacent North Atlantic waters. In 2014, the fishery accounted for 28,000 mt total allowable catches, which was well below the maximum sustainable yield of 31,680 mt.

The northern albacore fishery's interest in MSC certification began in the late 2000s, but did not move forward due to the status of the stocks. The fishery re-entered the certification process in 2014, when the stock was still recovering, and

---

[78] ICCAT (2008); OPAGAC & WWF (2016).

[79] Carleton et al. (2010).

[80] Ibid.; ICCAT (2016); Silva et al. (2016).

[81] ICCAT (2016); Silva et al. (2016).

[82] OPAGAC & WWF (2016).

[83] ICCAT (2015); OPAGAC & WWF (2016).

went into full assessment in 2015, when there was a near full recovery of the stock (97 per cent). The fishery was ultimately awarded certification in July 2016.[84] One of the five conditions set under the certification is to develop clear and well-defined HCRs, despite the fact that ICCAT's recommendation 11–13 for HCRs was in place and operational. The point raised by the MSC assessors was that these HCRs remain too poorly defined to assess North Atlantic albacore as a separate stock.[85] This indicated that progress toward agreeing on HCRs had been severely stalled while these were under development. In response, the MSC organized a pilot 'harmonization' meeting for ICCAT albacore fisheries; as a result, the northern albacore fishery has developed an action plan to adopt well-defined HCRs from the Commission by 2020, with the hope of integrating these into the existing strategic plan from 2015 to 2020.[86]

7.3.3.3 Pathways

In contrast to other RFMOs, what we observe in the context of ICCAT is multiple pathways converging to reinvigorate a stalled process of setting reference points and HCRs (see Figure 7.3). These pathways are separate in that they are set up in separate fisheries and regions. But they also differ in the order through which they seek to pressure ICCAT. The St. Helena fishery's assessment failed, but played an important advocacy role in highlighting that, when measured against an independent measure, the ICCAT process was failing. The OPAGAC fisheries improvement plan is based on a more traditional process of closing out conditions before certification, which in itself is a risky proposition because it is based on an ambition for certification rather than the risk of (very publicly) losing certification. The northern albacore certification may play a highly complementary role to the OPAGAC fisheries improvement plan because it is under a strict timeline to close out conditions, similar to what we see in the WCPFC and IOTC cases.

7.4 DISCUSSION AND CONCLUSION

The three cases in this chapter show how the incorporation of MSC certification into RFMOs can lead to dynamic and adaptive improvement pathways towards the fulfilment of precautionary transboundary management. But more fundamentally, these cases also demonstrate how smart regulatory mixes in transboundary resource management can contribute to improved cooperation between states, as well as between states and the industry actors they seek to regulate. We argue that the mixes of public and private rules and regulations in these cases are 'smart' because they

---

[84] Silva et al. (2016).
[85] Ibid.
[86] ICCAT (2015).

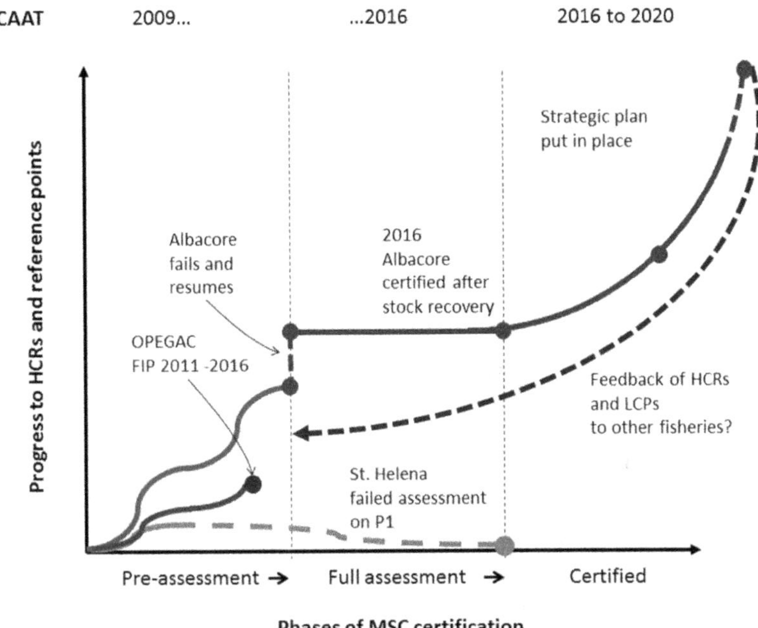

FIGURE 7.3 ICCAT-MSC improvement pathway

create sequenced interactions that contribute to the alignment of states and industry interests across multiple dimensions of borders, sectors and institutional levels. But, extending beyond the more instrumental view of Gunningham, Grabosky and Sinclair,[87] these regulatory mixes are also smart because they align the interests and strategies of actors for long-term strategic decision making in complex transboundary settings.

The pathways approach introduced in this chapter helps to illustrate the dynamic and nonlinear dimensions of smart regulatory mixes. In doing so, it goes beyond the notion of sequencing common to the smart regulation literature in two ways. First, whereas sequencing is used to describe a 'pyramid' of regulatory levels from voluntary to legal,[88] our pathways approach highlights the order of decisions and cooperative interactions between public and private actors involved in the application of a smart mix. Second, sequencing is not simply a matter of substituting or holding some tools in reserve;[89] it highlights the coordination and influence of one tool over actors involved in complex transboundary decision-making contexts. Based on these cases, we characterise the importance of these sequenced improvement pathways in

---

[87] Gunningham, Grabosky & Sinclair (1998).
[88] Braithwaite (2006).
[89] Gunningham & Sinclair (2017).

terms of dynamism, conditionality and experimentation, before turning to the consequences for RFMOs and the MSC moving forward.

First, the dynamic nature of the pathways illustrated in each of these cases demonstrates there is no single or first best sequencing of the RFMO-MSC smart mix. At each step of MSC engagement, from pre-assessment to full assessment and post-certification, member states are forced to make decisions either towards or away from the formulation of collective and cooperative rules (in this case HCRs). In some cases these decisions are highly programmed, in the sense that there is a clear strategy of preparation in the pre-assessment phase. As illustrated by the WCPFC case, this is attributable to a relatively high degree of coordination and control over the stock by the PNA as a subregional bloc. In contrast, the IOTC and ICCAT cases show that the MSC can align political and economic interests by providing a framework for states to objectively assess progress towards their obligation to establish harvest strategies. For instance, the failed MSC assessments of the St Helena fishery and improvements made in the pre-assessment of the Northern albacore fishery helped ICCAT create clarity over the consequences and need for cooperative decision making towards HCRs. In the IOTC, the ascension of the Maldives to membership in the Commission demonstrates the awareness of and incentive for regional cooperation.

Second, the different pathways show the importance of conditionality in setting timelines for demonstrating improvement. As exemplified by the 'law of least ambitious program', consensus-based international regimes tend towards extended (and sometimes perpetual) timelines for decisions that are disruptive to the interests and practices of the parties involved.[90] While not a solution to these extended timelines, all three cases of RFMO-MSC regulatory mixes demonstrate that MSC conditions can agitate states and private actors alike to come to consensus on the establishment of (amongst others) HCRs. But each case also highlights how these conditions can be applied at different stages of the certification process.[91] As a result, the RFMO-MSC interaction remains flexible enough to provide incentives for improvement at different stages of rule development in response to the strategies and interests of RFMO members. For example, the WCPFC highlights the role of the MSC pre-assessment phase in creating the ambition to establish international agreements on fishing access and effort allocation in preparation for full assessment. Similarly, the IOTC demonstrates the role that time-bound conditions for re-assessment in one fishery can play in creating change at the RFMO level.

Third, the multiple pathways of MSC-RFMO interaction demonstrate the value of experimentalist approaches to institutional change in international environmental regimes.[92] The various pathways observed not only highlight the value of certifying

---

[90] E.g. Pentz & Klenk (2017).
[91] As also demonstrated by Martin et al. (2012).
[92] Overdevest & Rickenbach (2006).

fisheries and setting conditions over time, but the value of parallel processes of, for example, failed and successful certification and fisheries improvement plans. We argue that even unsuccessful attempts at certification contribute pressure on RFMOs to change – largely because of the transparency and public engagement that they are exposed to through the certification process.[93] This is not to say that, ultimately, outcomes need to be seen in terms of the improving status of stocks; instead, it places greater appreciation on the multiple and various procedural changes that certification can foster.[94] For example, looking beyond the outcomes in a single fish stock, we see that conditions for establishing HCRs and reference points at the RFMO level also oblige all members to implement and comply, which in turn improves governance and management of transboundary tuna resources at all waters, regardless of where the fish is caught.

Seeing the RFMO-MSC smart regulatory mix as a dynamic and adaptive process of learning and change holds consequences for how we understand the role of certification as a tool for improving fisheries management. Our results, for instance, provide contrast to recent critiques of the role of conditions in promoting improvement through MSC certification. Christian et al.[95] argue, for instance, that the objections procedure of the MSC is flawed because objections that are lodged against fisheries rarely lead to a fishery being denied certification. Similarly WWF's review of MSC engagement in the Indian Ocean claims that 'the reliance on conditions only perpetuates a psychology that sub-standard fisheries should be embraced within "the MSC Program" in order to foster their improvement. Experience to date has largely discredited this notion'.[96] Both of these critiques ultimately argue that fisheries, in a very linear notion of improvement, should close out all conditions before being certified. This chapter has, on the one hand, noted that conditions set by certification to design HCRs in each of the RFMOs remain open. Nevertheless, the pathways outlined in each case show that the certification process, including failed assessments, provides a dynamic rather than linear vision of the impact these conditions have over cooperation and, ultimately, improvement in RFMOs.

Finally, our results also hold insights for understanding the apparent 'crisis' in RFMOs.[97] As international regimes mandated with managing trans-boundary fish stocks, their record may be less than perfect. But RFMOs are representative of international cooperative management bodies, faced with a complex political economy of competing public and private interests. Our results demonstrate that innovation and/or reform of RFMOs can be more fruitfully based on a dynamic understanding of smart regulatory mixes. Seen as such, RFMOs are not only the

---

[93] Cf. Auld & Guldbransen (2010).
[94] Green (2013).
[95] Christian et al. (2013).
[96] WWF (2016), at 2.
[97] See Cullis-Suzuki & Pauly (2010).

sum of cooperative international relationships; They are also the sum of transboundary interactions between public and private actors, structured through the interplay between private rules/norms and public regulation. Breaking through barriers to cooperation by understanding RFMOs as open, dynamic and adaptive international regimes can then help to better formulate strategies that make progressive change towards precautionary management of the world's transboundary tuna stocks.

REFERENCES

Allen, R. 2010. 'International management of tuna fisheries: Arrangements, challenges and a way forward', Technical Paper, FAO Fisheries and Aquaculture.

Anonymous. 2009. Report of the IOTC Performance Review Panel, IOTC: 56.

Aqorau, T. 2015. 'How Tuna is Shaping Regional Diplomacy'. In Fry, G. & Tarte, S. (eds.), *The New Pacific Diplomacy*. Canberra, ANU Press, 223–235.

Aranda, M., Murua, H. & de Bruyn, P. 2012. 'Managing Fishing Capacity in Tuna Regional Fisheries Management Organisations (RFMOs): Development and State of the Art'. *Marine Policy* 36(5), 985–992.

Auld, G. & Gulbrandsen, L.H. 2010. 'Transparency in Non-state Certification: Consequences for Accountability and Legitimacy'. *Global Environmental Politics* 10(3),97–119.

Bailey, M., Sumaila, U. & Martell, S. 2013. 'Can Cooperative Management of Tuna Fisheries in the Western Pacific Solve the Growth Overfishing Problem?' *Strategic Behavior and the Environment* 3, 31–66.

Banks, R. 2009. Pre-Assessment Report for the Unassociated and Associated Purse Seine Fisheries Operating in PNA Waters. Australia, Moody Marine Limited.

Banks, R., Clark, L., Huntington, T., Lewis, T. & Hough, A. 2011. MSC Assessment Report for PNA Western and Central Pacific Skipjack Tuna (*Katsuwonus pelamis*) Unassociated and Log Set Purse Seine Fishery. Intertek Moody Marine, Moody International, Version: 4 Final Report.

2012. MSC Assessment Report for PNA Western and Central Pacific Skipjack Tuna (Katsuwonus pelamis) Unassociated and Log Set Purse Seine Fishery. UK, Intertek Moody Marine.

Barclay, K. & Cartwright, I. 2007. 'Governance of Tuna Industries: The Key to Economic Viability and Sustainability in the Western and Central Pacific Ocean'. *Marine Policy* 31 348–358.

Bellchambers, L. M., Fisher, E. A., Harry, A. V. & Travaille, K. L. 2016. 'Identifying and Mitigating Potential Risks for Marine Stewardship Council Assessment and Certification'. *Fisheries Research* 182, 7–17.

Braithwaite, J. 2006. 'Responsive Regulation and Developing Economies'. *World Development*, 34(5), 884–898.

Brown, B. E. 2016. 'Regional Fishery Management Organizations and Large Marine Ecosystems'. *Environmental Development* 17, 202–210.

Carleton, C., Medley, P., Southall, T. & Gill, M. 2010. MSC Sustainable Fisheries Certification: St. Helen Pole & Line and Role & Line Tuna Fisheries for Albacore, Bigeye, Yellowfin and Skipjack Tuna (Final Report). Scotland, Food Certification International Ltd.

Cashore, B. 2002. 'Legitimacy and the Privatization of Environmental Governance: How Non-State Market-Driven (NSMD) Governance Systems Gain Rule-Making Authority'. *Governance* 15(4), 503–529.

Christian, C., Ainley, D., Bailey, M., et al. 2013. 'A Review of Formal Objections to Marine Stewardship Council Fisheries Certifications'. *Biological Conservation* 161, 10–17.

Cullis-Suzuki, S. & Pauly, D. 2010. 'Failing the High Seas: A Global Evaluation of Regional Fisheries Management Organizations'. *Marine Policy* 34(5), 1036–1042.

Daume, S. & Morison, A. 2014. The PNA Western and Central Pacific Unassociated Purse Seine Skipjack Tuna - 2014 3rd Annual Surveillance USA. Emeryville, CA, SCS Global Services.

2016. The Parties to the Nauru Agreement Western and Central Pacific Unassociated Purse Seine Fishery: Yellowfin Tuna (Thunnus albacares) Expedited Principle 1 Assessment. Emeryville, CA, SCS Global Services.

FAO. 1945. Constitution of the United Nations Food and Agriculture Organization (FAO). Quebec, FAO.

Grafton, Q. R., Arnason, R., Bjørndal, T., et al. 2006. 'Incentive-based Approaches to Sustainable Fisheries'. *Canadian Journal of Fisheries and Aquatic Sciences* 63(3), 699–710.

Green, J. F. 2013. *Rethinking Private Authority: Agents and Entrepreneurs in Global Environmental Governance.* Princeton, Princeton University Press.

Gulbrandsen, L. H. 2010. *Transnational Environmental Governance: The Emergence and Effects of the Certification of Forests and Fisheries.* Cheltenham, Edward Elgar.

Gunningham, N. & Sinclair, D. 2017. 'Smart Regulation'. In Peter, P. (ed.), *Regulatory Theory: Foundations and Applications.* Canberra, ANU Press, 133–148.

1999. 'Regulatory Pluralism: Designing Policy Mixes for Environmental Protection'. *Law & Policy* 21(1), 49–76.

Gunningham, N., Grabosky, P.N. & Sinclair, D. 1998. *Smart Regulation: Designing Environmental Policy.* New York, Clarendon Press.

Hampton, J., Berger, A. M., Harley, S., Pilling, G. M. & Davies, N. 2012. *Introduction to Harvest Control Rules for WCPO Tuna Fisheries,* Manila, WCPFC.MOW1-IP/06.

Harley, S., Davies, N., Hampton, J. & McKechnie, S. 2014a. Stock Assessment of Bigeye Tuna in the Western and Central Pacific Ocean. Scientific Committee - Tenth Regular Session. Majuro, Marshall Islands. Western and Central Pacific Fisheries Commission. WCPFC-SC10–2014/SA-WP-01.

Harley, S., Williams, P., Nicol, S. & Hampton, J. 2014b. The Western and Central Pacific Tuna Fishery: 2013 Overview and Status of Stocks. Western and Central Pacific Fisheries Commission.

Havice, E. 2010. 'The Structure of Tuna Access Agreements in the Western and Central Pacific Ocean: Lessons for Vessel Day Scheme Planning'. *Marine Policy* 34(5), 979–987.

2013. 'Rights-based Management in the Western and Central Pacific Ocean Tuna Fishery: Economic and Environmental Change under the Vessel Day Scheme'. *Marine Policy* 42, 259–267.

Havice, E. & Campling, L. 2010. 'Shifting Tides in the Western and Central Pacific Ocean Tuna Fishery: The Political Economy of Regulation and Industry Responses'. *Global Environmental Politics* 10(1), 89–114.

Hilborn, R., Orensanz, J. M. & Parma, A. M. 2005. 'Institutions, Incentives and the Future of Fisheries'. *Philosophical Transactions of the Royal Society B: Biological Sciences* 360 (1453), 47–57.

Huntington, T., Anderson, C., Macfadyen, G. & Powers, J. 2009. MSC Assessment Report for Pole and Line Skipjack Fishery in the Maldives. Maldives, Moody Marine Ltd., Version 3.

ICCAT. 2016. Report of the Standing Committee on Research and Statistics (SCRS). Madrid, ICCAT, 429.

2015. Strategic Plan Report: 2015–2020 SCRS Scientific Strategic Plan. Madrid, ICCAT, 21.

2008. Report of the Independent Review. International Commission for the Conservation of Atlantic Tunas (ICCAT). Madrid, ICCAT. PLE-106/2008.

2016. Compendium of Active Conservation and Management Measures for the Indian Ocean Tuna Commission (Updated November 2016). Seychelles, IOTC, 226.

2016. Report of the 2nd IOTC Performance Review. Seychelles, IOTC, 86. IOTC-2016-PRIOTCO2.

2015. Report of the 19th Session of the Indian Ocean Tuna Commission. Busan, Rep. of Korea, IOTC, 155.

2012. Report of the Fifteenth Session of the IOTC Scientific Committee. Seychelles, IOTC, 288.

2009. Report of the IOTC Performance Review Panel. Seychelles, IOTC, 56.

1993. Agreement for the Establishment of the Indian Ocean Tuna Commission. Rome, IOTC.

Martin, S. M., Cambridge, T. A., Grieve, C., Nimmo, F. M. & Agnew, D. J. 2012. 'An evaluation of environmental changes within fisheries involved in the Marine Stewardship Council certification scheme'. *Reviews in Fisheries Science* 20(2), 61–69.

Merino, G., Murua, H., Arrizabalaga, H., et al. 2016. Establishment of reference points and harvest control rules in the Framework of the International Commission for the Conservation of Atlantic Tunas (ICCAT). Specific Contract No. 8 under FRAMEWORK CONTRACT - MARE/2012/21, Brussels, European Commission, 98.

Miller, A. M., Bush, S. R. & van Zwieten, P. A. 2014. 'Sub-regionalisation of Fisheries Governance: The Case of the Western and Central Pacific Ocean Tuna Fisheries'. *Maritime Studies* 13(1), 1–20.

Nicks, P., Ambrosio, L., Keathinge, M. & DeAlteris, J. 2015. MSC Sustainable Fisheries Certification: Echebastar Indian Ocean Purse Seine Skipjack, Yellowfin, and Bigeye Tuna Fishery, F. Department. Edinburgh, Acoura Marine Ltd., 283.

OPAGAC and WWF. 2016. Fishery Improvement Project Work Plan [online] http://awsassets.wwf.es/downloads/opagac_fip_work_plan_final_1.pdf.

Overdevest, C. & Zeitlin, J. 2014. 'Assembling an Experimentalist Regime: Transnational Governance Interactions in the Forest Sector', *Regulation & Governance* 8, 22–48.

Overdevest, C. & Rickenbach, M. G. 2006. 'Forest certification and institutional governance: An empirical study of forest stewardship council certificate holders in the United States'. *Forest Policy and Economics* 9(1), 93–102.

Parris, H. 2010. 'Tuna Dreams and Tuna Realities: Defining the Term "Maximising Economic Returns from the Tuna Fisheries" in Six Pacific Island States'. *Marine Policy* 34(1), 105–113.

Pentz, B. & Klenk, N. 2017. 'The "Responsiveness Gap" in RFMOs: The Critical Role of Decision-Making Policies in the Fisheries Management Response to Climate Change'. *Ocean & Coastal Management* 145, 44–51.

Peters, B. G., Pierre, J. & King, D. S. 2005. 'The Politics of Path Dependency: Political Conflict in Historical Institutionalism'. *The Journal of Politics* 67(4), 1275–1300.

PNA. 2010. *Koror Declaration.* Koror, Palau, PNA.

Rochette, J., Billé, R., Molenaar, E. J., Drankier, P. & Chabason, L. 2015. 'Regional Oceans Governance Mechanisms: A Review'. *Marine Policy* 60, 9–19.

Saunders, P. & Haward, M. 2016. 'Politics, Science, and Species Protection Law: A Comparative Consideration of Southern and Atlantic Bluefin Tuna'. *Ocean Development & International Law* 47(4), 348–367.

Scott, I. & Stokes, K. 2014. Expedited P1 Assessment - The Pole & Line Yellowfin Fishery in the Maldives. UK, Intertek Fisheries Certification.

Silva, M., Garcia, D., Maguire, J. J., Blazquez, L. & Povedano, V. 2016. North Atlantic Albacore Artisanal Fishery. Final Report. Madrid, MSC, 250.

Squires, D., Maunder, M., Allen, et al. 2016. 'Effort Rights-based Management'. *Fish and Fisheries* 18(3), 440–465.

Tsamenyi, M., Rajkumar, S. & Manarangi-Trott, L. 2004. 'The International Legal Regime for Fisheries Management'. Paper presented at the UNEP Workshop on Fisheries Subsidies and Sustainable Fisheries Management, University of Wollongong.

Underdal, A. 1980. *The Politics of International Fisheries Management*. Oslo, Universitetsforlaget.

WCPFC. 2015a. 'Conservation and Management Measure on the Target Reference Point for WCPO Skipjack Tuna'. Commission 12th Regular Session, Bali, Indonesia, WCPFC. CMM 2015–06.

2015b. 'Information Paper: Data summaries in support of discussions on the CMMs on tropical tunas (CMM 2013–01 and CMM 2014–01)'. Commission 12th Regular Session, Bali, Indonesia, WCPFC.

WWF. 2016. WWF Retrospective on Indian Ocean Tuna Harvest Control Rules, Unpublished, World Wide Fund for Nature.

Yeeting, A. D., Weikard, H. -P., Bailey, M., Ram-Bidesi, V. & Bush, S. R. 2018. 'Stabilising cooperation through pragmatic tolerance: The case of the Parties to the Nauru Agreement (PNA) tuna fishery'. *Regional Environmental Change* 18(3), 885–897. DOI: 10.1007/s10113-017-1219-0

8

Smart Mixes in Forest Governance[1]

Jing Liu

8.1 INTRODUCTION

Forests play vital and diversified roles in supporting ecosystems and human welfare, including sustaining the livelihood of forest-dwelling species, producing goods, controlling climate and conserving water and soil.[2] An alarmingly high rate of forest conversion and degradation, however, threatens the health of forests. Around 13 million out of the global total of 4 billion hectares of forest coverage disappears every year.[3] Deforestation is not dispersed evenly across the globe. It is more prominent in developing countries, especially countries with rich tropical forests.[4] While forest coverage has been expanding in Europe and has been stable in North America,[5] forest degradation also remains a concern in developed countries. In North America, the health of riparian forests is a concern,[6] while in Europe, biodiversity is an important concern, with very few primary forests remaining apart from those in Russia.[7]

Existing scholarship identifies multiple drivers that cause forest degradation or deforestation. Commercial and subsistence agriculture, mining, infrastructure and urban expansion are common drivers of deforestation. The drivers of forest degradation include timber logging, uncontrolled fires, and livestock grazing in the forest

---

[1] This chapter is extending the research done for the book, Liu, Faure & Mascini (2018).

[2] Millennium Ecosystem Assessment (2005).

[3] FAO (2010), at 10.

[4] On the international level, South America and Africa have the highest deforestation rates, with annual losses of 4 and 3.4 million hectares, respectively, from 2000 to 2010. At the national level, the largest annual net loss can be found in Brazil, Australia, Indonesia, Nigeria, Tanzania, Zimbabwe, the Democratic Republic of the Congo, Myanmar, Bolivia and Venezuela. See FAO (2010), at xvi, 21.

[5] Ibid., at xvi.

[6] McDermott, Cashore & Kanowski (2010), at 95; Sweeney et al. (2004), at 14132; Verry & Dolloff (2000).

[7] Paillet et al. (2010), at 101.

and fuelwood use.[8] In addition to those activities which directly influence the status of forests, indirectly, the economic (poverty, market, global demand), cultural, demographic, technological and institutional factors may also contribute to deforestation and forest degradation through their effects on the above-mentioned human activities.[9]

Apart from these causes, which have been identified as leading to forest degradation and deforestation, there is one characteristic of forests that is an underlying factor to these problems, which is that forests are common pool resources. This means that the exclusion of other users is very costly if not impossible (non-excludability) and the exploitation of one user diminishes the availability for others (subtractability).[10] Because exclusion is costly, when no institutions exist to regulate and restrict access, resource users tend to overexploit the resources. In this case, 'the tragedy of the commons' scenario materialises.[11]

Scholars have identified three basic types of property rights to restrict access to common pool resources: public, private and communal.[12] Extensive academic attention has been given to how common pool resources are protected under these different property rights and to the conditions for each property rights to function effectively.[13] The effective functioning of property rights means that the overexploitation of common pool resources is prevented through excluding users so that the subtraction of resources does not exceed sustainability levels, as well as making sure that present users do not expand their activities to exceed these levels.

Reliance on property rights, however, raises two sets of problems. On the one hand, property rights are not established without costs. Property rights need to be defined, properties need to be policed and in case of disputes, some dispute settlement mechanisms are necessary. These various activities associated with the establishment and maintenance of property rights result in transaction costs,[14] a concept which is often only vaguely defined.[15] The property rights literature usually uses it in a broad sense to incorporate 'costs associated with the transfer, capture and protection of rights'[16] or 'the resources used to establish and maintain property rights'.[17] Cole differentiates between two types of transaction costs: exclusion costs

---

[8] Hosonuma et al. (2012), at 3–4.

[9] Gupta, Van der Grijp & Kuik (2012), at 30.

[10] Feeny et al. (1990).

[11] Hardin (1968), at 1243.

[12] Feeny et al. (1990); Ostrom (2010).

[13] For example, Ostrom has identified eight design principles for the success of communal property rights regimes. The limitations of private property rights and public property rights have been discussed extensively in law and economics' literature. See Ostrom (2010); Cole (2010); Anderson & Leal (1991); Stroup & Baden (1983).

[14] Allen (2000), at 898–899; Allen (1991), at 1–18.

[15] Allen (2000), at 898–899. For such examples, see Barzel (1985), at 8; Alchian & Woodward (1988), at 66.

[16] Barzel (1989), at 122.

[17] Allen (1991).

and coordination costs. According to Cole, 'exclusion costs are the costs of drawing and enforcing boundaries to restrict access to and use of the resource to the owner(s) of the property'.[18] Coordination costs are the costs associated with solving collective action problems.[19] Reducing exclusion costs requires a clear definition of property rights and the capacity to enforce the rights, including policing, dispute settlement and sanctioning. For private property, decision-making power is vested in individuals. Sometimes, however, the individual owners need to negotiate with external parties in using and managing resources. For both communal and public property, collective decision-making institutions are established and coordination takes place inside these institutions. These three types of transaction costs (definition, enforcement and coordination) are connected to two other factors: information and scale. Collecting and distributing information is crucial in both exclusion and coordination activities. Scale concerns the level at which property rights are managed and possibly the links between the units that manage property rights at different levels. The scale of governance can, therefore, influence both the enforcement of property rights and the coordination between stakeholders. These five factors set the necessary preconditions for property rights to effectively address common resource problems such as clear definition of property rights, enforcement, coordination, information creation and sharing, and scaling. In fact, these elements can be conceived of as conditions that need to be fulfilled for the proper functioning of property rights.

Even if problems of transaction costs would be solved, the establishment of property rights does not mean that all problems related to common pool resources are solved. For example, while the market value of timber is internalized through the establishment of forest property rights, nonmarket values, such as the ecological and esthetical value of the forest, are not considered by self-interested private actors. In other words, the right holder may still use his or her property in a way that creates externalities for third parties. This results in market failure, which may justify external intervention, including public and private regulations.

Both public and private regulations influence the functioning of property rights. For example, public regulation can affirm private ownership of forests through registration or the creation of (quasi-)private property rights[20] on state-owned forests,

---

[18] Cole (2002), at 131.

[19] Ibid.

[20] It is debatable whether such rights should be classified as 'property rights' or 'quasi-property rights' in strict legal terms. Property rights carry different meanings in legal and economics literature. As a legal term, property rights are narrowly defined as rights *in rem*, rights to exclude others from use of the assets, and are limited by the numerous clauses principle. However, economists use the term more broadly to include rights and privileges in accessing and using resources as well. Some of these rights or privileges may not be qualified as property rights. For example, whether the rights to use and manage collectively owned forests established by contracts are property rights in law is hotly debated in China. To avoid the complexity in the legal debate, this chapter uses the broad definition of the economics literature to denote 'a set

such as through the granting of concessions and influencing the definition of property rights. Besides, private regulatory regimes, such as forest certification, can apply to resources that are managed under different property rights, adding another layer of enforcement. In addition, public and private regulations are also closely linked. Therefore, property rights, along with public and private regulations, constitute an intricate and interactive system that governs common pool resources such as forests.

Against this background, this chapter focuses on the influence of institutional factors on the conditions for a proper functioning of property rights, more particularly on public and private regulation. In other words, this chapter examines the extent to which public and private regulation interact in facilitating the necessary preconditions for effectively addressing common pool resources by clearly defining property rights, enhancing enforcement capacity, smoothing coordination, generating and sharing information and operating the right scale of governance. In so doing, this chapter explores when such interaction would constitute 'smart mixes' in relation to the governance of forests.

Measuring the effectiveness and efficiency of regime and instrument mixes in the governance of common pool resources is empirically difficult. This chapter therefore examines the smartness of mixes in an indirect way. Firstly, it examines whether public and private regulation interacts in a complementary or counteracting way. Secondly, it discusses the effect of public and private regulation on the five preconditions for the establishment and maintenance of property rights, as well as their potential to overcome the externalities created by property rights. Following this Introduction, the chapter is divided into three main sections. Section 8.2 examines property rights and public and private regulation for forests respectively. Sections 8.3 and 8.4 explore whether the combinations of these regulatory schemes can be considered in terms of smartness; the former focusing on the interaction between public and private regulation and the latter discussing the influence of their interaction on the functioning of property rights. Section 8.5 concludes the chapter.

8.2 FOREST GOVERNANCE: PROPERTY RIGHTS, PUBLIC AND PRIVATE REGULATION

Deforestation in rainforests started to trigger global attention in the 1960s. Since the 1970s, several international environmental conventions have started to address forest protection.[21] However, a legally binding agreement specifically

of rights to control assets', or in other words, the form of power where 'a sanction and authority for decision-making' over resources has been established. See Cole (2010); Denman (1978); Dasgupta (1982).

[21] Such as the 1973 Convention on the International Trade in Endangered Species, the 1975 World Heritage Convention and the Ramsar Convention on Wetlands of International

dealing with protection of forests has not been concluded.[22] The forest-related rules are still dispersed among various non-forest–focused international environmental conventions. This chapter leaves aside these conventions, and focuses on domestic law of the selected countries in relation to forest management and protection.

In selecting countries, a number of factors were considered: notably, the prominence of forestry problems, the stage of development, the prevailing property rights regime and the presence of an institutionalised certification system. Based on these factors, five countries were chosen – Bolivia, Canada, Indonesia, Sweden and the United States of America. All of these countries have substantial forest coverage: 51.4 per cent in Indonesia, 52.2 per cent in Bolivia, 34.1 per cent in Canada, 33.3 per cent in the United States and 69.2 per cent in Sweden.[23] Bolivia and Indonesia are two developing countries suffering from rapid deforestation. They both have adopted diverse types of property rights: public, private and communal. The property rights are often less clear and less secure than in many developed countries. This ambiguity has led to conflicts between forest users and has contributed to high deforestation rates. Although fewer forests have been certified in Indonesia and Bolivia than in the three developed countries, given the low penetration rate of forest certification in developing countries, they represent examples of developing countries with comparatively well institutionalised certification schemes.

Canada, Sweden and the United States are developed countries with stabilized forest coverage, but still face forest degradation problems, such as the loss of ecological function caused by insufficient protection of riparian forests and the loss of biodiversity. While most forests in Canada are publicly owned, private forests are more popular in the United States and Sweden.[24] Community-owned or managed forests do exist in these three countries (such as those owned or managed by Native American tribes in the United States, First Nations in Canada and Sami people in Sweden), but only on a small scale.[25] Forest certification started to develop from these countries and has a wide coverage there: around 36 per cent of forest areas in North America are certified[26] and more than half of the forests are certified in Sweden.[27]

Importance, the 1992 United Nations Framework Convention on Climate Change and the 1992 United Nations Convention on Biological Diversity.

[22] Instead, the United Nations Forum on Forests adopted a non-legally binding instrument on all types of forest in 2007.

[23] See data at http://data.worldbank.org/indicator/AG.LND.FRST.ZS.

[24] McDermott, Cashore & Kanowski (2010), at 74, 80, 136.

[25] Sunderlin, Hatcher & Liddle (2008).

[26] www.unece.org/fileadmin/DAM/timber/publications/FPAMR-2014-final_01.pdf, at 17.

[27] https://se.fsc.org/preview.the-contribution-of-fsc-certification-to-biodiversity-in-swedish-forests.a-661.pdf.

8.2.1 *Property Rights on Forests*

Most property regimes governing environmental goods are 'mixtures of individual private ownership, private (non-state) common property management, state ownership and management (i.e., regulation)'.[28] Both ownership and less complete property rights can thus be established in relation to forests. Owners have broad rights over their property (either forests or the land where forests grow), including access, harvest, management, exclusion and alienation. Sometimes, less complete rights are established for forests. For example, private parties can gain rights to access, harvest and manage a specific part of publicly owned forests through permits, concessions or contracts.[29] In these cases, ownership and less complete users' rights coexist on the same forests: ownership is a public property right, while users' rights are private.

The Food and Agriculture Organisation (FAO) collected data from 188 countries (accounting for 99 per cent of total forest area) on forest ownership by the year 2005.[30] The FAO defines forests owned by individuals, corporate/institutions and communities as private forests. According to FAO, public forests compose the largest part of the global forests, amounting to 80 per cent of all forest ownership. In comparison, 18 per cent of forests are privately owned, and the remaining 2 per cent of forests are classified as 'other' (including unknown and disputed ownership).[31] Public ownership was predominant in most regions with private ownership being common in Europe excluding Russian (more than 50 per cent), Central America (46 per cent), Oceania (37 per cent), East Asia (33 per cent) and North America (31 per cent).[32] For private forests, data concerning the type of forest owner is only available for 55 per cent of global forest area. In these areas, '59 percent of private forests were owned by individuals, 19 percent by private corporations and institutions, and the remaining 21 percent by local communities and indigenous people'. This suggests that communally managed forests are only a small fraction of global forest areas. Such communal rights are most common in Africa (excluding Northern Africa) and Central America.[33]

An interesting phenomenon is the shift towards state ownership and control of property rights in developing countries since colonisation. Although historically forests were exploited by local communities in many developing countries, as a result of the influence of colonisation and decolonisation, state ownership and control over natural resources became common. One example is the case of

---

[28] Cole (2010), at 235.

[29] This is the main approach to manage public forest in Canada. Permits/concessions have also been established on public forests in Indonesia and Bolivia.

[30] FAO has been monitoring forests since 1946 and has produced the Global Forest Resources Assessments every five years. The recent assessment report was produced in 2010, which provides information on forest ownership in 2005.

[31] FAO (2010), at 122.

[32] Ibid., at 122–123.

[33] Ibid., at 123.

Indonesia, where forests were traditionally managed by local communities. During the colonial period, the colonial government claimed ownership over resources and controlled the land in Java.[34] The post-colonial government of Indonesia retained state ownership of land and forests.[35] During the Suharto period, the authority for managing the forests was concentrated in the hands of the central government, which claimed control over forests in the whole country; however, the government could only conduct direct control over forests on Java, and it outsourced forests in other areas to big timber companies with good political connections.[36] The rights of smaller loggers and customary communities were not recognised. It was not until the *Reformasi* period that the rights of these marginalized actors were recognised again to a limited extent.[37] The Basic Forestry Law of 1999 stipulates that 'forest control by the state shall respect customary laws, as long as it exists and its existence is recognised and not contradicting national interests'.[38] The law requires government to establish regulation on the process to recognise traditional (*adat*) communities; to date, however, no such regulation has been passed.[39] In practice, the existence of the *adat* community depends on the recognition of the government, and the *adat* rights can be revoked if the existence of *adat* is no longer recognised or if it is determined to contradict national interests.[40] Hence, the *adat* system has rarely enjoyed 'more than minimal force in practice'.[41] Private rights to forests were also permitted in Indonesia by the Basic Agrarian Law (1960) and Basic Forestry Law (1967). The definition of private forests, however, was not clear in these laws[42] and low land registration rates threaten the security of private forests.[43] Given the lack of clarity related to the definition of forest tenure, land conflicts are common in Indonesia.[44]

A shift from customary forests to public forests, similar to Indonesia's, also occurred in Bolivia. The first Forestry Law of Bolivia in 1974 declared state ownership of all forests, and required users of forests on both public and private land to

---

[34] Gold & Zuckerman (2014), at 3.

[35] Ibid., at 79.

[36] Arnold (2008), at 80.

[37] This is believed to be a result of the demise of the military's role in social affairs and the emergence of civil society. Under such circumstances, communities began to demand their rights and the conflicts between them and transmigrants/concessions operators came to rise. See McCarthy et al. (2006); Arnold (2008), at 81.

[38] English copy see http://theredddesk.org/sites/default/files/uu41_99_en.pdf; www.elaw .org/node/2644.

[39] Article 67(2)(3).

[40] Arnold (2008), at 86.

[41] Bartley (2011), at 530.

[42] Safiti (2010), at 88.

[43] Gold & Zuckerman (2014), at 13.

[44] It is estimated that 22.5–24.4 million out of 98.6 million hectares of forests in Indonesia are subject to conflicts. See Indrarto et al. (2012), at 13.

obtain permits from the state.[45] Permits were granted only to registered enterprises, and local and indigenous communities were excluded. Land and forest law reform started in the mid-1990s. The state still owned all forests in Bolivia, but the land on which forests grows could be owned by the state, private parties and indigenous groups (in the latter case through Tierras Communitarias de Origen-TCOs).[46] In this regard, three types of forest tenure were established on the public land: long-term contracting, concessions by private parties and concessions assigned to local communities (ASLs).[47] Reforms that have been implemented since 2006 have further strengthened the rights of indigenous people and have put a halt to issuing new concessions to timber companies. The highly theoretical clear forest tenure system, however, led to some issues in practice, especially in cases of private and communal rights.[48] Without the clear forest tenure system, forest users have little incentive to protect their forests, leading to alarming deforestation rates, which amount to 0.7 per cent of forest cover per year in Indonesia[49] and 0.53 per cent per year in Bolivia[50] in the late half of the first decade of the twenty-first century.

In developed countries, comparative clear forest tenure systems have been established. Some countries are dominated by public forests, including Canada, where over 90 per cent of forests are publicly owned.[51] But private parties can gain rights to harvest public forests either through area or volume-based licenses.[52] The licensees have an important influence over the forests through being responsible for the preparation of the forest management plans.[53] Privately owned forests have a more central place in the property systems in some other countries, such as the United States of America and Sweden. In the United States, 58 percent of forests are privately owned.[54] In Sweden, 48 percent of forests are owned by individuals, 24 percent by private sector companies; and another 6 percent by other private owners.[55] Clear forest tenure has been established in those countries, providing incentives to rights holders to conserve the forests. Therefore, deforestation is no longer a major problem in these developed countries. In spite of the establishment of clear forest tenure, however, some externalities remain. Because the rights holders cannot capture the full benefits of maintaining biodiversity, conserving the soil and water, these issues are often neglected. Therefore, many developed countries still

---

[45] Benneker (2008).
[46] Boscolo & Vargas Rios (2007), at 192.
[47] Müller, Pacheco & Montero (2014), at 29.
[48] Pacheco et al. (2008).
[49] FAO (2010).
[50] Müller, Pacheco & Montero (2014).
[51] McDermott, Pacheco & Kanowski (2010), at 74.
[52] Ibid., at 76.
[53] Ibid.
[54] Ibid., at 80.
[55] Skogsstyrelsen 2018, Strukturstatistik: Statistik om skogsägande 2017, at 6. Available at: https://www.skogsstyrelsen.se/globalassets/om-oss/publikationer/2018/rapport-2018-12-strukturstatistik-statistik-om-skogsagande-2017.pdf.

face forest degradation problems, such as biodiversity loss in Sweden[56] and unhealthy riparian buffer zones in the United States and Canada.[57]

8.2.2 *Public Forest Regulation*

Most countries have adopted public regulatory instruments relating to forests into their domestic law, including spatial planning, permits, tax, subsidy, soft targets, information and education, but such regulations vary.[58] This section focuses on three commonly used instruments, which have important influence over the functioning of property rights regimes: spatial planning, permits and concessions and other technical standards.

8.2.2.1 Spatial Planning

Because the same methods of resource exploitation – such as timber harvest (either by large companies and private owners or indigenous communities), food collection, creation of plantations and agriculture – may be used by different actors in the same forests, the coexistence of these multiple activities can potentially lead to spatial conflicts. In Indonesia, for example, conflicts between commercial timber companies and indigenous peoples are widespread.[59] In addition, the fast expansion of palm oil plantations, mining and agriculture has led to the conversion of some forests in this country.[60] In Bolivia, the competition between timber harvest, agriculture and cattle ranching has also led to rapid deforestation.[61]

By deciding which activities are allowed in specific areas, spatial planning is an important tool for addressing such conflicts. The coordination of different interests in spatial planning, however, is not easy. The case of Indonesia illustrates this phenomenon, as both the central and regional governments have authority to engage in spatial planning at their respective levels; this often results in conflicting plans.[62] In addition to the Ministry of Forestry, which is responsible for determining the boundaries for forestry activities, sector-base legislation is used by non-forestry agencies to determine the areas for their related activities, such as agriculture, mining and the establishment of eco-regions.[63] Although the Spatial Planning law requires coordination among the actors involved in land-use planning, this has never

---

[56] Sahlin (2011), at 2.
[57] McDermott, Pacheco & Kanowski (2010), at 95.
[58] For an overview of the regulatory instruments, see Gupta, Van der Grijp & Kuik (2012), at 34–44; McDermott, Cashore & Kanowski (2010).
[59] Indrarto et al. (2012), at 13.
[60] Ibid., at 4–9.
[61] Müller, Pacheco & Montero (2014), at 10.
[62] Gupta, Van der Grijp & Kuik (2012), at 128.
[63] Arnold (2008), at 95.

actually been implemented.[64] As a result, the conflicts between the different levels of government and different departments have led to incoherent spatial planning in Indonesia.

8.2.2.2 Permits and Concessions

Permits and concessions are commonly used by governments to restrict access to forests. In Indonesia, the Ministry of Forestry is responsible for the issuing of most logging permits. Local governments can issue some 'lesser' permits,[65] and other authorities can issue permits for mining and agricultural plantations. Based on conflicting spatial planning, these different agencies sometimes issue conflicting permits, thereby accelerating deforestation.[66] In Bolivia, concessions have been used to authorize private actors (including twenty-year logging contracts and forty-year concessions) and indigenous peoples (through the Bolivia Asociaciones Sociales de Lugar [ASLs]) to access public forests since 1996. Authorization is also needed to access and harvest private and communal forests (in terms of Tierras Communitarias de Origen (TCOs)). Similar to the case in Indonesia, land tenure and forests have been regulated by different agencies authorized by different laws in Bolivia. For example, while TCOs are decided by the National Institute for Agrarian Reform,[67] ASLs are granted by municipalities.[68] In the fear that claims for TCOs would endanger the availability of municipal forest reserves for ASLs, municipalities sometimes oppose the granting of TCOs, further complicating the process of regularising land ownership and solving the land-related conflicts.[69]

8.2.2.3 Other Technical Standards

Spatial planning and permits are closely related to, and necessary for, the establishment of effective property rights. In addition, public regulation can also determine how forests are to be managed directly, and therefore can influence the use of property rights. They can also help to address the remaining externalities of property rights, such as the critical condition of riparian buffer zones in Canada and the United States and biodiversity loss in Sweden. The technical standards that could be set by public regulation to address externalities include, for example, minimum logging cycles, a minimum cut diameter by tree species and so on. The examples of Sweden and Bolivia show that the design of such standards is usually based on

---

[64] Indrato et al. (2012), at 21.
[65] Such as timber extraction licenses for non-commercial purposes and non-timber forest product extraction permits. Barr et al. (2006).
[66] Indrarto et al. (2012), at 31.
[67] Boscolo & Vargas Rios (2007), at 192.
[68] Pacheco (2004), at 91.
[69] Real (2002), at 79–81.

Western science, which is more suitable for large-scale resource management by commercial parties than for small-scale forestry, which is often artisanal, low-impact, low capital-intensive and often lacks information. Using the same standards for commercial forestry and small-scale ones has created challenges for the latter.[70]

8.2.3 *Private Forest Regulation*

Due to the failure of the international community to reach a legally binding agreement on forest governance despite the apparent inability of existing international law and organizations to provide satisfactory solutions for forestry problems, since the 1990s private and hybrid regimes have started to develop to fill this gap. The earliest and most influential one of these private regimes relates to forest certification, represented by the Forest Stewardship Council (FSC) and the Programme for the Endorsement of Forest Certification (the PEFC). The forest certification schemes set standards for forest management as well as for the supply chain management of timber production. It requires an accredited, independent third-party certifier to evaluate and audit the production processes or methods according to predefined environmental and social sustainability standards.[71] Whereas the standards are defined by an independent governing body, the audits are conducted by private actors.

The FSC was initiated by environmental NGOs in 1993.[72] The FSC has developed a global standard for good forest management, called the FSC Principles and Criteria.[73] To tailor the general standards to the local situation in different jurisdictions, the FSC delegates the authority to elaborate them to its national affiliates.[74] Many countries have published their own FSC standards, such as the Bolivian Council for Voluntary Certified Forest Management Standards (Boliviano Certificación Forestal Voluntaria, CFV), FSC Sweden standards, as well as regional FSC standards in the United States and Canada. In countries where national standards have not been developed, forests are evaluated by certification bodies against their own standards adopted according to the FSC Principles and Criteria, as in Indonesia. The stringent standards of FSC in some cases led to caution among the forest industry, which responded by establishing their own industry-led national certification schemes, such as the Canadian Standards Association (CSA), the Sustainable Forest Initiative (SFI) in both Canada and the United States and the industry-led certification scheme in Sweden (PEFC Sweden). These national schemes were later endorsed by the PEFC. Together, the FSC and PEFC by

---

[70] Pacheco et al. (2008), at 40–44.
[71] Steering Committee of the State-of-Knowledge Assessment of Standards and Certification (2012).
[72] Gulbrandsen (2010), at 52.
[73] See https://ic.fsc.org/en/certification/principles-and-criteria.
[74] Gulbrandsen (2010), at 55.

2014 covered 11 per cent of global forests.[75] In addition to FSC and PEFC, some countries have developed other certification schemes, such as the Indonesian Eco-label Institute certification (LEI).[76] Most certified forests are, however, in North America and Europe.

In spite of the differences between national standards, they have to comply with some general principles, such as the FSC Principles and Criteria and the PEFC International Standards. These general principles set standards regarding environmental, economic and social issues. Clearly defined forest tenure and respecting indigenous peoples' rights are required under both FSC and PEFC standards.[77] Although the FSC standards are generally argued to be more stringent than PEFC standards, the competition between the two has led to the ratcheting up and convergence of standards.[78]

8.3 THE INTERACTION BETWEEN PUBLIC AND PRIVATE REGULATION

As discussed in Section 8.1, good governance of common pool resources relies on the proper functioning of property rights, which are influenced by institutional factors such as public and private regulation. Though forest certification started to develop due to dissatisfaction with public regulation, they do not function separately. Rather, public and private regulation is usually closely linked. A smart mix of these regimes/instruments can, therefore, be analysed from two perspectives: how public regulation interacts with private, and how that interaction influences the proper functioning of property rights regimes. This section focuses on the first. It argues, on the one hand, that public and private regulation can operate complementarily, and, on the other hand, that the differences between them demonstrate that they cannot replace each other.

8.3.1 *Complementarities between Public and Private Regulation*

Both public regulation and private regulation set performance standards for forest activities. Coherence of the standards can strengthen their capacity to realize protection goals. For example, in Bolivia, the indigenous movement, combined with deteriorating economic conditions and growing concerns about deforestation, triggered the forest law reform and the development of forest certification in the late 1990s, which means that the standards under two schemes (public and private) are

---

[75] Georgia-Pacific, *Forest Certification Around the World*, available at: http://www.fao.org/resources/infographics/infographics-details/en/c/325836/.

[76] Klassen, Romero & Putz (2014), at 256.

[77] Principles 2 and 3, FSC Principle and Criteria. PEFC International Standard, Sustainable Forest Management Requirement, criterion 6.

[78] Overdevest (2010).

now similar.[79] The coherence of standards led to the quick spread of forest certification in the early 2000s.[80] The same similarity and coherence of standards are also found in the United States and Canada regarding riparian forests protection. These are examples of how public and private regulation have joined forces in enforcement.[81] In Sweden, domestic law only provides a framework for forest regulation and includes minimum requirements, and thus relies largely on certification to concretise its standards.

The coordination between public and private regulators can strengthen the enforcement of forest protection and its associated rights. In North America, processes have been established for information sharing to enable public regulators to rely on the audits conducted by certification schemes. This saves the resources of public regulators, which can then be allocated for other purposes.[82] In Sweden, public regulation relies on a soft steering model and has a low coercion level.[83] In this case, the widely adopted certification provides a very useful complement to the oversight mechanism.[84]

Good public regulation often serves as a basis for private regulation. Most certified forests are located in developed countries, which generally score high on good governance and public regulation. Moreover, certified forests in developing countries are usually already regulated well before certification. For example, two field studies on certified forests in Indonesia show that most of these forests were already well-managed due to prior public regulation.[85] However, since developing countries generally still face substantial challenges in achieving good regulation,[86] it is more difficult for the largest part of forests in these countries to become certified.

When public regulation is lacking, certification has the potential to stimulate the creation of public regulation. For example, in Sweden, the legislation in the early 1990s did not require harvesters to leave retention trees in harvested forests, which is an important measure to maintain biodiversity. Attempts to regulate were unsuccessful due to objections from forest owners. A consensus was achieved in the FSC negotiation, which resulted in the requirement of leaving retention trees. Such

---

[79] Müller, Pacheco & Montero (2014), at 24–25.

[80] Espinoza & Dockry (2014), at 82.

[81] Anecdotal evidence shows that certification may have helped to improve the legal compliance of forest operators. See Carraway et al. (2002), at 26.

[82] Such as the arrangement between the Forest Practices Board in British Columbia and SFI; between Ontario's Ministry of Nature Resources and CSA as well as FSC. See Wood (2009), at 94.

[83] Johansson & Keskitalo (2014), at 121–123.

[84] Hysing & Olsson (2005), at 512.

[85] Hinrichs, Muhtaman & Irianto (2008), at 43; Harada & Wiyono (2014).

[86] The general performance of public regulation in different countries is ranked in some projects. For example, the World Bank has the Worldwide Governance Indicators project, ranking countries' performance in good governance. The indicators include, for example, voice and accountability, political stability and absence of violence, government effectiveness, regulatory quality, rule of law and control of corruption.

requirements were later incorporated into domestic law.[87] In addition, when public regulation is already in place, certification accredits public regulation to a certain degree.[88]

8.3.2 *Differences between Public and Private Regulation*

Although public and private regulation can operate in a complementary way, one cannot replace the other. Firstly, standards under public and private regulation are not always coherent. For example, while certification standards in North America and Bolivia closely mimic public regulation,[89] Indonesia provides an example of contradictory norms in public and private regulation regarding indigenous people's property rights. On the one hand, the certification schemes require clear forest tenure and the protection of indigenous rights over forests;[90] on the other hand, the state only gives minimum recognition to customary rights. The content of such rights and the procedure to formalize these rights, however, remain unclear. In other words, in Indonesia, private regulation supports the forest tenure of customary communities, while public regulation contradicts this. In Sweden, even though public and private regulation are largely complementary, sometimes the standards are not fully coherent, leading to confusion among forest operators.[91]

Secondly, private regulation and public regulation operate in different ways. Whereas public regulation applies at the national or regional level, certification applies to certified forests and operates at the management unit level. Moreover, certification control is based on company documentation and selected site visits and hence is selective rather than comprehensive.[92] In summary, both positive and negative interactions exist between public and private regulation. Private regulation has an important role to play to supplement public regulation, but it is not able to completely replace public regulation.

---

[87] Lister (2012), at 193–194.

[88] The existence of a good management system is a criterion in certification schemes. See Principle 7, FSC Principle & Criteria, FSC-STD-01–001; Section 4, PEFC International Standard, Sustainable Forest Management Requirements, PEFC ST 1003: 2010. Public regulations are a crucial part of the management system. Therefore, the evaluation process inevitably also involves the assessment of public regulation. This phenomenon is even more obvious with fishery certification, where governmental agencies sometimes act as clients of certification themselves. In this case, the agencies manage the fisheries themselves and certifying the fisheries are to a large extent accrediting the public regulatory system. Gulbrandsen (2010), at 125.

[89] McDermott, Noah & Cashore (2008), at 47.

[90] Principle 3, FSC Principle and Criteria; LEI 5000–1; Principles 2, 3 and 4, Standards FSC-STD-BOL-04–2000.

[91] For example, the requirements about setting aside forests are not fully aligned in certificate schemes and law. See Lister (2012), at 191–193.

[92] Keskitalo et al. (2009); Johansson (2014).

8.4 THE INFLUENCE OF PUBLIC-PRIVATE INTERACTION ON THE FUNCTIONING OF PROPERTY RIGHTS

This section discusses how the interaction between public and private regulation influences the five preconditions for the proper functioning of property rights, as well as its capacity to address the remaining externalities of property rights.

8.4.1 *Defining Property Rights*

Clear definitions of property rights and completeness in property rights are crucial to incentivize sustainable behaviour. Conflicting property rights do not only relate to inconsistent property law, but may also be aggravated by public regulation. As discussed in Section 8.2, spatial planning is crucial in reconciling the conflicting interests of different land users, since it plays a role in influencing to whom property rights are granted. Conflicting spatial plans of multiple public authorities in Indonesia, therefore, aggravate the conflicting uses among stakeholders in the forests. The overlapping authority of agencies to grant permits and concessions in the forests also adds to the conflicts of forest tenure in Indonesia and Bolivia.

The importance of completeness in property rights is also noted. Schlager and Ostrom argue that at the very least, communities need to have rights to access, harvest and manage the resources in forests in order to establish effective communal management.[93] The contents of property rights are usually defined in property statutes and customary law. Public regulation, however, has the potential to influence the contents as well. For example, in Indonesia, forest tenures of traditional communities can take two forms: ownership (customary forests) or less complete rights granted through community-based forest management (CBFM) policies. Under the CBFM policies, communities can participate in the management and have access to state or private forests, either through licensing or through agreements with companies or conservation offices.[94] In the licensing models, the rights granted to local communities are usually limited, sometimes prohibiting the transfer of rights, or the use of these rights as collateral for credit. Usually, limited participation of communities is allowed in the licensing process, during which the duration of the rights is determined.[95] The limited rights awarded to local communities fail to incentivise responsible behaviour.[96]

Institutional arrangements also play an important role in determining the contents of property rights as well as their functioning in practice. Institutional design is closely linked with public and private regulation. For example, under the customary law in Indonesia, customary communities manage their forests through their own

---

[93] Schlager & Ostrom (1999), at 105.
[94] Sari (2013), at 121.
[95] Ibid., at 293–295.
[96] Ibid., at 295–296.

traditional governance institutions (*adat*). The way in which the *adat* are organized varies greatly from one region of the country to another. The Village Governance Law, however, restructured local villages according to the Javanese model.[97] Therefore, traditional *adat* institutions lost their role in governing forests and were replaced by village heads, who were responsible for implementing policies top down, rather than designing and implementing rules that fitted the customary systems. In Bolivia, the traditional institutions were not formally abandoned, but the law also required the establishment of a new institution, the TCO, on top of traditional self-governing institutions (usually the villages) as a precondition to confirm the communities' property rights. Geographically separated and culturally different villages have been joined by the TCO, and this has weakened the basis of self-governance. The coexistence of different levels of governance has also weakened the capacity of public institutions to conduct effective control in Indonesia and Bolivia.[98] Therefore, unsuitable institutions can impede the definition of clear property rights. The introduction of private regulatory schemes may further complicate the institutional arrangements. Private certification schemes are usually market-based mechanisms, which require institutions to help forest operators to secure markets, and gain market power through joint bargaining processing and so on. Customary institutions and government authorities are usually not suited to conduct such tasks, and hence new institutions such as cooperatives are often established to fulfil the requirements related to certification.[99] In some developing countries, the cooperatives are subject to poor governance and community members are more reluctant to engage in the management of cooperatives than in traditional governing institutions.[100]

In addition, public and private regulation may favour one type of property rights over another. Both public and private regulations include technical standards regarding how forests should be managed. Such standards are often based on Western science, which is more suited for large-scale resource management by commercial parties than for small-scale customary forests. Such standards create challenges for customary communities.[101]

8.4.2 *Enforcement*

Property rights do not only need to be clearly defined, but also need to be enforced. Substantial costs may be involved in 'drawing and enforcing boundaries to restrict access to and use of the resource' to someone's property. The enforcement can either be conducted by rights holders themselves, who monitor the behaviour

---

[97] Arnold (2008), at 80.
[98] Pacheco (2011), at 10.
[99] Hinrichs, Muhtaman & Irianto (2008), at 50.
[100] Ibid., at 51.
[101] Pacheco et al. (2008), at 32–33.

influencing their forest tenure and solve the disputes through formal or informal mechanisms. In addition to the efforts of rights holders themselves, state and public regulators can also play a crucial role in monitoring and ensuring compliance.

The enforcement capacity of public regulators is related to the availability of staff, budget, expertise, vulnerability to corruption, the authority provided to a public agency and the relationship between different regulatory bodies. Indonesia and Bolivia provide examples indicating that conflicts between different departments and levels of government can lead to conflicting enforcement activities.[102] In Sweden, a deregulation process took place in forest regulation in the 1990s, leading to minimum public regulation and forest owners and companies' substantial discretion to choose the proper measures.[103] Therefore, the coercion of public regulation is not strong, and noncompliance by forest owners and operators is usually left unpunished.[104] After deregulation, the size and budget of the Sweden Forestry Agency have been reduced; hence the use of persuasive instruments has also been reduced.[105]

Certification can provide additional monitoring and assurance to the public enforcement capacity. In the United States, anecdotal evidence shows that certification has contributed to the improvement of compliance with public regulation regarding riparian buffer zone protection.[106] Certification does not, however, automatically improve monitoring and oversight. Certifiers may be captured by their customers in order to secure income. For example, in Sweden, objections brought by NGOs to certification have seldom led to the withdrawal of certificates.[107]

Public regulation, self-governing institutions and private regulation rely on different ways of monitoring: self-governing institutions rely on local users who conduct daily management activities and have easy access to information regarding the resources and exploitation activities. Public regulators rely on monitoring by government officials and the private management of certification schemes requires the expertise of forest professionals. The latter actors may have advantages in terms of economies of scale or expertise, but are further removed from the daily activities in forests, and monitoring can therefore be more costly. It is thus crucial to figure out whether these different approaches of monitoring complement or replace/weaken each other. Indonesia provides an example of a phenomenon where traditional self-governing institutions have been dismantled by law and self-enforcement has been

---

[102] See Section 8.2.2, this chapter.

[103] Johansson & Keskitalo (2014), at 121.

[104] Ibid., at 124; Hysing (2009), at 656.

[105] Hysing (2009), at 656.

[106] For example, a study on the forest operations and compliance of public regulation in Texas shows that those operations which transport timber to SFI mills have a higher rate of compliance than those transporting timber to noncertified mills. Carraway et al. (2002), at 26.

[107] NGOs complain that certification bodies sometimes delay in responding to formal complaints and are reluctant in requiring the suspension or revocation of certificates. The first partial suspension of certificate was only made in 2014. Sahlin (2013), at 8. Stora Enso (2014), at 38.

replaced by weak public enforcement.[108] In Bolivia, the traditional governing institutions have not been abolished, but an additional level of governing institution has been added to the local management, de facto weakening their functioning.[109] Certification may also require the replacement of traditional governing institutions by new governing institutions such as cooperatives or enterprises, adding to the complexity of enforcement.[110]

8.4.3 *Coordination*

Many different actors are involved in the exploitation of forest resources. Coordination between the various parties, both among rights holders or between forest rights holders and other parties, is therefore crucial for the proper functioning of property rights. The different types of property rights vary in their capacity to coordinate the behaviour of the different actors that are involved. Under public property rights, the state or other levels of government holds the rights, and consequently decides the allocation of user rights through acting as the coordinator. Under private property rights, rights holders need to coordinate with other stakeholders whose behaviour can influence the forests, such as the mining industry, plantation operators and farmers, as well as visitor services providers who are operating in the same areas. As far as communal property rights are concerned, self-governing institutions play an important role in coordinating the behaviour of their members. These self-governing institutions decide who has the rights to access and to harvest the resources, and how both of these will be done.

Both public and private regulation can influence the coordination of forest management, however; the coordination among governmental agencies can sometimes be problematic. For instance, in Indonesia and Bolivia, the departmentalisation and the struggling decentralization processes have led to conflicting decisions and regulatory gaps among different sectors or different levels of government.

Certification can also provide a platform for multiple stakeholders to collaborate since the standards of development and governance of the certification schemes usually involve industry, rights holders, social groups and other environmental interests. The governance structure of certification schemes, however, influences the extent of public involvement. For instance, the FSC scheme was initiated by NGOs, while industry generally plays a more important rule in PEFC schemes. Although the FSC is usually better received by NGOs, in recent years, it is also causing some concerns. For example, in the case of Sweden, dissatisfaction with certification has led some environmental groups to withdraw their support from the

---

[108] See Section 8.4.1, this chapter.
[109] Ibid.
[110] Hinrichs, Muhtaman & Irianto (2008), at 50.

certification schemes. This misalignment is problematic, since environmental groups are usually an important impetus for increasing the demand of certified products.[111]

The coexistence of different regulatory schemes may also add complexity to forest governance and cause some confusion among regulated actors. For example, multiple institutions have been created to manage community forests in Indonesia and Bolivia: traditional self-governing institutions, institutions reorganised or created by the government (e.g. the TCO in Bolivia) as well as cooperatives/enterprises created to get certification. These complexities have added to the cost of coordination. The example of Sweden also shows that having multiple layers of regulation often leads to confusion among forest owners in deciding their forest management activities.[112]

8.4.4 *Information*

The availability of information regarding resource status, forest activities and their interaction is crucial for resources management. Local users, government institutions and certification assessment bodies can all act together as information generators. These different sources of information all have advantages and disadvantages. Local actors are well informed about local issues, such as the status of the resources, local livelihood demands and behaviours. The indigenous exploitation of forests in Indonesia is an example. The uses of fallow periods, fires and open land for agriculture are based on local knowledge and respect for ecological limits. The replacement of customs by commercial exploitation and agriculture by outsiders, however, has had devastating effects on the tropical forests.[113]

Compared to local knowledge, 'the state has a regional and national advantage as well as a repertoire of tools and techniques which are not normally available to local institutions'.[114] However, obtaining information about vast forest areas remains a challenge, especially when, as in the cases of Indonesia and Bolivia, the capacity of the regulatory agencies is weak. Even in countries with comparatively strong governmental capacity, information insufficiency is not uncommon. This is especially true for issues with high scientific uncertainties, such as forest biodiversity protection. Such uncertainties may cause confusion for forest owners and managers in determining how to protect the forests. The inconsistencies and ambiguities of public and private regulation in Sweden aggravate the differences of opinions between forest owners/managers and environmental groups. Moreover, a recent study reviewing scientific literature on environmental standards shows that the

---

[111] Gulbrandsen & Auld (2016).

[112] Uggla, Forsberg & Larsson (2016), at 5.

[113] Arnold (2008), at 97.

[114] Berkes (2009), at 1694.

legislative and certification standards regarding many environmental issues do not reach the threshold recommended in scientific literature.[115] Therefore, the scientific uncertainty has not been well accommodated under either the public or the private regulatory system.

Certification can also play an important role in information generation and spreading. Certification schemes have rules concerning the documentation of processes related to forest management. Information is made available during the certification process. Assessment teams gather information from existing documentation and literature produced by clients, scientists, government authorities and other stakeholders. The publication of final reports and of surveillance reports also increases the transparency of information. Assessment reports also identify gaps in knowledge and issue conditions or recommendations accordingly, which provide drivers for information production.[116]

For effective resources management, it is beneficial to build a robust bridge between these complementary types of knowledge. For example, in North America, public and private regulators coordinate in sharing information and rely on the information provided by the others.[117]

8.4.5 *Scale*

Forest management requires considerations at different scales. Issues such as the retention of trees can be addressed at a local scale, but issues like biodiversity and riparian forests protection require landscape considerations. Regulatory approaches vary in their capacity to address different scales. Certification is conducted at management unit level and is therefore particularly tailored to local situations. The coordination between different units is therefore essential in order to incorporate higher-level considerations. Public regulation can address issues on both the local level and a larger scale. Research, however, shows that even the public regulation of riparian zone protection in Canada and the United States has not incorporated landscape and watershed-scale consideration sufficiently.[118] In Sweden, many forest areas are set aside from harvest in accordance with certification requirements. These areas, however, are often fragmented. Poor connectivity reduces their capacity to promote biodiversity protection. Public regulation, however, is subject to legal and financial limits. To set aside private forests from harvesting, public regulators need to compensate the private owners for lost production. Lack of funding is a major barrier to reaching formal protection targets.[119]

---

[115] Johansson et al. (2013), at 107–108.
[116] Cano Chacón (2013).
[117] See Section 8.3.1, this chapter.
[118] Richardson, Naiman & Bisson (2012), at 236.
[119] Lister (2012), at 192.

Decentralization is often recommended in the regulation and management of natural resources, the argument being that it brings the decision-making process closer to local users and consequently contributes to effective resource management. However, examples of forest governance show that when the system is captured and corrupted and there is a lack of coordination between government agencies and different levels of government – including when there are aggravating factors such as obscure and conflicting legislation – the decentralization process itself can become a driver of environmental problems. For example, in Indonesia, conflicting legislation concerning decentralization and recentralization reflects a tug-of-war between different levels of government.[120] Multiple agencies often conduct conflicting spatial planning and issue overlapping permits on the same piece of land. The self-interested motives of bureaucrats behind these conflicting behaviours have not yet been overcome. In Bolivia, conflicting interests among different levels of governments and different departments also exist, as indicated by the reluctance of municipalities to support TCOs and the nesting of the land regularization process under the agricultural model.[121]

8.4.6 *Externalities*

An important goal of property rights in addressing common pool resources problems is to internalize externalities and to limit the incentives of free riding. Property rights, however, have their own limitations. Establishing property rights over forests can help to internalise the economic values of the resources, and therefore can contribute to solving some aspects of the environmental problems, such as deforestation. However, property rights are less useful in internalizing the nonmarket values of resources. For example, the supporting value of forests for other species, as well as its functioning in soil and water conservation, will not be internalized by establishing property rights. Indeed, even in places like North America and Sweden, where clear forest tenure has been established, forest degradation still occurs. This has been one of the reasons for creating riparian zone and biodiversity protection through regulation.

8.5 CONCLUDING REMARKS

This chapter analyses some of the issues related to the protection and management of forests from the perspective of common pool resources. Because of the non-excludability and subtractability of such resources, users tend to over-exploit the resources and free ride on each other, leading to 'the tragedy of the commons'; This chapter discussed three types of property rights used to overcome the problems

---

[120] Singer (2009).
[121] See Section 8.2.2.2, this chapter.

associated with the governance of forest resources: public, private and communal property rights. A smart mix of regimes and instruments should be able to promote the good functioning of these property rights and overcome the limitations of these property rights. The discussion in this chapter shows some examples of smart (or not-so-smart) mixes of public regulation, private regulation and property rights. Although forest certification evolved due to the slow progress of international forest law, it interacts closely with public regulation. A smart mix is therefore needed to strengthen the complementarities between public and private regulation in order to offer effective governance, especially, since one system cannot replace the other. This suggests that the limitations of public and private regulation need to be addressed within the specific systems themselves. Moreover, the interaction between public and private regulation can also influence the functioning of property rights through defining property rights, influencing the enforcement of property rights and the coordination between stakeholders, providing information and addressing scale issues. The interaction between public and private regulation can also help to address the remaining externalities.

The tentative thoughts expressed in this chapter on which types of mixes are smart or not in addressing forest governance problems are mainly based on the examples in five countries, namely, Bolivia, Canada, Indonesia, Sweden and the United States. One must be very careful with generalisations, since the case studies in this chapter have shown that the effectiveness of the institutional design with respect to the interaction between public and private regulation, and the resulting effect on the way in which property rights protect the common pool resource, is highly context-specific. There is, therefore, not just one 'smart mix'; under the specific conditions resulting from the country context, some interactions may work better than others. It is important to take this context specificity into account when drawing normative conclusions from the results of this chapter.

REFERENCES

Alchian, A. & Woodward, S. L. 1988. 'The Firm is Dead; Long Live the Firm: A Review of Oliver E. Williamson's "The Economic Institutions of Capitalism: Firms, Markets, Relational Contracting"'. *Journal of Economic Literature* 26(1), 65–79.

Allen, D. W. 2000. 'Transaction Costs'. In Bouckaert, B. & De Geest, G. (eds.), *Encyclopedia of Law and Economics*. Cheltenham, Edward Elgar, 893–926.

1991. 'What are Transaction Costs?'. *Research in Law and Economics* 14, 1–18.

Anderson, T. L. & Leal, D. R. 1991. *Free Market Environmentalism*. London, Palgrave Macmillan.

Arnold, L. L. 2008. 'Deforestation in Decentralised Indonesia: What's Law Got to Do with It?'. *Law, Environment & Development Journal* 4(2), 75–101.

Barr, C. M. et al. 2006. *Decentralization of Forest Administration in Indonesia: Implications for Forest Sustainability, Economic Development and Community Livelihoods*. Jakarta, CIFOR.

Bartley, T. 2011. 'Transnational Governance as the Layering of Rules: Intersections of Public and Private Standards'. *Theoretical Inquiries in Law* 12(2), 517–542.

Barzel, Y. 1989. *Economic Analysis of Property Rights*. Cambridge, Cambridge University Press.
    1985. 'Transaction Costs: Are They Just Costs?'. *Journal of Institutional and Theoretical Economics* 141(1), 4–16.
Benneker, C. 2008. Dealing with the State, the Market and NGOs. The Impact of Institutions on the Constitution and Performance of Community Forest Enterprises in the Lowlands of Bolivia, Doctoral thesis, Wageningen University.
Berkes, F. 2009. 'Evolution of Co-Management: Role of Knowledge Generation, Bridging Organizations and Social Learning'. *Journal of Environmental Management* 90(5), 1692–1702.
Boscolo, M. & Vargas Rios, M. T. 2007. 'Forest Law Enforcement and Rural Livelihoods in Bolivia'. In Tacconi, L. (ed.), *Illegal Logging: Law Enforcement, Livelihoods and the Timber Trade*. London, Earthscan, 191–217.
Cano Chacón, M. 2013. The Role of Information of the Marine Stewardship Council Certification Process in Developing Countries: A Case Study of Two MSC Fisheries Certified in Mexico. Dissertation, Dalhousie University.
Carraway, B. et al. 2002. Voluntary Implementation of Forestry Best Management Practices in East Texas. Texas Forest Service. Available at: http://texasforestservice.tamu.edu/uploa dedFiles/Sustainable/bmp/round5.pdf.
Cole, D. H. 2002. *Pollution and Property: Comparing Ownership Institutions for Environmental Protection*. Cambridge, Cambridge University Press.
    2010. 'New Forms of Private Property: Property Rights in Environmental Goods'. In Bouckaert, B. (ed.), *Property Law and Economics*. Cheltenham, Edward Elgar. 225–269.
Dasgupta, P. 1982. *The Control of Resources*. Cambridge, MA, Harvard University Press.
Denman, D. R. 1978. *The Place of Property: A New Recognition of the Function and Form of Property Rights in Land*, Berkhamsted, Geographical Publications.
Espinoza, O. & Dockry, M. J. 2014. 'Forest Certification in Bolivia: A Status Report and Analysis of Stakeholder Perspectives'. *Forest Products Journal* 64(3–4), 80–89.
FAO 2010. *Global Forest Resources Assessment 2010*. Rome, FAO.
Feeny, D., Berkes, F., McCay, B. & Acheson, J. M. 1990. 'The Tragedy of the Commons: Twenty-Two Years Later'. *Human Ecology* 18(1), 1–19.
Gold, M. E. & Zuckerman, R. B. 2014. 'Indonesian Land Rights and Development'. *Columbia Journal of Asian Law* 28(1), 41–70.
Gulbrandsen, L. H. 2010. *Transnational Environmental Governance: The Emergence and Effects of the Certification of Forest and Fisheries*. Cheltenham, Edward Elgar Publishing.
Gulbrandsen, L. H. & Auld, G. 2016. 'Contested Accountability Logics in Evolving Nonstate Certification for Fisheries Sustainability'. *Global Environmental Politics* 16(2), 42–60.
Gupta, J., Van der Grijp, N. & Kuik, O. 2012. *Climate Change, Forests and REDD: Lessons for Institutional Design*. London, Routledge.
Harada, K. & Wiyono. 2014. 'Certification of a Community-Based Forest Enterprise for Improving Institutional Management and Household Income: A Case from Southeast Sulawesi, Indonesia'. *Small-scale Forestry* 13(1), 47–64.
Hardin, G. 1968. 'The Tragedy of the Commons'. *Science* 162(3859), 1243–1248.
Hinrichs, A., Muhtaman, D. R. & Irianto, N. 2008. *Forest Certification on Community Land in Indonesia*. Eschborn, Deutsche Gesellschaft für Technische Zusammenarbeit.
Hosonuma, N., Herold, M., De Sy, V., et al. 2012. 'An Assessment of Deforestation and Forest Degradation Drivers in Developing Countries'. *Environmental Research Letters* 7(4), 1–13.
Hysing, E. 2009. 'From Government to Governance? A Comparison of Environmental Governing in Swedish Forestry and Transport'. *Governance* 22(4), 647–672.
Hysing, E. & Olsson, J. 2005. 'Sustainability through Good Advice? Assessing the Governance of Swedish Forest Biodiversity'. *Environmental Politics* 14(4), 510–526.

Indrarto, G. B., Murharjanti, P., Khatarina, J., et al. 2012. The Context of REDD+ in Indonesia: Drivers, Agents and Institutions. CIFOR Working Paper No. 92.

Johansson, J. 2014. 'Towards Democratic and Effective Forest Governance? The Discursive Legitimation of Forest Certification in Northern Sweden'. *Local Environment* 19(7), 803–819.

Johansson, J. & Keskitalo, E. C. H. 2014. 'Coordinating and Implementing Multiple Systems for Forest Management: Implications of the Regulatory Framework for Sustainable Forestry in Sweden'. *Journal of Natural Resources Policy Research* 6(3), 117–133.

Johansson, T., Hjältén, J., De Jong, J. & von Stedingk, H. 2013. 'Environmental Considerations from Legislation and Certification in Managed Forest Stands: A Review of Their Importance for Biodiversity'. *Forest Ecology and Management* 303, 98–112.

Keskitalo, E. C. H., Sandstrom, C., Tysiachniouk, M. S. & Johansson, J. 2009. 'Local Consequences of Applying International Norms: Differences in the Application of Forest Certification in Northern Sweden, Northern Finland, and Northwest Russia'. *Ecology and Society* 14(2), 1–14.

Klassen, A., Romero, C. & Putz, F. E. (eds.). 2014. *Forest Stewardship Council Certification of Natural Forest Management in Indonesia: Required Improvements, Costs, Incentives, and Barriers.* Vienna, IUFRO.

Lister, J. 2012. *Corporate Social Responsibility and the State: International Approaches to Forest Co-regulation.* Vancouver, UBC Press.

Liu, J., Faure, M. & Mascini, P. 2018. *Environmental Governance of Common Pool Resources: A Comparison of Fishery and Forestry.* London, Routledge.

McCarthy, J., Barr, C., Resosudarmo, I. & Dermawan, A. 2006. 'Origins and Scope of Indonesia's Decentralization Laws'. In Barr, C., Resosudarno, I., Dermawan, A., McCarthy, J. F., Moeliniono, M. & Setiono, B., (eds.), *Decentralization of Forest Administration in Indonesia: Implications for Forest Sustainability, Economic Development and Community Livelihoods.* Bogor, CIFOR, 31–57.

McDermott, C., Cashore, B. & Kanowski, P. 2010. *Global Environmental Forest Policies: An International Comparison.* London, Earthscan.

McDermott, C., Noah, C. & Cashore, B. 2008. 'Differences that "Matter"? A Framework for Comparing Environmental Certification Standards and Government Policies'. *Journal of Environmental Policy and Planning* 10(1), 47–70.

Millennium Ecosystem Assessment. 2005. *Ecosystem and Human Well-being: Synthesis.* Washington, DC, Island Press.

Müller, R., Pacheco, P. & Montero, J. C. 2014. *The Context of Deforestation and Forest Degradation in Bolivia: Drivers, Agents and Institutions.* CIFOR Occasional Paper No. 108.

Ostrom, E. 2010. 'Beyond Markets and States: Polycentric Governance of Complex Economic Systems'. *American Economic Review* 100(3), 641–672.

Overdevest, C. 2010. 'Comparing Forest Certification Schemes: The Case of Ratcheting Standards in the Forest Sector'. *Socio-Economic Review* 8(1), 47–76.

Pacheco, P. 2004. 'What Lies behind Decentralisation? Forest, Powers and Actors in Lowland Bolivia'. *The European Journal of Development Research* 16(1), 90–109.

2011. 'Land Tenure, Forest and Political Reforms: A Look at their Implications for Common-Property Forests in Lowland Bolivia'. Conference Paper: Sustaining Commons: Sustaining Our Future, the Thirteenth Biennial Conference of the International Association for the Study of the Commons.

Pacheco, P., Barry, D., Cronkleton, P. & Larson, A. M. 2008. *The Role of Informal Institutions in the Use of Forest Resources in Latin-America.* Bogor, Center for International Forestry Research.

Paillet, Y., Bergès, L., Hjältén, J., et al. 2010. 'Biodiversity Differences between Managed and Unmanaged Forests: Meta-Analysis of Species Richness in Europe'. *Conservation Biology* 24(1), 101–112.

Real, B. 2002. 'Legal Reforms in Bolivia in the 1990s: Challenges and Opportunities for Decentralization, Indigenous Rights and Forest Management'. Doctoral thesis, University of Florida.

Richardson, J. S., Naiman, R. J. & Bisson, P. A. 2012. 'How Did Fixed-width Buffers Become Standard Practice for Protecting Freshwaters and Their Riparian Areas from Forest Harvest Practices?'. *Freshwater Science* 31(1), 232–238.

Safiti, M. 2010. 'Forest Tenure in Indonesia: The Socio-legal Challenges of Securing Communities' Rights'. Doctoral thesis, Leiden University.

Sahlin, M. 2013. *Credibility at Stake: How FSC Sweden Fails to Safeguard Forest Biodiversity.* Stockholm, Swedish Society for Nature Conservation.

2011. *Under the Cover of the Swedish Forestry Model.* Stockholm, Swedish Society for Nature Conservation.

Sari, I. M. 2013. 'Community Forests at a Crossroads: Lessons Learned from Lubuk Beringin Village Forest and Guguk Customary Forests in Jambi Province, Sumatra, Indonesia'. Doctoral thesis, University of Oslo.

Schlager, E. & Ostrom, E. 1999. 'Property Rights Regimes and Coastal Fisheries: An Empirical Analysis'. In McGinnis, M. D. (ed.), *Polycentric Governance and Development: Readings from the Workshop in Political Theory and Policy Analysis.* Ann Arbor, MI, University of Michigan Press. 87–113.

Singer, B. 2009. 'Indonesian Forest-Related Policies: A Multisectoral Overview of Public Policies in Indonesia's Forests since 1965'. Doctoral thesis, Institut d'Etudes Politiques and CIRAD, France. Available at: http://b-singer.fr/pdf/Forest_policies_in_Indonesia .pdf.

Steering Committee of the State-of-Knowledge Assessment of Standards and Certification. 2012. Toward Sustainability: The Roles and Limitations of Certification, Washington, DC, RESOLVE Inc. Available at: https://www.rainforest-alliance.org/sites/default/files/ 2016-08/toward-sustainability.pdf.

Stora Enso. 2014. 'Global Responsibility Performance'. Part of Stora Enso's Annual Report 2014. Available at: http://assets.storaenso.com/se/com/DownloadCenter-Documents/ Global_responsibility_Performance_2014.pdf.

Stroup, R. L. & Baden, J. A. 1983. *Natural Resources: Bureaucratic Myths and Environmental Management.* Cambridge, Ballinger Publishing Company.

Sunderlin, W. D., Hatcher, J. & Liddle, M. 2008. *From Exclusion to Ownership? Challenges and Opportunities in Advancing Forest Tenure Reform.* Washington, DC, Rights and Resources Initiative.

Sweeney, B. W., Bott, T. L., Jackson, J. K., et al. 2004. 'Riparian Deforestation, Stream Narrowing, and Loss of Stream Ecosystem Services'. *Proceedings of the National Academy of Sciences of the United States of America* 101(39), 14132–14137.

Uggla, Y., Forsberg, M. & Larsson, S. 2016. 'Dissimilar Framings of Forest Biodiversity Preservation: Uncertainty and Legal Ambiguity as Contributing Factors'. *Forest Policy and Economics* 62, 36–42.

Verry, E. S. & Dolloff, C. A. 2000. 'The Challenge of Managing for Healthy Riparian Areas'. In Verry, E. S., Hornbeck, J. W. & Dolloff, C. A. (eds.), *Riparian Management in Forests of the Continental Eastern United States.* Boca Raton, FL, CRC Press LLC. 1–22.

Wood, P. J. 2009. 'Public Forests, Private Governance: The Role of Provincial Governments in FSC Forest Certification'. Doctoral thesis, University of Toronto.

9

Governing Forest Supply Chains

Ratcheting up or Squeezing out?[*]

Constance L. McDermott

9.1 INTRODUCTION

This chapter employs inductive reasoning to assess two key initiatives of EU foreign forest policy from two contrasting theoretical perspectives, referred to as "Trading Up" and "Ecologically Unequal Exchange" (EUE) respectively. Trading Up is focused on the impacts of global trade on the content of environmental regulation, while EUE examines the inequalities of global production and consumption. The purpose of this comparative analysis is not to definitively evaluate EU policy from either perspective, but rather to illustrate how different assumptions about the roles of policy and trade in shaping environmental and social outcomes may drive different policy choices. Awareness of these different policy logics can then inform the design of a policy mix that accommodates insights from both perspectives.

In recent years the EU has engaged in a growing suite of "global forest policy," i.e. initiatives designed to influence the governance and management of forests outside the EU. Central among these, and the focus of this chapter, are policies to implement the 2003 Forest Law Enforcement, Governance and Trade (FLEGT) Action Plan, and EU engagement with Reducing Emissions from Deforestation and Forest Degradation (REDD+) under the UN Framework Convention on Climate Change (UNFCCC). FLEGT aims to eliminate imports of illegally produced timber into the EU, while REDD+ provides economic incentives to developing countries to reduce forest loss.

While both FLEGT and REDD+ are based on intergovernmental agreements and include strong state participation, they are also accompanied by a wide range of multilateral and bilateral regulatory, financing and support arrangements, national and subnational government policies and private market initiatives such as legality

[*] This paper draws on work funded by the European Union's 7th Framework Program for Research (FP-7) under Grant Agreement No FP7–282887 (the INTEGRAL project).

195

and forest carbon certification schemes. In other words, FLEGT and REDD+ have become prime examples of multiscale policy "mixes," as described by Gunningham, Grabosky and Sinclair[1] and discussed in Chapter 1 of this book. The complexity of these policies complicates evaluations of their "effectiveness", or "smartness", of their policy mix. But it also offers two important analytical opportunities: to consider 1) how the core logics of FLEGT and REDD+ shape the types of policy instruments they generate, and 2) how those instruments nevertheless vary in their assumptions, priorities, and strategies for environmental and social impacts.

In regard to overarching logics, a core rationale for the EU's engagement in both FLEGT and REDD+ is to address the region's role as a world-leading producer and consumer. There is little dispute that the EU has played a significant role in driving global forest change. For example, EU countries are major investors in, and importers of, palm oil and other agricultural products whose production has spurred the conversion of forestland to other uses. In 2012 the EU-27 was the world's third-largest importer of palm oil, the leading driver of deforestation in Indonesia.[2] Likewise the EU is the second-largest importer of tropical sawnwood and plywood, both contributors to deforestation in SE Asia.[3] While China leads the world in the import of tropical logs, a large fraction of this material is manufactured as furniture and re-exported to the EU.[4]

Increasing awareness of this interconnectedness has spurred a wide array of research, revealing differences in how the challenge of resource globalization is framed and evaluated. On the one hand, David Vogel's concept of "Trading Up"[5] has helped to frame comparative research on environmental policy that focuses on differences among countries in the stringency of their environmental regulations, policies and standards.[6] The questions posed by these studies revolve around where, and under what conditions, global trade favors a "race to the bottom" or a "race to the top" in the stringency of environmental standards. Often inherent in this argument is the association of stringent standards and regulations with "high" environmental performance. From this perspective, the EU and member states are considered environmental leaders due to relatively stringent and well enforced environmental regulations. In contrast, developing countries are environmental laggards due to inadequate regulatory frameworks or lack of enforcement.[7] A possible conclusion to draw from these findings is that the EU should leverage its dominant position in global markets to ratchet up environmental standards and squeeze out producers who do not meet them.

---

[1]   Gunningham, Grabosky & Sinclair (1998).
[2]   Oosterveer (2014); Rudel et al. (2009).
[3]   ITTO (2011).
[4]   Xiufang & Canby (2011).
[5]   Vogel (1997).
[6]   Cashore & Stone (2014).
[7]   McDermott, Cashore & Kanowski (2010).

On the other hand, work inspired by Hornborg's theory of "Ecologically Unequal Exchange" (EUE) has highlighted the inequality of global resource distribution.[8] This perspective explains forest loss and other environmental problems as the result of excessive consumption in the developed world and displacement of environmental harms elsewhere. According to this theory, poor countries suffer these harms because much of their population is unable to accumulate the economic and social capital necessary to safeguard their local welfare and natural resources.[9] The solution, from this perspective, is to decrease consumption in the developed world while improving the equity of resource distribution in the developing world.

The following analysis begins with a further elaboration of the Trading Up and Ecologically Unequal Exchange (EUE) perspectives, and their implications for EU global forest policy design. It then assesses the EU's engagement with two core strategies, i.e. FLEGT and REDD+, from each of these vantage points, and in light of the following research questions: With which perspective is the design of EU global forest policies most closely aligned? How might policy implementation be evaluated from these two perspectives, and with what implications for sustainability? The latter question is further broken into two subquestions: Is the stringency of environmental standards actually a good proxy for environmental performance? How might raising formal standards to meet the demands of international trade impact equity, and hence the relative capacity of developing countries to tackle their own environmental problems? Finally, this chapter considers how the priorities and impacts of FLEGT and REDD+ might be deliberately altered through changes in their "mix" of actors and policy instruments.

9.2 CONTRASTING PERSPECTIVES

9.2.1 *Trading Up*

David Vogel, in his seminal work on "Trading Up", argues that international trade – a core driver of economic growth – can, under certain circumstances, have a "spillover" effect that leads to a global ratcheting up of environmental policies.[10] The reason for this is that firms operating in jurisdictions with relatively high environmental standards will seek competitive advantage by supporting a "ratcheting up" of national or global environmental standards in order to create a "level playing field" where all firms must meet similar requirements. Vogel has applied this theory, for example, to understanding changes in auto emissions standards. He argues how firms operating in California state, a state with relatively high per capita wealth and stringent environmental requirements, have pressured for federal regulations to level

---

[8] Hornborg (1998).
[9] Rice (2007).
[10] Vogel (1995).

the playing field across other states. California has also been an attractive market for German automobile manufacturers, who in turn supported the European Commission's adoption of American emissions standards.

There is a growing literature on the conditions under which this "ratcheting up" will, or will not, happen. For example, Cashore and others have applied Trading Up theory to the forest sector, examining the role of international forest certification and the verification of the legal origin of wood products as vehicles to promote improved forest standards.[11] They argue that ratcheting up requires support from large firms with sufficient influence in the industry, and that such support is most likely to be gained through initially modest environmental requirements that do not impose much additional cost for high-performing firms. These firms will then join forces with environmentalists to exclude producers who do not meet these standards.

In sum, the Trading Up perspective would support using global trade as a lever to ratchet up environmental standards and practices. Furthermore, Trading Up theory argues that decision-makers can facilitate this spread by promoting the use of international environmental standards that are supported by large influential firms. These large firms will compete with each other to become environmental leaders as a means to gaining competitive advantage over other firms that are environmental laggards, leading to a continual ratcheting up of environmental performance over time.

9.2.2 *Ecologically Unequal Exchange*

Alf Hornborg's theory of Ecologically Unequal Exchange (EUE), in contrast, focuses not on the presence or absence of environmental policy, but on the global drivers and distribution of production and consumption. In particular, EUE theory highlights major inequalities in the appropriation of natural resources by developed countries and the displacement of environmental pressures or environmental "cost shifting" to lesser-developed countries and regions. For example, when international trade is taken into account, the average EU citizen in 2004 induced more than twice the greenhouse gas emissions, appropriated roughly twice the land area, and consumed 10 percent more ground and surface water than the average global citizen. Furthermore, these percentages vary among EU countries and across scales. For example, the UK is the largest displacer of environmental pressure in intra-EU trade, while others such as Poland are net receptors of some environmental burdens.[12]

In general, research suggests that environmental footprints increase with income and, as a country's GDP grows, its footprints are increasingly displaced elsewhere.[13] For example, in most countries recently undergoing a forest transition, as in a

[11] Cashore et al. (2007).
[12] Steen-Olsen et al. (2012).
[13] Meyfroidt et al. (2013); Wiedmann et al. (2013).

transition from net forest loss to net forest gain, displacement of land use abroad has accompanied local gains in forest cover.[14] Understanding the full extent of this displacement therefore requires consideration of the full material flows involved in product production. According to Wiedmann et al, the amount of nondomestic resources used by a country is roughly three times greater than the physical quantity of the goods actually traded.[15] Furthermore, many of these goods come from countries with lower environmental performance. Thus, while the EU and other developed world regions appear to be ratcheting up environmental performance within their borders, this is supported by environmental degradation elsewhere.

Hornborg's EUE framework explains the reasons for these patterns by drawing on Marxist "world systems" analysis.[16] World systems theory views social, political and economic phenomena as embedded in a global system of economic exchange that pivots around competition for surplus capital. This has led to a tripartite of "core" countries that hold a dominant position in world trade through the exploitation of "peripheral" countries, with "semiperipheral" countries situated in the middle and playing the role of political mediators.[17] This dynamic of core and periphery also applies to the quality of governance. According to Wallerstein, the strength of the state in core countries is a function of the weakness of other states.

While world systems theory has historically focused on the accumulation of surplus economic capital, EUE employs similar arguments for the flow of natural resources or "natural capital," and the ways in which core countries benefit from the extraction of natural capital from the periphery. Global trade, Hornborg reasons, enables those in a strong position in the world trading system (i.e. developed countries) to maintain economic growth without exhausting their resources by exploiting the resources of those in the periphery (i.e. least developed countries). The peripheral countries are unable to accumulate the capital needed for their own economic and political development, and hence remain "stuck at the bottom" without a sufficiently affluent civil society to demand environmentally and socially sustainable resource management.[18]

Environmental harms may be displaced across sectors as well as countries. For example, commercial agriculture, not wood production, is the primary driver of forest loss in developing countries. Analysis of the deforestation embedded in agricultural production reveals that a significant portion of this production is destined for developed countries in the EU and elsewhere.[19]

The issue of displacement is a concern even within sectors where developed countries appear to produce as much as they consume. For example, the EU is

[14] Meyfroidt, Rudel & Lambin (2010).
[15] Wiedmann et al. (2013).
[16] Hornborg (1998); Rice (2007).
[17] Wallerstein (1974).
[18] Hornborg (1998); Rice (2007).
[19] EC (2013b).

estimated to consume only about one third of the amount of wood its forests are able to produce, and the quantity (by volume and weight) of its wood product production is nearly equal to the quantity of its apparent consumption.[20] But this does not mean that all of the wood the EU consumes is sustainably produced, or that the costs and benefits of production and consumption are equally distributed. For example, there has been rapid growth of intra-EU trade from new member states in Central and Eastern Europe, regions known to have problems with illegal logging and unsustainable harvest practices. Likewise, the EU is an important importer of tropical wood. For example, it is the largest importer of sawnwood, by volume, from Cameroon – a country where there are high rates of "illegal" logging, and where even the legal rates of harvest intensity exceed estimates of maximum sustainable yield.[21]

Thus, the problem is not simply the quantity of the EU's consumption of wood and agricultural products associated with deforestation, but the "quality" of this EU footprint. From an EUE perspective, therefore, effective environmental governance would require not just reducing the overall quantity of EU consumption, but also ensuring a more globally equitable distribution of resource production and consumption. In contrast to Trading Up, an EUE perspective would not support policies that favor large, multinational companies over local and small-scale producers. Rather, it hypothesizes that capturing more benefits locally is necessary to incentivize and enable local populations to practice environmental stewardship.

9.2.3 A Comparative Assessment Framework

The Trading Up and EUE theories are based on significant empirical study. However, the consideration of both theories together highlights differences in their framing that obscure some challenges while emphasizing others. The EUE theory does not address the potential role of global trade in redistributing environmental and social goods as well as harms. The Trading Up perspective, on the other hand, overlooks the issues of consumption and displacement, i.e. the degree to which developed countries have achieved growth by displacing environmental and social harms elsewhere. A focus on overall consumption and displacement, in turn, highlights the finite nature of global resources and calls into question the ability of poorer countries to emulate the development pathways previously experienced by developed countries.

The Trading Up and EUE perspectives are both focused on the dynamics of self-interest and economic competition, but predict different outcomes. Trading Up explains how powerful interests may support a ratcheting up of global environmental

---

[20] McDermott et al. (2014). "Apparent consumption" is calculated as production plus imports minus exports, consistent with the methodology of the Global Footprint Network, Ewing et al. (2010).

[21] Cerutti et al. (2011).

standards as a means to gain advantage over those with lesser capacities, and implies that the net result of such competition will be good for the environment. From an EUE perspective, however, such strategies reinforce global inequities and leave poor countries and actors "stuck at the bottom" in a downward spiral of environmental degradation.

The following analysis of key EU global forest policies therefore compares and contrasts policy logics from both perspectives. In regard to Trading Up theory, it considers whether or not the approach uses global trade as a lever to ratchet up environmental policies. For EUE theory, it considers whether a core logic of the initiative is focused on reducing consumption and/or enhancing the sharing of benefits from global trade across actors and countries. It then briefly reviews the evidence of how policy instruments associated with the initiative have been designed and implemented thus far, and consistent with what logic. Finally, it considers the implications of the mix of actors and instruments for overall policy effectiveness.

9.3 ANALYSIS OF KEY EU POLICIES

9.3.1 *The EU FLEGT Action Plan*

The EU FLEGT Action Plan was launched in 2003 to support developing countries to combat illegal logging by blocking illegally harvested timber from entering the EU. According to the Plan, illegal logging is associated with corruption and violent conflict, undermines the competitiveness of "legitimate" industry operations, costs governments "vast sums of money" in lost revenue and causes "enormous environmental damage and loss of biodiversity."[22] Stopping illegal logging has therefore become a central pillar of the EU's efforts to promote sustainable forest management globally.

As is evident in the "T" in "FLEGT," this Action Plan aims to leverage the power of EU trade to achieve its objectives. Specifically, the EU proposes to sign voluntary partnership agreements (VPAs) with developing countries to enact legality licensing schemes. Once these schemes are in place, wood products coming from VPA countries must be licensed as legal before they are allowed into EU markets.[23]

The core logic of the FLEGT Action Plan appears most coherent with a Trading Up perspective. That is, its focus is on creating a "level playing field" whereby all forest producers will be required to adhere to the laws of the country if they are to access EU markets. However, the Action Plan also includes language consistent with an EUE perspective. Specifically, the Plan states a commitment to "promoting

[22] EC (2003).
[23] Ibid.

equitable and just solutions to the illegal logging problem that do not have an adverse impact on poor people."[24]

Currently, twenty-six countries have engaged in various degrees with the EU VPA process. Six have signed VPA agreements, nine are in active negotiation, and another eleven have initiated discussions.[25] Of these, Ghana and Indonesia are arguably the closest to full VPA implementation. We therefore draw on case studies of these two countries for our analysis of VPA implementation.

In regard to "ratcheting up," there is more than one way to interpret changes in forest governance in Indonesia and Ghana. As part of their VPA negotiations, both countries have engaged in multi-stakeholder processes to improve the coherence and consistency of their legal frameworks governing timber harvest. Both have also developed systems for independent monitoring and legality licensing.[26] Assuming these policy instruments lead to more consistent enforcement, this could be viewed as a type of "ratcheting up."[27] Given past contradictions in forest laws and policies in both countries, however, it is difficult to determine whether their previous legal content was more or less environmentally stringent. Employing inductive logic, it would seem the enforcement of legality could provide incentives to reduce legal requirements to make legality more easily attainable – at least in the short term. Thus, for our purposes we note potential for a ratcheting-up effect in terms of compliance but uncertain effect in terms of legal content.

In regard to EUE theory, there is little evidence to suggest that the FLEGT VPAs will lead to a marked reduction in the EU's overall consumption of wood products. While Indonesia is a leading source of tropical timber for the EU, tropical timber contributes only a very small percentage to overall EU consumption and there are many substitute products available.[28]

The effects on the distribution of benefits may be more significant. EU member states have invested aid money to support civil society engagement in the negotiation of VPA processes in both Indonesia and Ghana. There is some evidence that these processes, which were thus incentivized through nonmarket channels, may have enhanced local participation in decision-making.[29] At the same time, both processes seem to be decreasing local access to forest products. It has been estimated that roughly 80 percent of wood production in both Indonesia and Ghana is produced for domestic markets, mostly by small-scale producers and chainsaw millers.[30] Yet in both countries the VPAs cover domestic as well as exported timber. Many of the smallholders who produce for domestic markets lack formal land or tree tenure

---

[24] Ibid.
[25] EFI (2014).
[26] Lesniewska & McDermott (2014).
[27] Cashore & Stone (2012).
[28] Koulelis & McDermott (2019, in press).
[29] Overdevest & Zeitlin (2014).
[30] Lesniewska & McDermott (2014).

rights, and many of those who hold such rights lack capacity to navigate the complex wood harvest permitting processes. The VPAs add to these challenges by requiring third-party verification of legality. In Indonesia, where the policy instrument used to verify compliance is private, third-party certification, producers are expected to pay for the costs of private auditing. These additional requirements have created disproportionate barriers for smallholders and domestic producers, while favouring large-scale, high-capacity forest producers, and less heavily regulated large-scale, intensive plantation production.[31]

Meanwhile, until November 2016,[32] none of the VPA countries had FLEGT legality licenses recognized by the EU. Nevertheless, the EU continued apace in implementing another key piece of legislation to implement the FLEGT Action Plan – The EU Timber Regulation (EU TR) 2010.[33] The EU TR came into force in March 2013 and is a policy instrument designed to complement FLEGT by requiring due diligence in assuring the legality of internationally traded timber. It therefore follows a logic similar to that of the VPA processes but takes the form of a unilateral EU regulation, without the emphasis on negotiated agreements with developing countries or on local stakeholder participation. The EU TR was intended to support the FLEGT VPA process by recognizing FLEGT licenses as sufficient proof of due diligence. However, until the EU approves the legality licensing schemes of VPA countries, importing timber from these countries faces the same due diligence requirements as elsewhere.

There is as yet a lack of published research on the net impacts of the EU TR either inside or outside the EU. Furthermore, assessing impacts on this scale presents major challenges of attribution, since it is difficult to determine the degree to which changes in forest governance and trade can be attributed to the EU TR as opposed to a myriad of other factors. However, it is possible to observe a certain coherence of the EU TR with a Trading Up perspective, in that the EU TR is designed to enforce an international standard of legality and exclude producers who are unable to demonstrate compliance. At the same time, the exclusive focus on legality offers no apparent incentive to raise environmental standards and could actually have the effect of driving standards down to make compliance less difficult. Furthermore, there is a risk that supplier countries will respond by shifting their exports to other countries and regions without legality requirements.

In regards to EUE, the EU TR further weakens the position of developing countries in the world trading system. This could exacerbate inequalities with the result of weakening rather than strengthening forest governance among peripheral trading partners. From the EUE perspective, the problem of weak governance is not

---

[31] Obidzinski et al. (2014); Setyowati & McDermott (2017).
[32] 15 November 2016 marks the first shipments of FLEGT licensed products from Indonesia. See: www.flegtlicence.org/flegt-licensed-products-from-indonesia.
[33] EC (2010).

simply a problem of inadequate enforcement, but rather a systemic problem relating to inequalities in the accumulation of capital and the ability of countries and producers to develop both economically and politically. This perspective would call for replacing or modifying the current mix of FLEGT policies, with their heavy focus on state regulation and the licensing of wood for export, with a mix of policies expressly designed to support governance reform that enables local communities and domestically oriented enterprises to capture a greater share of the benefits from forest resources.

9.3.2 *The EU and REDD+*

Reducing Emissions from Deforestation and Degradation (REDD+) is a mechanism under the United Nations Framework Convention on Climate Change (UNFCCC).[34] The core logic of REDD+ is to provide financial and technical support to developing countries to reduce their carbon emissions from forest loss and enhance forest carbon storage. The UNFCCC's emphasis on economic incentives as the central driver of REDD+ is similar to the EU's framing of FLEGT, and is consistent with a Trading Up perspective. However, as the EU has done with FLEGT, the UNFCCC has also integrated elements of an EUE perspective into its REDD+ text. Many REDD+ stakeholders have expressed concern that a market-based approach to REDD+ might spur the conversion of natural forests to more intensively managed plantations to maximize carbon uptake, or lead to dispossession and loss of traditional livelihoods among indigenous and local communities.[35] As a result, the UNFCCC has developed a set of environmental and social safeguards.[36] The UNFCCC REDD+ safeguards state that REDD+ interventions should ensure effective and transparent governance, full and effective participation of relevant stakeholders (including indigenous and local communities) and enhanced environmental and social benefits. These safeguards have since been further defined in different ways by international aid agencies and financial institutions, including EU and member-state funding bodies, nonstate certification schemes and other policy stakeholders. A comparison of the content of these various state and nonstate-based safeguards reveals considerable variation in their content and stringency.[37]

The ways in which one might assess the effectiveness of these diverse REDD+ safeguards would vary if one adopted a Trading Up or an EUE perspective. A proponent of Trading Up might measure changes in state-based or private environmental and social regulation. As of the 2013 UNFCCC COP 19 in Warsaw, it has

---

[34] UNFCCC/AWGLCA (2011).
[35] McDermott et al. (2012).
[36] UNFCCC/AWGLCA (2011).
[37] McDermott et al. (2012).

been decided that countries must provide a summary of how the REDD+ safeguards are being addressed and respected before receiving results-based payments.[38] This emphasis on making REDD+-related environmental and social policies internationally transparent could lead to norm diffusion and a ratcheting-up effect.

From an EUE perspective, however, the more important questions would be whether REDD+ will foster reduced consumption and facilitate more equitable benefit sharing. Analysis of the evidence on consumption is not encouraging. The amount of finance currently available for REDD+ is inadequate to cover the opportunity costs of converting forests to palm oil or cattle.[39] Even if such finance were to be made available, this might not necessarily reduce consumption. For example, the recent dramatic decline of forest loss in Brazil was achieved in part by the intensification of cattle production, with no net loss of production.[40] This suggests that reducing forest loss did little to address consumption.

The ability of REDD+ to promote equitable resource distribution is also uncertain. The UNFCCC empowers country parties to develop nationally appropriate REDD+ strategies and safeguards. While it is still too early to tell how these strategies will materialize at the country level, there is a growing body of research on REDD+ preparation activities at the national and project levels. Some countries, such as Indonesia and Mexico, have developed REDD+ strategies or visions that situate REDD+ within the broader goals of sustainable forest management, participatory decision-making and social equity.[41] While reduction of emissions from forest loss is one objective, this is to be achieved only in concert with other sustainable development goals. Furthermore, participatory REDD+ processes in some cases appear to have enhanced civil society participation in forest governance more generally.[42] Importantly from an EU perspective, EU donor countries such as Germany and the UK have helped to encourage, as well as financially support, this enhanced participation. However, it is unclear whether current priorities for public participation will be sustained if REDD+ transitions into a purely "performance-based" policy instrument as envisioned under the UNFCCC.[43] In the absence of sustained finance for safeguard activities, and without associated policy instruments to support safeguards monitoring and implementation that are proportionately equivalent to investments in the monitoring and sale of carbon, the EU would have little leverage or legitimacy to insist on rigorous safeguards implementation.

At the project level, case study evidence suggests that REDD+ carbon payments have favoured relatively large landowners with secure tenure over smaller

[38] UNFCCC (2014).
[39] Borrego & Skutsch (2014); Butler, Koh & Ghazoul (2009).
[40] Nepstad et al. (2014).
[41] CONAFOR (2010), at 57; Ituarte-Lima, McDermott & Mulyani (2014).
[42] Ituarte-Lima, McDermott & Mulyani (2014); Mulyani & Jepson (2013).
[43] Barron & McDermott (2015).

landholders and the landless.[44] However, REDD+ projects, which are frequently implemented by a mix of private sector and civil society actors and draw on voluntary certification, have in some cases provided important noncarbon benefits, including the strengthening of tenure rights for some farmers.[45]

In sum, the effect of REDD+ on benefit sharing will be variable at both national and project levels. At the same time, the influence of the EU on REDD+ priorities will depend on the particular mix of REDD+-related actors and state and nonstate policy instruments that the EU and its member states support, and the degree to which this promotes participatory processes and non-carbon benefits.

9.4 SUMMARY AND CONCLUSION

The EU's global forest policy goals, as currently stated in its 2013 Forest Strategy and associated action plans, are resonant with two contrasting theoretical perspectives: Trading Up and EUE. The Trading Up perspective frames the EU as an environmental leader whose role is to leverage its influence on global trade to bring the rest of the world up to standard. The EUE perspective, in contrast, acknowledges the EU's role as a leading consumer, and views its dominant position in world trade as driving a downward spiral of inequality and environmental degradation. From this perspective, effective policies must lead to a reduction in EU consumption and foster a more equitable distribution of the world's resources, thereby enabling developing countries to shape their own strategies for environmental stewardship.

The assessment of EU global forest policy according to both the Trading Up and EUE perspectives yields mixed results (see Table 9.1). The FLEGT VPAs, EU TR and REDD+ all place heavy emphasis on verification and international reporting as a means to incentivize desired behavior. In the case of the VPAs and TR the focus is on compliance with forest law, while the focus of REDD+ is on carbon accounting and reporting on safeguards compliance. From a Trading Up perspective, this emphasis on internationally transparent compliance with laws and standards could foster a type of "ratcheting up" of environmental enforcement by squeezing out those actors unable or unwilling to meet reporting requirements. However, it could also incentivize a weakening of the content of environmental and social policies in order to make compliance easier to achieve.

None of these policies address the role of net EU consumption. In regards to benefit distribution, the negotiation of FLEGT VPAs and REDD+ actions to date have created new platforms for previously marginalized actors to participate in forest-related decision-making. These platforms, furthermore, were realized via a variety of policy mechanisms, some of which emerged outside of the formal EU FLEGT and UNFCCC

---

[44] Peskett, Schreckenberg & Brown (2011).
[45] Osborne (2011).

TABLE 9.1 *An Evaluation of FLEGT and REDD+ from the Trading Up and Ecologically Unequal Exchange (EUE) Perspectives*

Policy Instrument	Trading Up	Ecologically Unequal Exchange	
Outcome	**Ratchet up Environmental Standards?**	**Reduce Consumption?**	**Redistribute Benefits?**
FLEGT VPAs	**Mixed.** Uncertain effect on environmental policies, displacement of timber trade, but may increase compliance.	**No.** Likely to hasten intensification of timber production.	**Mixed.** May enhance public participation. Has created disproportionate barriers for domestic/small-scale production.
EU Timber Regulation	**Mixed.** Uncertain effect on environmental policies, displacement of timber trade, but may increase compliance.	**No.** Likely to favor developed countries, large industry.	**No.** Likely to favor developed countries, large industry.
UNFCCC (REDD+)	**Mixed.** Increased emphasis on transparency of forest policies and safeguards may encourage ratcheting up of standards.	**No.** Likely to support intensification of timber and agricultural production.	**Mixed.** Depends on relative investment in safeguards and noncarbon benefits.

REDD+ governance structures – such as bilaterally funded multistakeholder platforms and voluntary legality and forest carbon certification schemes. However, it is unclear which actors and instruments will remain active, and how much priority either the VPAs or REDD+ will continue to place on safeguarding local participation or welfare, once they become fully operational and performance-based. Meanwhile, the procedural benefits gained through both the VPA and REDD+ projects have not necessarily enhanced local benefit capture: the VPAs have created disproportionate market barriers for small-scale and domestic-oriented forest producers, and REDD+ carbon payments have favored relatively large landowners with clear tenure.

In conclusion, there is a need to more carefully consider the evidence offered by both Trading Up and EUE perspectives in order to improve the likely impacts of EU global forest strategies. From a Trading Up perspective, there is a need to transition international attention beyond legality to assessing how current strategies impact the content – and social equity – of environmental and social rules and standards and their on-the-ground outcomes. From an EUE perspective, much greater attention needs to be placed on reducing the size, and improving the quality, of the EU's global footprint, supporting small-scale and domestic producers and promoting local benefit sharing.

It is also important to recognize that the Trading Up and EUE perspectives are more contradictory than complementary, and that theoretical and political consensus on their relative value is unlikely. Nevertheless, achieving some degree of balance across the two perspectives may be necessary – in terms of both the overarching logics used to design policy initiatives and the mix of policy instruments employed for their implementation – if the EU is to balance its goals for global leadership with its goals for equity, justice and a new green economy.

REFERENCES

Barron, D. P. & McDermott, C. L. 2015. 'Private Funder Perspectives on Local Social and Environmental Impacts in "Reducing Emissions from Deforestation and Degradation+"'. *Journal of Environmental Policy & Planning* 17(2), 277–293.

Borrego, A. & Skutsch, M. 2014. 'Estimating the Opportunity Costs of Activities that Cause Degradation in Tropical Dry Forest: Implications for REDD+'. *Ecological Economics* 101, 1–9.

Butler, R. A., Koh, L. P. & Ghazoul, J. 2009. 'REDD in the Red: Palm Oil Could Undermine Carbon Payment Schemes'. *Conservation Letters* 2(2), 67–73.

Cashore, B. & Stone, M. W. 2012. 'Can Legality Verification Rescue Global Forest Governance? Analyzing the Potential of Public and Private Policy Intersection to Ameliorate Forest Challenges in Southeast Asia'. *Forest Policy and Economics* 18, 13–22.

Cashore, B. W. & Stone, M. W. 2014. 'Does California Need Delaware? Explaining Indonesian, Chinese, and United States Support for Legality Compliance of Internationally Traded Products'. *Regulation & Governance* 8(1), 49–73.

Cashore, B., Auld, G., Bernstein, S. & McDermott, C. 2007. 'Can Non-state Governance 'Ratchet Up' Global Environmental Standards? Lessons from the Forest Sector'. *Review of European Community and International Environmental Law* 16(2), 158–172.

Cerutti, P. O., Tacconi, L., Nasi, R. & Lescuyer, G. 2011. 'Legal vs. Certified Timber: Preliminary Impacts of Forest Certification in Cameroon'. *Forest Policy and Economics* 13(3), 184–190.

CONAFOR. 2010. *Visión de México sobre REDD+, CONAFOR, SEMARNAT.* Mexico City, Gobierno Federal.

EC. 2013a. 'A New EU Forest Strategy: For Forests and the Forest-based Sector', Communication from the Commission to the European Parliament, the Council, the European Economic and Social Committe and the Committee of the Regions, Brussels, European Commission, 1–17.

2013b. The impact of EU consumption on deforestation: Comprehensive analysis of the impact of EU consumption on deforestation. Study funded by the European Commission, DG #NV, and undertaken by VITO, IIASA, HIVA and IUCN NL. Views or opinions expressed in this report do not necessarily represent those of IIASA or its National Member Organizations. Brussels, European Commission, 1–348.

2010. Regulation (EU) No 995/2010 of the European Parliament and of the Council of 20 October 2010 laying down the obligations of operators who place timber and timber products on the market (Text with EEA relevance). Brussels, European Commission, 1–12.

2003. Forest Law Enforcement, Governance and Trade (FLEGT) Proposal for an EU Action Plan. Brussels, European Commission, 1–32.

EFI. 2014. *EU FLEGT Facility > Voluntary Partnership Agreements*, European Forest Institute. Available at: http://www.euflegt.efi.int/vpa.

Ewing, B., Moore, D., Goldfinger, S., Oursler, A., Reed, A. & Wackernagel, M. 2010. *Ecological Footprint Atlas 2010*. Oakland, Global Footprint Network, 1–113.

Gunningham, N., Grabosky, P. & Sinclair, D. 1998. *Smart Regulation: Designing Environmental Policy*. Oxford, Oxford University Press.

Hornborg, A. 1998. 'Towards an Ecological Theory of Unequal Exchange: Articulating World System Theory and Ecological Economics'. *Ecological Economics* 25(1), 127–136.

ITTO. 2011. *Annual Review and Assessment of the World Timber Situation 2011*. Yokohama, International Tropical Timber Organization (ITTO), 1–206.

Ituarte-Lima, C., McDermott, C. L. & Mulyani, M. 2014. 'Assessing Equity in National Legal Frameworks for REDD+: The Case of Indonesia'. *Environmental Science and Policy* 44, 291–300.

Koulelis, P. & McDermott, C. 2019 (in press). 'Incorporating A Global Perspective into Future-Oriented Forest Management Scenarios: The Role of Forest Footprint Analysis'. *International Journal of Agricultural and Environmental Information Systems* 10(1), 21.

Lesniewska, F. & McDermott, C. L. 2014. 'FLEGT VPAs: Laying a Pathway to Sustainability via Legality: Lessons from Ghana and Indonesia'. *Forest Policy and Economics* 48, 16–23.

McDermott, C., Koulelis, P., Kubo, K. & Barron, D. 2014. 'Integrating Global Forest Footprints into Forest Management Scenarios'. In Hinterseer, T., Koulelis, P., Jonsson, R., et al. (eds.), Synthesis Report on Integrated Forest Management Scenarios in Europe Including the National Case Study Reports and the Report on the Role of EU Commodity Consumption. Uppsala, INTEGRAL, EU FP7 Programme.

McDermott, C. L., Cashore, B. & Kanowski, P. 2010. *Global Environmental Forest Policies: An International Comparison*, London, Earthscan.

McDermott, C. L., Coad, L., Helfgott, A. & Schroeder, H. 2012. 'Operationalizing Social Safeguards in REDD+: Actors, Interests and Ideas'. *Environmental Science and Policy* 21, 63–72.

Meyfroidt, P., Lambin, E. F., Erb, K. H. & Hertel, T. W. 2013. 'Globalization of Land Use: Distant Drivers of Land Change and Geographic Displacement of Land Use'. *Current Opinion in Environmental Sustainability* 5(5), 438–444.

Meyfroidt, P., Rudel, T. K. & Lambin, E. F. 2010. 'Forest Transitions, Trade, and the Global Displacement of Land Use'. *Proceedings of the National Academy of Sciences* 107(49), 20917–20922.

Mulyani, M. & Jepson, P. 2013. 'REDD+ and Forest Governance in Indonesia: A Multistakeholder Study of Perceived Challenges and Opportunities'. *The Journal of Environment and Development* 22(3), 261–283.

Nepstad, D., McGrath, D., Stickler, C., et al. 2014. 'Slowing Amazon Deforestation through Public Policy and Interventions in Beef and Soy Supply Chains'. *Science* 344(6188), 1118–1123.

Obidzinski, K., Dermawan, A., Andrianto, A., Komarudin, H. & Hernawan, D. 2014. 'The Timber Legality Verification System and the Voluntary Partnership Agreement (VPA) in Indonesia: Challenges for the Small-scale Forestry Sector'. *Forest Policy and Economics* 48, 24–32.

Oosterveer, P. 2014. 'Promoting Sustainable Palm Oil: Viewed from a Global Networks and Flows Perspective'. *Journal of Cleaner Production* 107, 146–153.

Osborne, T. M. 2011. 'Carbon Forestry and Agrarian Change: Access and Land Control in a Mexican Rainforest'. *The Journal of Peasant Studies* 38(4), 859–883.

Overdevest, C. & Zeitlin, J. 2014. 'Constructing a Transnational Timber Legality Assurance Regime: Architecture, Accomplishments, Challenges'. *Forest Policy and Economics* 48, 6–15.

Peskett, L., Schreckenberg, K. & Brown, J. 2011. 'Institutional Approaches for Carbon Financing in the Forest Sector: Learning Lessons for REDD+ from Forest Carbon Projects in Uganda'. *Environmental Science & Policy* 14(2), 216–229.

Rice, J. 2007. 'Ecological Unequal Exchange: Consumption, Equity, and Unsustainable Structural Relationships within the Global Economy'. *International Journal of Comparative Sociology* 48(1), 43–72.

Rudel, T., DeFries, R., Asner, G. P. & Laurance, W. 2009. 'Changing Drivers of Deforestation and New Opportunities for Conservation'. *Conservation Biology* 23(6), 1396–1405.

Setyowati, A. & McDermott, C. L. 2017. 'Commodifying Legality? Who and What Counts as Legal in the Indonesian Wood Trade'. *Society & Natural Resources* 30, 750–764.

Steen-Olsen, K., Weinzettel, J., Cranston, G., Ercin, A. E. & Hertwich, E. G. 2012. 'Carbon, Land, and Water Footprint Accounts for the European Union: Consumption, Production, and Displacements through International Trade'. *Environmental Science & Technology* 46(20), 10883–10891.

UNFCCC. 2014. Report of the Conference of the Parties on its nineteenth session, held in Warsaw from 11 to 23 November 2013: Part Two. Bonn, UNFCCC, 1–43.

UNFCCC/AWGLCA. 2011. Report of the Conference of the Parties on its sixteenth session, held in Cancun from 29 November to 10 December 2010. United Nations Framework Convention on Climate Change/Ad hoc Working Group on Long-term Cooperative Action.

Vogel, D. 1995. *Trading Up: Consumer and Environmental Regulation in a Global Economy.* Cambridge, MA, Harvard University Press.

1997. 'Trading up and Governing across: Transnational Governance and Environmental Protection'. *Journal of European Public Policy* 4(4), 556–571.

Wallerstein, I. 1974. 'The Rise and Future Demise of the World Capitalist System: Concepts for Comparative Analysis'. *Comparative Studies in Society and History* 16(4), 387–415.

Wiedmann, T. O., Schandl, H., Lenzen, M., et al. 2013. 'The Material Footprint of Nations'. *Proceedings of the National Academy of Sciences* 112(20), 6271–6276.

Xiufang, S. & Canby, K. 2011. *FLEGT Asia, Baseline Study 1, China: Overview of Forest Governance, Markets and Trade,* Washington DC, EFI-FLEGT Asia Regional Office, Forest Trends, 1–52.

10

Public Sector Engagement with Private Governance Programmes

Interactions and Evolutionary Effects in Forest and Fisheries Certification

Lars H. Gulbrandsen

10.1 INTRODUCTION

Over the past few decades, private standard setting has become a prominent mode of governing the practices of transnational production, distribution, and consumption in a globalized world economy.[1] From the self-regulation of businesses and codes of conduct to performance-based standards requiring third-party auditing, private regulatory efforts vary greatly in the demands they place on corporations. Certification and eco-labelling schemes – created in sectors such as forestry, fisheries, and agriculture – arguably represent the most demanding types of these efforts, as they typically involve mandatory, third-party verification of compliance with performance-based standards. These programmes establish environmental standards and standards for socially responsible production that often go beyond government regulations.

The proliferation of private regulatory efforts is widely understood as a response to the failure of governments to ameliorate pressing social and environmental problems related to global production, distribution and consumption.[2] Globalized supply chains and multinational corporations have become focal targets of new institutional arrangements that often seek to fill governance gaps resulting partly from liberalized trade and lack of state capacity, ability or willingness to regulate transboundary environmental problems. Frustrated with the lack of government intervention, nongovernmental organizations (NGOs) have increasingly targeted companies and their supply chains directly, using a range of strategies to persuade them to support or participate in certification programmes.

Because private regulatory programmes are market-based and operated by non-state actors, they are sometimes portrayed as advanced cases of what Rosenau and

---

[1] This chapter is a significantly revised version of Gulbrandsen (2014). I am grateful to André Nollkaemper for constructive and detailed comments on an earlier version of the chapter.

[2] E.g. Cashore (2002); Bartley (2007); Vogel (2010); Auld & Gulbrandsen (2014).

Czempiel[3] have termed 'governance without government'. However, studies of nonstate certification have increasingly recognized the close interconnections between private and public governance efforts.[4] These studies share the understanding of certification programmes as advanced governance systems interacting with public authorities at several levels and in several ways. Some scholars have advanced terms such as 'co-regulation'[5] and 'hybrid forms of governance'[6] in place of private and public governance, arguing that those terms seem better adapted to social and environmental certification schemes.

Notwithstanding recent research advances, the role of states in nonstate governance schemes remains an understudied area of contemporary global governance. This chapter examines the role and influence of states in nonstate governance programmes by comparing forest and fisheries certification programmes. These cases are interesting because of their different approaches to the management of forests and fisheries. While most of the world's forests are domestic resources managed by private owners or companies with logging concessions, coastal and open-ocean fish stocks are common-pool resources managed by international, regional and national fisheries management regimes. Unlike the forestry sector, there is a long history of international governance in the fisheries sector, and a wide range of legally binding international rules. Did these differences in the management of forest and fisheries result in divergent state responses to the emergence of forest and fisheries certification? How did states' responses to forest and fisheries certification programmes influence the subsequent development of these programmes?

Given that little is known about the processes, outcomes and underlying mechanisms of public-private interaction dynamics, an inductive approach seemed the appropriate design for this study. After having reviewed the establishment of forest and fisheries certification programmes, I proceed in three analytical steps to investigate interaction dynamics. First, I examine how states have responded to the emergence of nonstate certification in the forest and fisheries sectors. Research shows that states can influence nonstate certification programmes at all stages of the regulatory process, from agenda setting to negotiation of standards and on to implementation, monitoring and enforcement.[7] Less attention has been paid to the development and effects of public procurement policies. In recent years, many countries have developed timber procurement policies and recognized certification under credible schemes as assurance of legally and/or sustainably sourced timber.

---

3   Rosenau & Czempiel (1992).
4   E.g. Boström (2003); Meidinger (2006); Tollefson, Gale & Haley (2008); Abbott & Snidal (2009); Lister (2011); Vogel (2010).
5   Tollefson, Gale & Haley (2008); Lister 2011.
6   Gale & Haward (2011).
7   See Boström (2003); Tollefson, Gale & Haley (2008); Gale & Haward (2011); Lister (2011).

Hence, for the forestry case, I investigate documentation from countries with timber procurement policies. Investigating public procurement policies is less relevant in the case of fisheries, given the general absence of such policies in that sector. Here I focus instead on the engagement of the UN Food and Agriculture Organization (FAO), which has taken a rather active role in the case of seafood eco-labelling, and how states have otherwise acted to enhance or restrict the rule-making authority and legitimacy of certification programmes in the fisheries sector.

Second, I examine what I call the 'evolutionary effects' of state responses to certification programmes by investigating two indicators: producer and market adoption of certification and changes in certification standards, rules, and procedures. Concerning *producer and market adoption*, governments can influence the adoption decisions of producers and companies along global supply chains by enacting timber procurement policies, seafood policy strategies and other policy mechanisms. Whereas governments can facilitate the adoption of certification by expressing their support of certain schemes or labels, they can also impede adoption by rejecting particular schemes or labels. For example, if governments favour one certification scheme over another in public procurement policies, it would be a strong signal to producers that are considering the various certification options. Concerning the *regulatory effects* of government engagement with certification programmes, public procurement policies can influence the standards, rules and procedures of competing certification programmes through public comparisons and benchmarking. For example, public comparisons of competing programmes can put pressure on poorly performing programmes to improve standards. Intergovernmental engagement with fisheries certification, seafood policy strategies and other government policies can have similar effects on fisheries certification programmes. Here I pay particular attention to the effects of FAO engagement on fisheries certification standards, rules and procedures.

Third, from the empirical examination of the forestry and fisheries cases, I identify types of public–private governance interaction. Specifically, I draw upon these cases to identify pathways of interaction between public regulation and private standards and the causal mechanisms that shed light on the dynamics of interaction. I also identify conditions under which state involvement is likely to result in either the strengthening or weakening of nonstate programmes. The purpose of this inductively paired comparison is to identify mechanisms of governance interaction and generate hypotheses that can be examined across a larger number of cases in future research.

The chapter draws on primary research into certification initiatives in the forest and fisheries sectors conducted over more than a decade. I have investigated policy documents of government agencies, certification programmes, NGOs, and other stakeholders, as well as secondary sources. In addition, I draw on a number of interviews with representatives of forest and fisheries certification programmes, environmental NGOs, industry associations, and government agencies. The interviews were not conducted specifically for the purpose of investigating governance interactions, but the topic emerged nevertheless as a key issue in many interviews.

10.2 THE EMERGENCE OF FOREST AND FISHERIES CERTIFICATION

Forest certification emerged in response to growing concern about deforestation, global forest degradation and failed intergovernmental efforts to develop a legally binding agreement on forests. According to FAO's latest *Global Forest Resources Assessment*, there was a net loss of some 129 million hectares of forest – about the size of South Africa – between 1990 and 2015.[8] Most of the forest lost in this period was in the tropical domain, which has shown losses in every measurement period since 1990. The lack of a global forest convention or other intergovernmental instruments able to address the problems of deforestation and forest degradation prompted NGOs and other organizations concerned about forest destruction to seek alternative solutions.

In 1993, primarily at the initiative of the World Wide Fund for Nature (WWF), a broad coalition of environmental NGOs, social groups, retailers, forest companies, and professional certifiers founded the Forest Stewardship Council (FSC).[9] The idea was to audit and verify sustainable forestry practices and encourage retailers to support such practices by sourcing certified timber and timber products. To promote 'environmentally appropriate, socially beneficial and economically viable' forest management, FSC developed global principles and criteria for its definition of 'well-managed forests', including tenure and use rights and responsibilities, indigenous peoples' and workers' rights; use of forest products and services to maximize economic viability and environmental and social benefits; maintenance of forests with high conservation value; environmental impact; monitoring and assessment; and planning and management of plantations.[10] These principles and criteria are elaborated and specified for each country or region by national or regional working groups through a process in which ecological, economic and social interests in principle take part on an equal footing. The members of FSC make up a General Assembly, which serves as its highest decision-making body and has a tripartite structure that includes social, environmental and economic chambers. Because FSC was an attempt to work outside intergovernmental forest processes, governments and state agencies are not allowed to hold membership; as of 2002, however, government-owned and -controlled companies may apply for membership in the economic chamber.

---

[8] FAO (2015).
[9] See www.fsc.org/.
[10] Ibid.

Forest industry and landowner associations in many countries responded quickly to the emergence of FSC by creating 'industry-friendly', national-level certification programmes with more flexible and discretionary standards.[11] FSC's governance structure and relatively prescriptive standards motivated forest industry and landowner associations to establish schemes that gave forest owners greater influence over rule-making processes, and standards that, at least initially, paid less attention to environmental and social criteria for sustainable forestry, and more to economic criteria. These national-level programmes eventually formed an umbrella organization, known since 2003 as the Programme for the Endorsement of Forest Certification (PEFC). The modus operandi of this scheme is somewhat different to the FSC's, in that it is an umbrella certification scheme that facilitates the mutual recognition of national certification schemes and provides an 'internationally credible framework', with a common logo and ecolabel for such schemes. Although a few other programmes have emerged, FSC and PEFC remain the two major forest certification schemes; the former enjoying the strong support of the WWF and other NGOs, the latter backed by forest industry and landowner associations in many countries.

Like forest certification, fisheries certification emerged partly in response to perceived governance failures and gaps.[12] Driven by the rapid growth of the world's fishing fleet, world marine capture fisheries production quadrupled between 1950 and 1990, reaching a production peak of 86.4 million tonnes in 1996. Since then, production has exhibited a general declining trend.[13] Decades of overfishing have had dire consequences. The share of fish stocks within biologically sustainable levels declined from 90 per cent in 1974 to 68.6 per cent in 2013.[14] Hence, almost one-third of fish stocks were judged to have been fished at biologically unsustainable levels and therefore merited the designation 'overfished'. Despite progress in some areas, the overall state of the world's fish stocks has not improved.[15] The global imbalance between fish resources and harvesting capacity has been – and remains – the most serious problem in the fisheries sector.

In response to increasing concerns over the inability or unwillingness of governments to resolve the challenges of fisheries management on their own – and inspired by FSC's success – the WWF exported the certification and labelling idea to the fisheries sector. The Marine Stewardship Council (MSC) was formally established in 1997 as a nonprofit organization.[16] The programme's three main fisheries standards – known as the MSC principles and criteria – focus on the health of the target

---

[11] Cashore, Auld & Newsom (2004); Gulbrandsen (2004); Auld, Gulbrandsen & McDermott (2008).

[12] Gulbrandsen (2010).

[13] FAO (2016), at 38.

[14] Ibid.

[15] FAO (2016).

[16] See www.msc.org.

fish stocks, the ecosystem impacts of the fishery, and the performance of the fishery management regime. Accredited certification bodies conduct the MSC assessment process. These certifiers appoint expert assessment teams that evaluate whether applicant fisheries meet the MSC principles and criteria.

Unlike FSC, the MSC standards and assessment process build on international agreements and guidelines, particularly the 1995 FAO Code of Conduct for Responsible Fisheries.[17] Governments have created rather elaborate international and regional fisheries regimes and a wide range of legally binding rules, centred in recent decades on the 1982 Law of the Sea Convention.[18] The MSC is therefore embedded in a regulatory field that has strongly influenced state responses to the programme and the subsequent evolution of its standards and procedures.

10.3 STATE RESPONSES TO THE EMERGENCE OF CERTIFICATION

10.3.1 *Forest Certification*

Several countries have considered restricting imports of tropical timber to address the problems of deforestation and forest degradation in the tropics. Similarly, the European Parliament has on several occasions called for European Union (EU) restrictions on the importation of tropical timber. However, a tropical timber import ban has never been enacted for fear of violating international trade law, and many countries have instead adopted public procurement policies aimed at purchasing timber only from legal and/or sustainable sources. While some governments have developed comprehensive legality and sustainability criteria for their timber procurements policies, others have simply decided that timber certified under a credible certification scheme can be accepted. Around thirty countries now have central government procurement policies aimed at ensuring that public buyers source only legal and/or sustainable timber and wood products. Here I examine how governments with such policies have assessed forest certification programmes.

Examination of timber procurement policies shows that although some differences remain, there has been substantial convergence in such policies over time. When European countries began to develop public procurement guidelines in the early 2000s, FSC quickly became the programme of choice. For example, Denmark declared in 2003 that FSC was the only certification scheme to provide adequate assurance of legally *and* sustainably sourced tropical wood.[19] Several other European countries expressed similar preferences for FSC-certified timber, and some

---

[17] FAO (1995).

[18] Full title: United Nations Convention on the Law of the Sea of 10 December 1982, UN doc. A/CONF.62/122; text in UNTS, Vol. 1833, at 3ff.

[19] Danish Ministry of the Environment (2003).

countries, notably Austria, the Netherlands, Switzerland, and the UK, even provided project funding to the FSC early on.[20]

A group of frontrunner countries – the UK, Denmark, the Netherlands, Belgium and Luxembourg – eventually developed their own comprehensive criteria for legally and sustainably sourced timber and assessed the degree to which certification programmes met those criteria.[21] The UK, for example, established a Central Point of Expertise on Timber (CPET) to assess certification schemes and advise the government on timber procurement. Similarly, the Netherlands established a Timber Procurement Assessment Committee (TPAC) to evaluate both international and national-level certification schemes. All the countries with comprehensive criteria for assessing timber legality and sustainability have learned from one another's experiences, and, in some cases, borrowed from one another's criteria and assessments.[22]

Although several countries initially signalled a preference for FSC-certified wood, most countries have accepted PEFC as a credible certificate as well. After having published its own comprehensive criteria in 2007,[23] Denmark declared that it accepted both FSC and PEFC as proof of legal *and* sustainable timber. Similarly, while PEFC was not acknowledged as evidence of sustainable timber in the first assessment of certification schemes by the UK government,[24] by the time the government implemented preferential treatment, PEFC had improved its standards sufficiently to meet the requirements for both legality and sustainability.[25] As mentioned previously, however, PEFC is different in nature from FSC; whereas FSC is a global certification scheme with one set of principles and criteria, PEFC is a mutual recognition framework that endorses national-level certification schemes on the basis of certain minimum requirements. Given the significant differences in the criteria and requirements of national member schemes, NGOs have repeatedly urged countries to assess each national-level scheme and accept only those that meet legality and sustainability criteria.[26] Because such assessments require considerable effort, most countries have adopted a less elaborate system, deciding that PEFC as an umbrella scheme meets public procurement criteria. The exception is the Netherlands, where TPAC has assessed and approved PEFC International and five member schemes – PEFC Belgium, Germany, Finland, Sweden, and Austria. These five national schemes account for about 90 per cent of the supply of PEFC timber on the Dutch market.

---

[20] Bartley (2003), at 448.
[21] Brack (2014).
[22] Ibid., at 9.
[23] Danish Ministry of the Environment (2007).
[24] CPET (2004).
[25] CPET (2006).
[26] E.g. FERN (2009).

Some countries have included social issues in their timber-procurement policies.[27] Of the countries with comprehensive legality and sustainability criteria, Denmark, the Netherlands and Belgium were frontrunners in incorporating social requirements in public procurement contracts. Since 2003, Denmark has included in its definition of sustainable forest management various social criteria relating to issues such as land tenure, rights of indigenous peoples, workers' rights, and community relations.[28] Similarly, the Dutch and Belgian governments have included social criteria as an integral element of their sustainable forest management policies.[29] However, the inclusion in public procurement policies of social requirements over and above those legislated for in the producer country itself remains controversial. The UK used to exclude social issues, which meant that social requirements above those legislated for in the producer country could not be included in public procurement contracts.[30] Since April 2010, however, all government procurement contracts for timber products in the UK must include social criteria as contract conditions.[31] Social issues recognized by Denmark, the Netherlands, Belgium and the UK include rights of indigenous peoples, rights of local communities, land tenure, recognition of customary rights, workers' rights, health and safety, fair prices (only the Netherlands), multiple functions of forests (only the Netherlands), participation, dispute resolution, and law enforcement.

A larger group of countries has adopted much simpler systems, accepting a wider range of certificates, eco-labels, legality verification schemes or even industry self-declarations as acceptable proofs of legality and/or sustainability.[32] These countries have not carried out elaborate assessments of certification programmes, but accept FSC or PEFC, or equivalent programmes. The exception is Norway, which since 2007 has banned the use of all tropical timber – including FSC- and PEFC-certified timber – in public-sector buildings and other construction works.[33] Although this policy is most likely in contravention of WTO procurement rules (WTO Agreement on Government Procurement),[34] it has not been challenged by any producer country, possibly because Norway has always been a rather minor importer of tropical timber.

In recent years, advances in EU timber legality law and licensing have made any EU member state's timber procurement policy aimed only at legality verification almost redundant, because timber products being placed on that country's market

---

[27] Brack (2008); FERN (2009).

[28] Danish Ministry of the Environment (2003).

[29] Brack (2008); Proforest (2010a, 2010b).

[30] Brack (2005).

[31] CPET (2010).

[32] Brack (2014), at 11.

[33] Norwegian Ministry of the Environment (2007). Available at: www.regjeringen.no/Upload/MD/Vedlegg/Planer/T-1467.pdf (accessed 27 October 2017).

[34] See Gulbrandsen & Fauchald (2015).

must be legal anyway.[35] The EU has developed an action plan to counter illegal logging, the Forest Law Enforcement, Governance and Trade (FLEGT) action plan. This action plan, approved in Council Conclusions in 2003, requires the EU to develop Voluntary Partnership Agreements (VPAs) with producer countries on timber licensing. Producer countries entering into such agreements with the EU commit to exporting to the EU only legally logged timber accompanied by a 'FLEGT license'.[36] Each VPA establishes a legality assurance system designed to identify legal timber products that can be licensed for export to the EU. To complement the timber licensing system, in 2008 the European Commission proposed a new, legally binding EU Timber Regulation.[37] This regulation, agreed in 2010 and applying in full since March 2013, prohibits the placing of illegally harvested timber and timber products on the EU market. The regulation also requires timber operators in the EU to implement systems of 'due diligence' to minimize the risk of their handling illegal timber. Since FLEGT-licensed timber and timber products are considered to have been legally harvested in accordance with the requirements of the regulation, the EU Timber Regulation provides an additional push for producer countries to complete VPA legality assurance systems and issue FLEGT licenses.[38]

To summarize, while one group of frontrunner countries developed comprehensive criteria for assessing legally and sustainably sourced timber, most other countries adopted simpler systems, deciding that certification under certain schemes or other assurance systems was proof of legality and sustainability. In practice, most countries with timber procurement policies eventually accepted both FSC and PEFC as acceptable proof. Although the EU Timber Regulation has made timber *legality* procurement policies less relevant in EU member states, forest certification still appears to be an important tool for verifying *sustainability* under the timber procurement policies of many EU member states; outside the EU, certification is used to verify both legality and sustainability. Hence, timber procurement policies in most countries continue to be closely linked to forest certification schemes.

10.3.2 *Fisheries Certification*

Given the long history of intergovernmental fisheries governance, some governments were deeply sceptical to the MSC from its inception, questioning the right of nongovernmental bodies to govern common-pool resources such as fish stocks. Seeing the MSC as an attempt to create a nonstate management regime beyond national jurisdiction, these governments argued that nongovernmental actors had

---

[35] Brack (2014), at 7.
[36] European Commission (2003).
[37] European Commission (2008).
[38] Brack (2014), at 7.

neither the necessary experience nor the mandate to govern fisheries. Like the FSC, the MSC allocates no preferred position to governments, which are treated like other stakeholders such as conservation organizations, fishing companies, industry associations and seafood buyers.

The Nordic Council of Ministers quickly became a central proponent of an FAO-endorsed labelling scheme, but neither meetings of the FAO Committee of Fisheries (COFI) nor consultations were sufficient to reach an agreement about how FAO should respond to the emergence of fisheries certification.[39] In 2003, at another COFI meeting, governments agreed to develop *voluntary guidelines* for the eco-labelling of fish and fisheries products from wild-capture fisheries. A set of such labelling guidelines was subsequently drafted during a series of FAO expert and technical consultations. In 2005, the FAO adopted these nonbinding guidelines, which addressed most aspects of fisheries certification. The FAO Guidelines for the Ecolabelling of Fish and Fishery Products from Marine Capture Fisheries (here-after: FAO Guidelines) emphasized the need to include in labelling programmes objective, third-party fisheries assessments using scientific evidence; transparent processes with extensive stakeholder consultation and opportunities for complaints and rules for adjudication; and standards based on the sustainability of target species, ecosystems and management practices.[40]

Despite initial mistrust of the MSC, many governments with large fishing industries have become increasingly supportive of the programme. As I discuss in Section 10.4.2 on evolutionary effects, this support seems to be related to FAO's engagement with fisheries certification and eco-labelling. Rather than seeing the MSC as a potential threat to states' sovereign authority to manage fisheries, these governments now view it as a helpful supplement to domestic, regional, and international management regimes. Government management authorities have in many cases provided financial support for MSC certification or have even been the clients for certification.[41] For instance, the Alaska Department of Fish and Game was the client for the certification of the Alaska salmon fishery, which re-entered assessment for its third, five-year MSC certification in July 2012. In Canada, where the government had been critical of the MSC, the federal Department of Fisheries and Oceans in 2007 made certification and eco-labelling a cornerstone of its Ocean to Plate policy strategy. This strategy was accompanied by presentations and government documents more or less officially endorsing the MSC.[42] A similar pattern of support is found across the Atlantic, where several European governments in the mid-2000s shifted from scepticism to actively supporting the MSC. The UK, Norway, Denmark, Sweden, and the Netherlands have all made certification part of their seafood

[39] Gulbrandsen (2009), at 656–657.
[40] FAO (2005).
[41] Gulbrandsen (2010); Gale & Haward (2011); Foley (2013).
[42] Foley (2013).

policy strategies and have in some cases provided funding for preassessment or full assessment of fisheries.

A few government-backed certification schemes for wild-capture fisheries have emerged, notably in Iceland, Japan, and the US state of Alaska. Dissatisfied with the MSC, these territorial schemes used the FAO Guidelines as a point of reference when they started developing their own standards. However, the MSC has continued to hold a virtual monopoly in the setting of widely adopted certification standards for wild-capture fisheries[43] and the government-backed territorial schemes do not challenge MSC's position.

To summarize, state responses to the emergence and development of the MSC have included establishing intergovernmental guidelines for fisheries certification, providing support for MSC certification, and endorsing certification and labelling in seafood policy strategies. Notwithstanding the emergence of a few government-backed schemes, many governments with a significant stake in seafood exports have become increasingly supportive of the MSC, facilitating industry uptake of certification within their jurisdictions – much as with the case of forest certification.

10.4 EVOLUTIONARY EFFECTS OF STATE RESPONSES TO CERTIFICATION

10.4.1 *Evolution of Forest Certification*

We have seen that timber procurement policies are closely linked to forest certification schemes. The key question to be examined here is what these policies have meant for the adoption of certification and the evolution of standards, rules, and procedures.

Beginning with adoption as of 2017, PEFC had a certified forest area of just over 300 million hectares in 35 countries, while FSC had certified nearly 200 million hectares of forests in 83 countries.[44] Correcting for forest certified under both schemes, the total global certified area in 2017 was estimated at 429 million hectares, or about 11 per cent of the global forest area. The distribution of certified forestland between the northern and southern hemispheres was extremely uneven: 85 per cent of the certified forest area worldwide was in the northern hemisphere, while Latin America, Africa, Asia and Oceania combined accounted for only 15 per cent of the global certified forest area.[45]

One effect of timber procurement policies has been to increase demand for certified products, particularly in the European market. Because there are no official

---

[43] Ponte (2012), at 301.
[44] UNECE/FAO (2017), at 17. Available at: www.unece.org/fileadmin/DAM/timber/publications/ FPAMR2017AdvanceDraft.pdf (accessed 28 October 2018).
[45] *Ibid.*, at 15.

statistics on public-sector timber consumption, it is difficult to estimate the percentage of total timber consumption covered by procurement policies that can be shown to engage with nonstate certification schemes. According to one estimate by the International Tropical Timber Organization (ITTO),[46] public procurement policies account for between 3 and 20 per cent of total timber consumption, depending on the importing country and market segment. In the EU, before the entry into force of the EU Timber Regulation, around 25 per cent of imported timber has been estimated to be subject to legality and sustainability verification requirements.[47] It is impossible to isolate the independent effects of public procurement, NGO pressure, and industry commitment to environmental responsibility on market uptake of certified timber, but according to one estimate, 'it seems likely that procurement policy has had the greatest single impact'.[48] A 2011 survey of the timber market in six EU countries (Denmark, France, Germany, Italy, the Netherlands, and the UK) concluded that the public sector and commercial big buyers were the main drivers generating demand for certified timber.[49]

Central government procurement policies not only facilitate public-sector uptake of certified products, they also help increase private-sector uptake. By approving specific certification schemes, governments signal that those schemes are credible governance systems on which private procurers can rely. Evidence suggests that suppliers' preference for relatively simple supply chains magnifies this effect: timber suppliers are increasingly switching to certified products for all their customers, not only public-sector ones.[50] This switch has also helped timber companies spread procurement policy to local government, which is not covered by central government policy.[51] Public procurement policies have thus had a much broader effect on the timber market than simply through the direct effect of central government purchases.

Turning to the evolution of certification standards, rules and procedures, the most interesting effect of public procurement policies to emerge from this case is the direct influence on certification schemes. Whereas governments with *simple* legality and sustainability policies accepted whatever criteria were in the certification schemes, the governments with *comprehensive* criteria assessed the certification schemes against the criteria. Since the establishment of the PEFC umbrella scheme, a controversial issue has been the consistency and robustness of an arrangement that has a single universal label but covers many national-level schemes with different standards. Unlike the case of the FSC, there was no global set of principles and criteria stipulating requirements to be met by forest operations. PEFC did have

---

[46] ITTO (2010).
[47] Oliver (2009).
[48] Brack & Buckrell (2011), at 11.
[49] Cited in Brack (2014), at 14.
[50] Brack & Buckrell (2011).
[51] Brack (2008), at 6.

a set of basic requirements for national-level certification programmes, first established in 2002 and revised many times since, but many of the specific requirements for forest operations were left to the discretion of national member programmes. For these reasons, the Dutch TPAC has, as noted, assessed not only PEFC International but also several member schemes. Similarly, CPET, established by the UK government, has examined the consistency of member schemes in assessments of PEFC International. In 2004, CPET concluded that PEFC did not meet the public procurement requirements for sustainable forest management because of unbalanced governance, inadequate public consultation during the certification process, and lack of public disclosure of auditing outcomes.[52] The American Sustainable Forestry Initiative (SFI) – since 2005 a PEFC member scheme – was not approved because the chain-of-custody certificate did not specify the amount of uncertified material used in the end product.[53]

The schemes that did not pass the test were allowed six months, beginning in November 2004, to improve their standards and rules before the UK government implemented preferential treatment.[54] In response, PEFC took up each of the CPET issues at its 2004 General Assembly and at a second, especially scheduled meeting in April 2005. Subsequently, PEFC moved to adopt new conventions on balanced governance, public consultation, and transparency.[55] Similarly, in April 2005, the SFI adopted new chain-of-custody rules and requested that the scheme be submitted for reassessment against the UK government requirements.[56]

In August 2005, the UK government announced that PEFC and SFI had improved their standards sufficiently to meet the public procurement requirements; but they were put on probation until the end of 2005, when they would be reassessed by CPET.[57] Greenpeace, Friends of the Earth, and other NGOs immediately attacked the government's decision, saying in a joint statement that the UK government had approved two schemes that 'allow large-scale, unsustainable logging in ancient forest areas, the destruction of endangered species and the abuse of indigenous people's rights'.[58] Notwithstanding these NGO protests, CPET's reassessment confirmed that PEFC and SFI – like FSC – could be used as evidence of legal and sustainable timber.[59]

In 2010, the PEFC General Assembly adopted a new meta-standard for sustainable forest management to improve the international system and enhance consistency across member schemes.[60] The requirements in this standard document, which entered into force in May 2011, must be reflected in all forest management

---

[52] CPET (2004).
[53] Ibid.
[54] ENDS (2004).
[55] PEFC (2005); Overdevest (2010), at 67–68.
[56] CPET (2005).
[57] ENDS (2005).
[58] Ibid., at 22–23.
[59] CPET (2006).
[60] PEFC (2010).

standards submitted for PEFC endorsement. Member schemes endorsed by PEFC before May 2011 had to comply with the new standards by May 2013. Hence, public assessments have contributed to an upward harmonization of PEFC member schemes through the adoption of similar forest management rules and conventions on stakeholder participation, transparency, and balanced governance.

Finally, several certification schemes have developed legality assurance standards in response to the EU FLEGT and Timber Regulation as well as member-state procurement policies.[61] The FSC worked with government officials to ensure that its controlled wood standard (first approved in 2004, revised in 2006), developed for non-FSC certified material used in FSC Mix products, was consistent with these regulations and that FSC certification met the legality requirements in the EU. At the 2011 General Assembly, the FSC membership unanimously approved a motion to strengthen the standard. A main objective of this process was to 'bring the Controlled Wood standard in link with EUTR [EU Timber Regulation], Lacey Act [US] and other national legality legislation'.[62] Hence, while tracking systems developed in non-state certification programmes have proven central in advancing international efforts to eliminate trade in illegal wood,[63] international legality requirements have influenced certification standards and created opportunities for expansion into new areas of verification.

To conclude, the timber procurement policies of governments with *simple* criteria have contributed to increase the uptake of certification but have not had any impact on certification standards, rules and procedures. By contrast, the timber procurement policies of governments with *comprehensive* legality and sustainability criteria have forced the public exposure of information about certification standards, putting pressure on poorly performing programmes to improve their standards. Public comparisons with FSC's higher standards have pushed PEFC International and some of its member schemes to modify their standards and governance proced-ures and enhance participation from outside stakeholders in certification processes. As predicted by the Trading-Up hypothesis developed by Vogel,[64] certification programmes have thus converged upwards, at least in terms of procedural require-ments, to satisfy timber procurements requirements in European countries with comprehensive legality and sustainability criteria.

10.4.2 *Evolution of Fisheries Certification*

As in Section 10.4.1, the key question to be examined here is what state responses to certification have meant for the adoption of fisheries certification and the evolution of standards, rules and procedures. As explained in Section 10.1, I focus on how

---

[61] Proforest (2010b); Overdevest & Zeitlin (2014).
[62] See Gulbrandsen (2014), at 84.
[63] Auld et al. (2010).
[64] Vogel (1995).

FAO's engagement with fisheries certification influenced the uptake of the MSC and its standards and procedures.

Beginning with adoption, the growth of the MSC following the issuance of the 2005 FAO Guidelines is evident when we examine the number of certified fisheries and the quantity of all certified catches. In 2000, after fifteen months of assessment, Western Australia's rock lobster fishery became the first to be certified to the MSC standards. After a slow start, the MSC experienced rapid growth during the second half of its first decade of operations. The number of certified fisheries increased from 12 in 2005 to 122 by the end of 2010. By the end of 2016, 296 fisheries in 35 countries were certified to the MSC standard, and a further 67 fisheries were in assessment.[65] The certified fisheries record annual catches of about 12 per cent of the annual global harvest of wild-capture fisheries, a figure that has doubled since 2010.[66]

Given the many factors that may have contributed to the growth of the MSC, it is impossible to isolate the independent effect of FAO's engagement with fisheries certification on MSC uptake. Other government policies, market demand, NGO pressure, and a growing retailer commitment to source sustainable seafood are factors that all have influenced MSC uptake.[67] Indeed, a major breakthrough came in January 2006, less than one year after the issuance of the FAO Guidelines, when Walmart – the world's biggest retailer – announced a commitment to source all of its fresh and frozen seafood supplies in North America from MSC sources within five years. According to MSC chief executive Rupert Howes, the Walmart commitment pressured its suppliers to become involved in the MSC and spurred other retailers to look at their own commitments to the programme.[68] Similarly, other analysts have highlighted Walmart's importance in raising the number of certifications in the late 2000s.[69]

The evidence suggests, however, that the FAO's engagement in combination with other government policies and market dynamics have had a positive impact on the development and uptake of the programme. First, the MSC has actively employed the FAO Guidelines in efforts to enhance the credibility of the programme and attract new clients for certification. It has welcomed the guidelines, issued a statement of full compliance as soon as the necessary organizational and procedural changes were implemented, and has been careful to communicate that it is the only global certification programme consistent with them.[70] In short, the FAO Guidelines have represented an endorsement of fisheries certification,

---

[65] MSC (2017). Available at: www.msc.org/documents/environmental-benefits/global-impacts/ msc-global-impacts-report-2017 (accessed 27 October 2017).

[66] Ibid.

[67] Gulbrandsen (2010).

[68] Author's interview with Rupert Howes, 23 May 2006.

[69] E.g. Ponte (2012).

[70] See, for example, Gulbrandsen (2014), at 85.

enabling the MSC to attract clients that otherwise may have remained unconvinced by the programme.[71]

Second, the fact that the MSC could rapidly demonstrate full consistency with FAO Guidelines also helped convince several governments to support the programme. Recall that Nordic fisheries and their governments in the 1990s were highly doubtful of the MSC, and any other private standards in the fisheries sector, and had called for the establishment of an FAO-endorsed labelling scheme. When those efforts failed, the Nordic governments became key proponents of labelling guidelines issued by FAO. The subsequent establishment of the FAO Guidelines should be seen as an attempt to reassert authority in assessing credible private standards in the fisheries sector, not to weaken or undermine the MSC. When the MSC announced full consistency with the FAO Guidelines in 2006, the Nordic fisheries and their governments (with the exception of Iceland) eventually accepted it as a credible certification programme. But it also had to do with the evolution of the MSC from a minor experiment in the late 1990s to a full-fledged and elaborate certification programme by the time the FAO Guidelines were issued.

Third, while various aquaculture certification programmes have emerged, the MSC has remained the only *global* certification programme for wild-capture fisheries that is consistent with the FAO Guidelines. When Walmart and other big companies began to demand MSC-certified products, many governments, concerned about the market access of their seafood industries, decided to encourage and support certification. Government support for the MSC has expanded in recent years, as governments with significant stakes in the fishing industry have sought to secure, maintain and expand domestic seafood exports.

How, then, did FAO's engagement influence the MSC standards and procedures? The MSC responded swiftly to the issuance of FAO Guidelines by welcoming them, saying they represented 'a significant endorsement of eco-labelling as a tool to achieve the sustainable management of fisheries'.[72] As MSC chief executive Rupert Howes stated when the FAO Guidelines were issued: 'there is no other [certification] organization that is remotely near FAO compliance; the MSC is the only one'.[73] Importantly, the three main principles of fisheries assessment identified by FAO correspond to the MSC principles – health of the fish stock, impact of the fishery on the ecosystem, and performance of the fishery management system.[74] Moreover, the MSC could also claim conformance with the key sections of the FAO Guidelines, which stipulated, *inter alia*, that any credible fisheries eco-labelling programme must be independent, transparent and science-based, and have objections procedures and a high degree of stakeholder involvement.[75]

---

[71] Author's interview with MSC policy officer, 2 July 2012.
[72] MSC (2005).
[73] Author's interview with Rupert Howes, 23 May 2006.
[74] MSC (2002).
[75] FAO (2005).

Although they were nonbinding, the FAO Guidelines did exert normative influence over the MSC to comply with their provisions. In choosing to fully comply with the FAO Guidelines, the MSC separated the standard-setting and accreditation functions (consistent with section 69 of the Guidelines). Whereas independent certification bodies (certifiers) used to be accredited by the programme itself, the MSC thus outsourced accreditation decisions to Accreditation Services International – an independent organization that also accredits certifiers for the FSC. Similar to the FSC, MSC's approach to accreditation has therefore evolved from an in-house process to one controlled by a separate organization.

Furthermore, the MSC changed certain aspects of its objections procedure. Opportunities to lodge complaints have been in place since the MSC launched, but a formal objections procedure was first introduced in 2001 and has since undergone several revisions.[76] The first approach for complaints and dispute resolution gave the certifiers responsibility for specifying and implementing policy and procedures for handling disputes. An unresolved dispute could then be referred to the MSC. In 2001, an objection to the certification of the New Zealand hoki fishery – one of the first MSC-certified fisheries – exposed serious shortcomings in this approach, including the lack of a time limit for submitting a compliant; guidance regarding the scope and handling of a complaint; and a timetable for the processing and determination of a complaint.[77] In 2002, the MSC board of trustees adopted an objections procedure for use in all subsequent assessments. Any dispute a certifier could not resolve would be referred to the board of trustees, which would establish a dispute panel chaired by a board member with no interest in the fishery.

In order to fully comply with the FAO Guidelines (section 148), the MSC made its procedure for handling objections to fisheries assessments independent of the certification programme. This meant that the programme could no longer have a board member as chair, or even as a member, of the objections panel.[78] The MSC board of trustees therefore appointed an independent adjudicator (now a roster of adjudicators) to replace its objections panel. The adjudicator is a lawyer with experience in jurisprudence, fisheries law and dispute resolution or mediation related to natural resource management. If the adjudicator finds that the certifier committed a procedural or scoring error, the certifier must reconsider its final report and determination in light of the concerns raised by the objector. This process is repeated until the adjudicator accepts the changes made by the certifier or upholds the complaint.[79] Only stakeholders involved in or consulted during the fisheries assessment process may object to the certifier's decision, and the objector must cover

---

[76] Gulbrandsen & Auld (2016), at 51.
[77] Leadbitter & Ward (2003), at 81.
[78] Gulbrandsen & Auld (2016), at 53.
[79] MSC (2010).

the cost of an objection process.[80] The MSC reported that the changes to the objections procedure had been fully implemented by September 2006.[81]

In conclusion, the key normative influences on the MSC certification standards and procedures seem to be the 2005 FAO Guidelines on fisheries eco-labelling, as well as the 1995 FAO Code of Conduct for Responsible Fisheries and international fisheries agreements. Although the development of the FAO Guidelines could be seen as an intergovernmental attempt to regain control of an issue area dominated by non-state actors, the *effect* of these guidelines has been to strengthen MSC's position as the 'gold standard' of wild-capture fishery certification. Instead of presenting a challenge to MSC authority, the FAO Guidelines seem to have consolidated the MSC's position as the leading global certification programme in the fisheries sector. Hence, this case demonstrates the normative influence international 'soft law' can have on nonstate certification schemes seeking to enhance the credibility of their standards.

10.5 INTERACTION MECHANISMS AND EFFECTS

The evidence presented in this chapter shows how the interactions between private and public authority shape policy trajectories and policy outcomes. Four inductively generated interaction mechanisms and hypotheses follow from this analysis of forest and fisheries certification. The first mechanism is *comparison and benchmarking,* which occurs when competing private standards are measured against some public benchmark or standard. The evidence indicates that when certification programmes compete for public recognition, public procurement policies are more likely to put pressure on poorly performing programmes to improve their standards.[82] Public procurement policies have resulted in an upwards convergence of PEFC member schemes, narrowing the gap between their approach and that of the FSC. Hence, comparison and benchmarking can force the public exposure of information about certification standards and procedures, and pressure the poorer performers to improve their standards.

The second interaction mechanism is *collaboration,* which occurs when private and public rule-makers coordinate governance efforts, formally or informally, to achieve certain objectives. The evidence suggests that collaboration is more likely to occur when the state depends on the nonstate programme to help it implement public regulations or attain public policy objectives, and *vice versa.* In the case of forestry, states with timber procurement policies depend upon private certification schemes for sustainability verification. This form of collaboration can be referred to

---

[80] Since 2010, the MSC has capped the cost of filing objections to £5,000, and about 10 per cent of objectors have had the fee waived because of financial hardship. Brown, Agnew & Martin (2016).

[81] MSC (2006).

[82] See also Overdevest & Zeitlin (2014).

as conditional rule referencing: 'if you comply with X's rule, that will constitute compliance with mine'.[83] Several governments with timber procurement policies have made supplier eligibility contingent on compliance with forest certification standards. In the case of fisheries, many coastal states depend on the MSC for providing sustainability assurance to export markets for seafood products. In turn, nonstate programmes receive legitimacy and greater uptake from incorporation into public policies. Similarly, the expertise, competencies and 'social capital' (such as access to transnational networks) of nonstate programmes can lend credibility to state-based regulation. Symbolic resources such as a widely recognized logo and brand name associated with a particular organization (like FSC or MSC) can also bring credibility to government programmes. Generally, the evidence indicates that collaboration between public and private governance efforts occurs as they strive for legitimacy and policy relevance, carve out governance niches, and seek to obtain mutually beneficial outcomes.

The third mechanism is *co-optation*, which occurs when a stronger party imposes certain policies or practices on a weaker party. Co-optation will usually reflect an asymmetrical distribution of power between states and nonstate actors, where the state can rely on legal provisions and its sovereign rule-making authority to influence nonstate governance efforts. The comparison of state responses to forest and fisheries certification indicates that states are more likely to reassert authority in assessing credible standards when nonstate programmes emerge in densely regulated sectors. Given the exclusive authority of governments in fisheries governance and allocation decisions, many governments were initially sceptical of the MSC, questioning the right of nonstate certification programmes to play a role in fisheries governance. The fact that fisheries are managed by governments and international regimes can explain why governments responded to the MSC by mandating the FAO with the task of developing rules for fisheries certification and eco-labelling. The FAO Guidelines have become the global 'meta-governance' standard for fisheries certification schemes and political responses to them.[84] Similar to the MSC, territorial fisheries certification schemes in Iceland, Japan, the US state of Alaska, and elsewhere used the FAO Guidelines as a point of reference when they started developing their own standards. Indeed, all fisheries certification and labelling initiatives claim compliance with the FAO Guidelines and use them as a reference point. Although states do not force certification programmes to comply with the FAO Guidelines, it is likely that the Guidelines will continue to provide the global meta-regulatory point of reference for different fisheries certification initiatives.

The fourth mechanism at work here is *cognitive interaction*, which occurs when information, knowledge or ideas generated in one institution influence governance

---

[83] Eberlein et al. (2014), at 11.

[84] Foley (2013).

efforts in another.[85] The evidence suggests that policy learning across private and public governance efforts is more likely to occur when nonstate certification programmes disclose not only information about rules and procedures, but also information about outcomes and effects. As highlighted by Eberlein *et al.*,[86] the information nonstate governance schemes produce and disseminate may be among the most important sources of transnational governance interaction. One example is how tracking systems developed in forest certification schemes have become critical in advancing intergovernmental efforts to eliminate trade in illegal wood.[87] However, incomplete information disclosure across the field of non-state certification programmes presents problems for this collective model of learning and innovation.[88] While procedural transparency for auditing (the specific steps an audit will involve and information on when consultation and stakeholder engagement are to occur) has become a norm across all programmes in the forest and fisheries sectors, several PEFC-endorsed programmes disclose incomplete information about the outcomes of audits. Without full disclosure of the outcomes and effects of nonstate certification, collective learning to advance both public and private governance efforts will be difficult.[89]

To conclude, the "four Cs" – comparison, collaboration, co-optation, and cognitive interaction – are inductively generated mechanisms that describe types of interaction between public and private governance schemes. Table 10.1 summarizes these causal mechanisms, along with inductively generated hypotheses on interaction dynamics that can be tested in future research.

10.6 CONCLUSIONS

This chapter has demonstrated how differences in the management of forest and fisheries resulted in divergent state responses to forest and fisheries certification. In the case of forest certification, many producer countries responded to the FSC by supporting the development of industry-dominated schemes, while many consumer countries responded by developing procurement policies stipulating the purchase of 'FSC or equivalent' certificates. The effect of government support, either for FSC or industry-dominated competitors, was to legitimate forest certification as a major international forest policy tool embraced by decision-makers around the world.

In contrast, states responded to the emergence of MSC by mandating FAO with the task of developing guidelines for fisheries certification and eco-labelling. This attempt by governments, backed by fisheries associations in some countries, to regain control of an issue area of increasing importance in the fisheries sector is

[85] See also Gehring & Oberthür (2009).
[86] Eberlein et al. (2014).
[87] Auld et al. (2010).
[88] Auld & Gulbrandsen (2014).
[89] Ibid.

TABLE 10.1 *Public–Private Interaction Dynamics: Hypotheses, Mechanisms and Examples*

Hypotheses	Mechanisms	Examples
Public procurement policies are more likely to pressure poorly performing programmes to improve their standards when certification programmes compete for public recognition	Comparison and benchmarking	Upwards convergence of PEFC member schemes
Collaboration is more likely to occur when the state depends on the nonstate programme for help in implementing regulations or attaining public policy objectives.	Collaboration	Public recognition of private certification schemes in timber procurement policies and seafood policy strategies
States are more likely to reassert authority in assessing credible standards when nonstate programmes emerge in densely regulated sectors.	Co-optation	Development of FAO Guidelines for eco-labelling of fish and fisheries products
Policy learning across private and public governance efforts is more likely to occur when nonstate certification programmes disclose information about outcomes and effects.	Cognitive interaction	Tracking systems for legally and sustainably sourced wood

related to the way fisheries are managed. We have seen that, unlike the forestry sector, there is a long history of international governance in the fishery sector and a wide range of international, regional and national fisheries management regimes. I have argued, however, that the *effect* of the FAO Guidelines was to strengthen MSC's position as the leading global certification programme for wild-capture fisheries. The FAO Guidelines represented an endorsement of fisheries certification as a global policy tool to achieve sustainable fisheries management, much as in the case of government support for FSC or industry-dominated schemes. This outcome can partly explain why several governments discarded their initial scepticism of the MSC in favour of actively supporting the programme. Most governments with a significant stake in seafood exports have become increasingly supportive of MSC, encouraging, enabling and endorsing fisheries certification. As with the FSC, government support for the MSC has contributed significantly to its impressive growth in recent years. In both cases, state responses to the emergence of certification programmes have served to legitimate certification as a credible policy tool that companies and consumers can trust.

The evidence presented in this chapter demonstrates that certification has not supplanted public rules and regulations in the forestry and fisheries sectors. Although certification sometimes compensates for inadequate public governance systems, it has generally emerged as a supplement to public regulation, and in many countries it has been actively promoted by governments. In brief, states have influenced and continue to influence nonstate certification programmes in more ways, and more profoundly, than many previous studies imply. The emergence of private regulatory programmes reflects in many cases a desire of states and nonstate actors alike to experiment with new governance approaches. Policy changes in the forestry and fisheries sectors are not indicative of less government involvement, but rather of the ambition to develop a 'smart mix' of private and public policy instruments at multiple governance levels. These developments reflect a changing role of the state and suggest that states and nonstate actors are increasingly co-regulating natural resources. Rather than seeing private regulation as a challenge to government authority, states have generally accepted it as a useful supplement to – and sometimes a substitute for – traditional command-and-control regulation. Indeed, states have promoted, supported, and even required certification of producers and products in accordance with rules developed by nonstate actors. A critical area of study is therefore the intersection of private and public efforts to resolve environmental and social problems related to the use of natural resources. More research is needed to determine how public and private governance initiatives co-produce positive and negative environmental effects as well as intended and unintended outcomes in specific jurisdictions.

REFERENCES

Abbott, K. W. & Snidal, D. 2009. 'The Governance Triangle: Regulatory Standards Institutions and the Shadow of the State'. In Mattli, W. & Woods, N. (eds.), *The Politics of Global Regulation*. Princeton, NJ, Princeton University Press, 44–88.

Auld, G., Cashore, B., Balboa, C., Bozzi, L. & Renckens, S. 2010. 'Can Technological Innovations Improve Private Regulation in the Global Economy?'. *Business and Politics* 12(3), 1–39.

Auld, G., Gulbrandsen, L. H. & McDermott, C. L. 2008. 'Certification Schemes and the Impacts on Forests and Forestry'. *Annual Review of Environment and Resources* 33, 187–211.

Auld, G. & Gulbrandsen, L. H. 2014. 'Learning through Disclosure: The Evolving Importance of Transparency in the Practice of Nonstate Certification'. In Gupta, A. & Mason, M. (eds.), *Transparency in Global Environmental Governance: Critical Perspectives*. Cambridge, MA, MIT Press, 271–296.

Bartley, T. 2007. 'Institutional Emergence in an Era of Globalization: The Rise of Transnational Private Regulation of Labor and Environmental Conditions'. *American Journal of Sociology* 113(2), 297–351.

  2003. 'Certifying Forests and Factories: States, Social Movements, and the Rise of Private Regulation in the Apparel and Forest Products Fields'. *Politics and Society* 31(3), 433–464.

Boström, M. 2003. 'How State Dependent is a Non-State-Driven Rule-Making Project? The Case of Forest Certification in Sweden'. *Journal of Environmental Policy and Planning* 5(2), 165–180.

Brack, D. 2014. *Promoting Legal and Sustainable Timber: Using Public Procurement Policy*. London, Chatham House.

2008. *Social Issues in Timber Procurement Policies* (third draft). London, Chatham House.

2005. *Public Procurement of Timber: EU Member State Initiatives for Sourcing Legal and Sustainable Timber*. London, Chatham House.

Brack, D. & Buckrell, J. 2011. *Controlling Illegal Logging: Consumer-Country Measures*. London, Chatham House.

Brown, S., Agnew, D. J. & Martin, W. 2016. 'On the Road to Fisheries Certification: The Value of the Objections Procedure in Achieving the MSC Sustainability Standard'. *Fisheries Research* 182, 136–148.

Cashore, B. 2002. 'Legitimacy and the Privatization of Environmental Governance: How Non-State Market-Driven (NSMD) Governance Systems Gain Rule-Making Authority'. *Governance: An International Journal of Policy, Administration, and Institutions* 15(4), 503–529.

Cashore, B., Auld, G. & Newsom, D. 2004. *Governing Through Markets: Forest Certification and the Emergence of Non-State Authority*. New Haven, CT, Yale University Press.

CPET. 2010. *UK Government Timber Procurement Policy: Definition of Legal and Sustainable for Timber Procurements* (fourth edition). Oxford, CPET.

2006. *Evaluation of Category A Evidence: Review of Forest Certification Schemes: Results*, December. Oxford, CPET.

2005. *Evaluation of Category A Evidence: Assessment Results Sustainable Forestry Initiative*, September. Oxford, CPET.

2004. *UK Government Timber Procurement Policy: Assessment of Five Forest Certification Schemes*. CPET Phase 1 Final Report, November. Oxford, CPET.

Danish Ministry of the Environment. 2007. *Draft Criteria for Legal and Sustainable Timber and Assessment of Certification Schemes*. Copenhagen, Danish Ministry of the Environment.

2003. *Purchasing Tropical Timber. Environmental Guidelines*. Copenhagen, Danish Ministry of the Environment.

Eberlein, B., Abbott, K.W., Black, J., Meidinger, E. & Wood, S. 2014. 'Transnational Business Governance Interactions: Conceptualization and Framework for Analysis'. *Regulation and Governance* 8(1), 1–21.

ENDS (Environmental Data Services). 2005. 'DEFRA's Approval of Industry-Certified Timber Blasted by Green Groups'. *ENDS Report* 368 (September).

2004. 'PEFC Timber Scheme "Inadequate" Says DEFRA'. *ENDS Report* 358 (July).

European Commission. 2008. *Public Procurement for a Better Environment*. EU document COM (2008) 400/2. Brussels, European Commission.

2003. *Communication from the Commission to the Council and the European Parliament: Forest Law Enforcement, Governance and Trade (FLEGT), Proposal for an EU Action Plan*. EU document COM (2003) 251 final, 21 May. Brussels, European Commission.

FAO. 2016. *The State of World Fisheries and Aquaculture 2016*. Rome, FAO.

2015. *Global Forest Resources Assessment 2015*. Rome, FAO.

2005. *The FAO Guidelines for the Ecolabelling of Fish and Fisheries Products from Marine Capture Fisheries*. Rome, FAO.

1995. *Code of Conduct for Responsible Fisheries*. Rome, FAO.

FERN. 2009. *Buying a Sustainable Future? Timber Procurement Policies in Europe and Japan*. Moreton-in-Marsh (UK), FERN.

Foley, P. 2013. 'National Government Responses to Marine Stewardship Council (MSC) Fisheries Certification: Insights from Atlantic Canada'. *New Political Economy* 18(2), 284–307.

Gale, F. & Haward, M. 2011. *Global Commodity Governance: State Responses to Sustainable Forest and Fisheries Certification*. New York, Palgrave Macmillan.

Gehring, T. & Oberthür, S. 2009. 'The Causal Mechanisms of Interaction between International Institutions'. *European Journal of International Relations* 15(1), 125–156.

Gulbrandsen, L. H. 2014. 'Dynamic Governance Interactions: Evolutionary Effects of State Responses to Non-State Certification Programmes'. *Regulation & Governance* 8(1), 74–92.

    2010. *Transnational Environmental Governance: The Emergence and Effects of the Certification of Forests and Fisheries*. Cheltenham, Edward Elgar.

    2009. 'The Emergence and Effectiveness of the Marine Stewardship Council'. *Marine Policy* 33(4), 654–660.

    2004. 'Overlapping Public and Private Governance: Can Forest Certification Fill the Gaps in the Global Forest Regime?'. *Global Environmental Politics* 4(2), 75–99.

Gulbrandsen, L. H. & Auld, G. 2016. 'Contested Accountability Logics in Evolving Nonstate Certification for Fisheries Sustainability'. *Global Environmental Politics* 16(2), 42–60.

Gulbrandsen, L. H. & Fauchald, O. K. 2015. 'Assessing the New York Declaration on Forests from a Trade Perspective'. *BIORES: Analysis and News on Trade and Environment* 9(4), 4–7.

ITTO. 2010. *The Pros and Cons of Procurement. Development and Progress in Timber-Procurement Policies as Tools for Promoting the Sustainable Management of Tropical Forests*. Yokohama, ITTO.

Leadbitter, D. & Ward, T. 2003. 'Dispute Resolution and the MSC'. In Phillips, B., Ward, T. & Chaffee, C. (eds.), *Eco-Labelling in Fisheries: What Is It All About?*. Oxford, UK, Blackwell, 80–85.

Lister, J. 2011. *Corporate Social Responsibility and the State: International Approaches to Forest Co-Regulation*. Vancouver, BC, University of British Columbia Press.

Meidinger, E. 2006. 'The Administrative Law of Global Private-Public Regulation: The Case of Forestry'. *The European Journal of International Law* 17(1), 47–87.

MSC. 2017. *Global Impacts Report 2017*. 2017. London, UK, Marine Stewardship Council. Available at: www.msc.org/documents/environmental-benefits/global-impacts/msc-global-impacts-report-2017 (accessed 27 October 2017).

    2010. *TAB Directive 023: Revised Fisheries Certification Methodology Objections Procedure*. London, UK, Marine Stewardship Council.

    2006. *Leader in Fishery Certification and Eco-labelling Announces 100% Consistency with UN Guidelines*. Press release, 26 September. London, UK, Marine Stewardship Council.

    2005. *MSC Welcomes FAO Guidelines on Marine Eco-labelling*. Press release, 31 March. London, UK, Marine Stewardship Council.

    2002. *MSC Principles and Criteria for Sustainable Fishing*. London, UK, Marine Stewardship Council.

Norwegian Ministry of the Environment. 2007. *Miljø- og samfunnsansvar i offentlige anskaffelser. Handlingsplan 2007–2010*. Oslo, Norwegian Ministry of Environment. Available at: www.regjeringen.no/Upload/MD/Vedlegg/Planer/T-1467.pdf (accessed 27 October 2017).

Oliver, R. 2009. *EU Market Conditions for 'Verified Legal' and 'Verified Legal and Sustainable' Wood Products*. Prepared for the Timber Trade Federation and the Department for International Development by Forest Industries Intelligence Ltd, Settle (UK).

Overdevest, C. 2010. 'Comparing Forest Certification Schemes: The Case of Ratcheting Standards in the Forest Sector'. *Socio-Economic Review* 8, 47–76.

Overdevest, C. & Zeitlin, J. 2014. 'Assembling an Experimentalist Regime: Transnational Governance Interactions in the Forest Sector'. *Regulation and Governance* 8(1), 22–48.

PEFC. 2010. *PEFC International Standard: Requirements for Certification Schemes, PEFC ST 1003: 2010*. Geneva, PEFC Council.

　2005. *Assessment of PEFC Scheme against DEFRA/CPET Criteria*. Luxembourg, PEFC Council.

Ponte, S. 2012. 'The Marine Stewardship Council (MSC) and the Making of a Market for "Sustainable Fish"'. *Journal of Agrarian Change* 12(2–3), 300–315.

Proforest. 2010a. *Market Requirements for Legal and Sustainable Timber, and the Implications for Chinese Suppliers*. Oxford, Proforest.

　2010b. *FLEGT Licensed Timber and EU Member State Procurement Policies*. Oxford, Proforest.

Rosenau, J. N. & Czempiel, E. -O. (eds.). 1992. *Governance without Government: Order and Change in World Politics*. Cambridge, Cambridge University Press.

Tollefson, C., Gale, F. & Haley, D. 2008. *Setting the Standard: Certification, Governance, and the Forest Stewardship Council*. Vancouver, BC, UBC Press.

UNECE/FAO. 2017. *Forest Products Annual Market Review, 2016–2017*. Geneva, UN Economic Commission for Europe. Available at: www.unece.org/fileadmin/DAM/timber/publications/FPAMR2017AdvanceDraft.pdf (accessed 28 October 2018).

Vogel, D. 2010. 'The Private Regulation of Global Corporate Conduct: Achievements and Limitations'. *Business and Society* 49(1), 68–87.

　1995. *Trading Up: Consumer and Environmental Regulation in a Global Economy*. Cambridge, MA, Harvard University Press.

PART III

Climate Change and Oil

11

Smart Mixes, Non-State Governance and Climate Change

Neil Gunningham

11.1 INTRODUCTION

The concept of smart policy mixes goes back to at least the 1990s. One well-known version was advocated by Gunningham et al., in the form of 'smart regulation'.[1] Subsequently, the concept has been refined by Gunningham and Sinclair and others.[2] Smart regulation refers to a form of regulatory pluralism that embraces flexible and innovative forms of social control with an emphasis on harnessing not only governments but also business and third parties and designing complementary instrument mixes. For example, it encompasses self-regulation and co-regulation, using commercial interests and non-governmental organizations (such as peak bodies) as regulatory surrogates, together with improving the effectiveness and efficiency of more conventional forms of direct government regulation. The underlying rationale is that, in most circumstances, the use of multiple policy instruments (rather than a single policy), and a broader range of regulatory actors, will produce better regulation. As such, smart regulation envisages the implementation of complementary combinations of instruments and participants tailored to meet the imperatives of specific environmental issues.

To put smart regulation in context, it is important to note that, traditionally, regulation was thought of as a bipartite process involving government (as the regulator) and business (as the regulated entity). However, a substantial body of empirical research reveals that there is a plurality of regulatory forms, and that numerous actors influence the behaviour of regulated groups in a variety of complex and subtle ways. Crucially, informal mechanisms of social control often prove more important than formal ones. Accordingly, the smart regulation perspective suggests

---

[1] Gunningham, Grabosky & Sinclair (1998).
[2] Gunningham & Sinclair (2002); Bloor et al. (2006); Toffelson et al. (2008).

that those who design and implement regulatory policy should focus their attention on understanding such broader regulatory influences and interactions.

In terms of its intellectual history, smart regulation evolved in a period in which it had become apparent that neither traditional command-and-control regulation nor the free market provided satisfactory answers to the increasingly complex and serious environmental problems that confront the world. This led to a search for alternatives more capable of addressing the environmental challenge and to the exploration of a broader range of policy tools such as economic instruments, self-regulation and information-based strategies. Smart regulation also emerged in a period of comparative state weakness, in which the dominance of neoliberalism had resulted in the relative emasculation of formerly powerful environmental regulators and in which third parties such as NGOs and business were increasingly filling the 'regulatory space' formerly occupied by the state.

A substantial number of policy approaches are consistent with the precepts of smart regulation. In Canada, smart regulation became the principal focus of federal government–driven regulatory reform in the mid-2000s and continued under a different name for some time thereafter (Government of Canada (2005)).[3] There it has been characterized by 'a restructuring of the process of assessing, reforming and improving the regime in which regulations are developed, managed, enforced and measured'.[4] A few years later, the European Union (EU) promoted smart regulation as a vehicle to achieve a 'cleaner, fairer and more competitive Europe'.[5] Indeed, according to Toffelson et al., 'the smart-regulation critique has also resonated in potentially sweeping law and policy reforms in many other Western liberal democracies'.[6] For example, EU and Dutch regulators have invoked smart regulation to combat the global problem of e-waste, and as a vehicle for effective governance of the globalised shipping industry.[7] It has also been contemplated for other emerging globalizing industries.

A growing interest in the environmental challenges of globalization, and in global environmental governance more broadly, raises the question of whether, to what extent or in what circumstances 'smart mixes' (a better term in the international context[8]) might be invoked in *transnational*[9] governance rather than just as an

---

[3]   Leese (2005).

[4]   Wood & Johannson (2008), at 361.

[5]   Danish Ministry of Economic and Business Affairs et al. (2010).

[6]   Toffelson et al. (2008), at 257.

[7]   Bloor et al. (2006).

[8]   While 'smart regulation' is an appropriate term to describe (inter alia) how the nation-state might best harness multiple policy instruments and a broader range of regulatory actors to achieve preferred outcomes, 'smart mixes' is a more apposite term to capture a similar approach at international and transnational levels, given that the latter involve 'governance' rather than 'regulation' and may be engaged in by non-state actors even in the absence of state intervention.

[9]   In 'transnational governance', actors other than, or in addition to, states play significant roles that encompass action across geopolitical borders.

approach to instrument and policy design at a domestic level. Researchers exploring the prospects for smart mixes within the transnational sphere have been concerned primarily with understanding how international treaties can be supplemented with other instruments to better mitigate transboundary environmental harm, and with examining the interactions between international environmental agreements and private standards.

However, the previously mentioned forms of governance are far from being the whole picture. Certainly, various public-private partnerships and private standards (and their interactions with other instruments and actors), are important examples of initiatives that are consistent with market liberal (emphasizing the virtues of markets in stimulating innovation) and institutional (emphasizing the role of rules and institutions in incentivizing environmental protection) worldviews.[10] But they are in no way representative of what Bulkeley et al. term 'deep green' initiatives that emphasize confrontation and conflict rather than partnerships and cooperation.[11] The latter, which variously aspire to create norms, to set agendas and reframe environmental debates and to monitor and enforce behaviour through boycotts, shaming and other reputational penalties (in marked contrast to environmental partnerships), are far less studied but no less important.

This chapter is concerned with one 'deep green' transnational initiative – the global fossil fuel divestment movement – that has had a remarkable impact in shaping perceptions and norms concerning climate change, within a relatively short period and which also provides insights into the potential for 'smart mixes' in the transnational context. Section 11.2 describes the divestment movement's evolution, aims and strategies. Section 11.3 examines the transnational dimensions of climate governance and some of the lenses through which scholars have sought to understand it. Section 11.4 seeks to illuminate the divestment movement's role in climate governance, with a focus on its interactions with and impact on a cluster of other climate change actors. Section 11.5 asks the normative question: what smart mixes *should* be invoked in this context and whether, to what extent and in what ways, the insights of smart regulation in the domestic context are capable of extrapolation to transnational governance. Section 11.6 concludes.

Data concerning the divestment movement used in this article is drawn from the writer's ongoing research on that movement, published elsewhere.[12]

11.2 THE FOSSIL FUEL DIVESTMENT MOVEMENT

The commitments made by the 194 parties that signed the 2015 Paris Climate Change Agreement undoubtedly suggest a greater willingness on the part of states

---

[10] Bulkeley et al. (2014).
[11] Ibid.
[12] Gunningham (2017).

to engage in climate change mitigation. Yet climate change remains a 'super-wicked problem' involving global public goods and high risks of free-riding. The unwilling-ness of many governments to take effective action to date, and the Paris Agreement's heavy reliance on voluntary 'nationally determined contributions', on markets and on emergent and commercially untested technologies such as carbon capture and storage, should be sufficient reasons to look beyond government action *alone* to achieve effective climate change action. Indeed, the very reason that a plethora of transnational governance initiatives have emerged is largely to fill the governance space left by inadequate arrangements at the national and international levels. One such experiment that has generated considerable interest, but little analysis, is the fossil fuel divestment movement.

The movement's origins can be traced back to 2008, when environmental writer and activist Bill McKibben and six of his students established the climate activist group 350.org. The number 350 (parts per million) is regarded by climate scientists as the safe upper limit for atmospheric carbon dioxide, beyond which we risk 'dangerous climate change'.[13] Until 2012 the group participated in various campaigns with some success, but its ambition, reach, and direction changed dramatically following the publication in *Rolling Stone* of McKibben's article, 'Global Warming's Terrifying New Math'.[14] This set out the case for keeping most fossil fuel reserves in the ground to avoid catastrophic warming and led directly to what became 350.org's divestment campaign. Building on existing campus activism, 350. org catalysed multiple campus-based mini-campaigns aimed at pressuring individual universities to divest of their fossil fuel assets and to reinvest in low-carbon alterna-tives (to 'go fossil-free'). Universities (and subsequently other institutional investors vulnerable to moral appeals to divest), however, are merely convenient vehicles through which to engage with the movement's real targets – the fossil fuel extraction industry, whose behaviour it aims to stigmatize, and the public, whose norms and perceptions of a 'social problem' it seeks to re-shape.

The divestment movement is now at the forefront of civil society initiatives to raise public consciousness about the need for a 'fossil-free' future. While it has the highest profile in North America, parts of Western Europe and in Australasia, it is also becoming significant elsewhere. Realizing the opportunities that divestment pro-vided for political engagement, but lacking material resources, the movement has been adept at harnessing grassroots activists though multiple locally driven cam-paigns, and innovative and often disruptive forms of activism, coordinated nationally and internationally. These have sought to name and shame the fossil fuel industry and to threaten its social license to operate. Crucially, it has recognized the importance of cognitive framing and symbolic politics in gaining media interest, in persuading the public of the importance and legitimacy of its claims and in

[13] Hansen et al. (2008).
[14] McKibben (2012).

translating climate change science into language that is readily understood by target audiences. Finally, recognizing the embedded institutional and instrumental power of its adversary and its own weakness in these terms, it has managed to shift the contest, at least in part, onto a terrain where it may hold a comparative advantage: morality and expressive politics.[15]

The movement has rapidly evolved into a Transnational Activist Network, with multiple nodes working in concert across national borders.[16] By December 2016 it had been instrumental in persuading 688 institutions (including major banks, insurers and investment funds) and 58,399 individuals across 76 countries to commit to divest from fossil fuel companies based on climate change concerns. The value of assets represented by these institutions and individuals is conservatively estimated at US $5 trillion, up from $3.4 trillion the previous year (Arabella Advisors (2016)). As such, the movement's growth has been dramatic and it is becoming an increasingly significant climate change actor on the global stage and certainly 'the fastest growing [such] movement in history'.[17]

11.3 TRANSNATIONAL CLIMATE GOVERNANCE

The divestment movement's role as a climate policy actor, and more importantly, its capacity to amplify its impact by engaging in smart mixes, are best understood within the wider context of transnational climate governance.

While the divestment movement does not seek, at least in the short term, to directly shape law or regulation within the nation-state, it is becoming a significant actor in climate *governance* transnationally. Within the social science literature, 'governance' is variously conceived of as 'societal steering' and the 'management of the course of events within a social system'.[18] As such, 'governance' conceives of state law as simply one node among many in a world of diffused power and responsibility.

When a governance lens is applied to the sphere of climate change mitigation (hereafter 'climate governance') at the regional or global level, it becomes apparent that international relations are much messier and more complex than many traditional scholars in that field had conceived them to be. No longer is the international stage the predominant or exclusive domain of states and International Organizations (IOs), as realists and neo-realists had asserted. Rather, governance can be seen to take place in 'many rooms' and at many levels and to involve not just states (through hierarchical relationships), but also markets and networks. Certainly, states can and often do play leading roles, but governance scholarship recognizes that there are many other potentially influential climate change actors. These include

---

[15] Gunningham (2017).
[16] Keck & Sikkink (1999).
[17] Vaughn (2014).
[18] Burris, Drahos & Shearing (2005), at 30.

institutional investors and other financial market actors, business associations, corporations, cities, municipal networks, and various manifestations of civil society, including international NGOs such as 350.org, local community groups and many others who, individually or in combination, have the capacity to accelerate or constrain climate action at levels that range from the local to the global.[19]

As recognition of the transnational dimensions of climate governance has increased, researchers have found that this 'agency beyond the state' involving 'the capacity of individual and collective actors to change the course of events or the outcome of processes'[20] exists on a remarkably large scale.

They have also found climate governance's manifestations to be exceptionally complex, being not only fragmented but also decentralized, multi-level, and commonly *coproduced* by multiple governors, 'whose authority is contested and whose legitimacy is questionable'.[21] Theorizing to better take account of the multiple dimensions of climate change governance has evolved in tandem with the empirical work that has revealed its extent and complexity. The result is a shift away from the state and international relations–centric approaches that characterised traditional international relations scholarship and towards ones that take account of climate governance's non-state dimensions.[22]

For example, the 'regime complex' approach conceives of 'an array of partially overlapping and non-hierarchical institutions governing a particular issue area'.[23] For Abbott and Snidal, what they term 'the Transnational Regime Complex for Climate Change' (TRCCC) involves state, civil society and business actors who form a 'governance triangle'.[24] Within this triangle, they argue, one may find some actors engaged in standard setting, and others operating variously as activists, collectors and disseminators of information and lobbying. Other writers, coming from different perspectives, have also come to recognize that not only traditional top-down processes involving international negotiations between states, but also bottom-up processes involving limited groups of actors, including transnational NGOs, can play significant roles.[25]

For present purposes, what is striking about this literature is not only the extent to which it reveals the importance of non-state actors in climate governance but, by extension, the likely significance of smart mixes that involve such actors, sometimes indeed without any involvement by the state. This is a theme that will be

---

[19] Bulkeley et al. (2014).
[20] Pattburg & Stripple (2008), at 369, 374.
[21] Dellas, Pattberg & Betsill (2011).
[22] Biermann, Pattberg & Zelli (2012), at 16.
[23] Raustalia & Victor (2004), at 279; see also Keohane & Victor (2011).
[24] Abbott & Sindal (2009).
[25] Biermann, Pattberg & Zelli (2012); Bodansky (2016), at 305; Vandenbergh & Gilligan (2015); Ostrom (2009); Abbott, Levi-Faur & Snidal (2017).

revisited in the following sections, noting especially the need for the smart mix literature to 'de-centre the state'.

However, while students of transnational climate change governance (if not some theorists of smart mixes) have increasingly recognized its complexity and multidimensional nature, they have not subjected all these dimensions to detailed study. For the most part, attention has focused on a few high-profile initiatives (private standards and certification), to the relative neglect of other manifestations of this phenomenon. In short, we 'know a lot about a few cases'.[26] None of this work, however, extends to 'deep green' initiatives such as the divestment movement, which have very different characteristics and pursue very different goals. The divestment movement is also distinctive in recognizing the extent to which behaviour is 'pervasively a function of norms'[27] and that changing these must form a fundamental underpinning to effective climate change action by states. More specifically, it might be thought of as 'governing by discourse',[28] as a moral entrepreneur engaged in agenda setting and norm creation and doing so not in isolation but through broader 'webs of influence'. As will become apparent, it is these webs of influence that have particular significance for present purposes and out of which smart mixes may evolve.

11.4 NORMS, INTERACTIONS AND WEBS OF INFLUENCE

The following discussion explores the interconnections between the divestment movement and a cluster of others within a 'governance triangle' that is characterized by multiple potential interactions at multiple levels by multiple actors. It will do so in terms of a number or orienting concepts, namely: norms, interactions, mechanisms and webs of influence.

The importance of *social norms* (some of which ultimately become legal norms) in shaping climate outcomes may be considerable. Norms are not only pervasive but a fundamental underpinning to the functioning of individuals, organizations and societies. Norms may, for example, be the glue that connects potential allies and that makes weak ties stronger, or the battleground on which contests for legitimacy are fought. And it is precisely because the divestment movement is concerned with changing norms that its focus is on stigmatizing the fossil fuel industry and challenging the necessity for a fossil fuel–based economy.

The movement's role in the architecture of transnational climate governance is also distinctive in that rather than operating according to a 'logic of consequences' like most climate actors, it exhorts its targets to divest because it is the 'right thing to

---

[26] Macleod & Park (2011), at 80.
[27] Sunstein (1996), at 907.
[28] Fuchs (2005), at 791.

do' and in accordance with a 'logic of appropriateness'.[29] And rather than seeking authority and legitimacy through making rules or setting standards[30] it seeks to change norms and practices, seeking to embed such norms and increase their authority, as through the various 'legitimacy networks' in which it is engaged.[31]

The impact of a norm entrepreneur such as the divestment movement is likely to be complemented, resisted or otherwise shaped by the activities of other climate change actors. Indeed, it is often the *interactions* between actors that will shape who wins and who loses a contest, including those which are essentially contests of norms. Those interactions may take many forms, with multiple potential permutations, each with implications for governance outcomes. For example, these may involve catalysis, complementarity, mutual reinforcement, redundancy, synergy, incompatibility and conflict.[32] Recognizing that some actors are better placed to engage in some governance functions than others, the scope for harnessing complementarity may be considerable, while conversely, other interactions may be counter-productive.

Actors seek to achieve their goals by invoking a variety of *mechanisms*, including coercion, reputational rewards and sanctions, capacity building, forum-shifting and modelling.[33] Of course, not all mechanisms are available to all actors, but which mechanisms are available – and how they are harnessed in practice – will influence whether, to what extent and in what ways actors or coalitions of actors achieve their goals or indeed are capable of constraining the capacity of others to achieve theirs.

In conceptualizing how the above mechanisms and interactions shape climate governance beyond the state, the metaphor of *webs of influence*[34] is preferred to the lens of networks or that of a regime complex. The networks lens is unhelpful in the present context because the movement's influence extends well beyond its networks. And notwithstanding the virtues of the regime complex lens, it operates on a much broader canvas than the present analysis, and is more suited a macro than to a micro or meso-level analysis. In contrast, the 'webs' metaphor can accommodate to the fact that while some strands in the web are connected, this need not be by way of networks, and these connections may be weak or strong (and indeed, that webs themselves are available to the weak as well as the strong). Moreover, whether, to what extent or in what ways strands are connected will have implications for how societal steering takes place.

[29] March & Olsen (1989).
[30] Green (2014), at 6.
[31] Ayling (2017).
[32] Gunningham, Grabosky & Sinclair (1998), Ch. 3.
[33] Gunningham, Grabosky & Sinclair (1998); Braithwaite & Drahos (2000), at 15–17. Braithwaite and Drahos use this term specifically to explain events in terms of degrees of globalization of business regulation. As such, their study concerned a rather different issue from that which is the subject of this chapter.
[34] Braithwaite & Drahos (2000), at 550–553.

Against this backdrop, what roles does the divestment movement, as an emerging climate actor which aims to promote an explicit normative agenda, play within the architecture of climate governance beyond the state in conjunction with a mix – smart or otherwise – of other actors? What are its interactions with these other actors, what drives those interactions and how do they shape outcomes? What mechanisms are invoked by different actors, and to what effect? How are the different strands in the web of influence connected, and with what consequences?

A cluster of other climate change actors might variously contribute to (or constrain) the movement's progress through their interactions with it. Its relationship with the fossil fuel industry itself is predictably distant and adversarial. Since BP's widely mocked attempt to rebrand itself as 'Beyond Petroleum' was revealed to be little more than greenwash and quietly dropped, initiatives by fossil fuel companies to embrace renewable energy or to otherwise change their business models have been notable by their absence. Instead, an increasingly sophisticated fossil-fuel counter-movement has developed, which has attacked the divestment movement through extensive advertising, planted news 'stories', disinformation and counter-framing initiatives.[35] The industry has also called upon its allies in the media and in government for support, primarily through the mechanism of adverse publicity.

Arrayed against the fossil fuel industry and its allies, and interacting with the divestment movement in various ways and to varying degrees, are a cluster of other climate actors, the most significant of which will be examined in the following text.

The movement has various *close allies* comprising civil society organizations that have similar goals and values. There are obvious synergies and complementarities between 350.org and other 'deep green' groups such as Greenpeace and Friends of the Earth, which also engage in contentious politics, although its most important alliances are context specific. Various coalitions involving like-minded NGOs have coalesced around individual campaigns. Transitory alliances, which are rather looser in nature, sometimes involve unlikely bedfellows, such as farmers and indigenous groups. In terms of mechanisms, both types of alliance might be regarded as involving capacity and coalition building and networks of varying strength and duration. A rather different role, as conceptive ideologist[36] is played by organizations such as Carbon Tracker, whose considerable contribution was to reframe the risks of climate change in the language of 'stranded assets' and a 'carbon bubble', using concepts familiar to financial market actors unconventionally and thereby generating new ways of thinking.

Overall, alliances with deep green organisations and looser couplings with other civil society groups serve to amplify the impact of the divestment movement's activities. They do so variously by swelling the numbers at particular actions (in itself an indicator of impact) and by increasing the legitimacy, authority and

---

[35] Oreskes & Conway (2010).
[36] Cain (1983).

recognition of the movement through its association with other, perhaps more respected, civil society groups, such as farmers. Sometimes, network participants enhance each other's agency by providing mutual reputation benefits (as in alliances between NGOs which draw from different legitimacy communities). The relationship between the movement and conceptive ideologists is also positive, and essentially complementary: Carbon Tracker, notwithstanding the power of its conceptualization, was getting limited traction until 350.org effectively translated and communicated its message to a broad audience.

Also important to the movement's impact are *elite allies* in the form of supportive international organisations (IOs) and key individuals within them, who have, respectively, institutional and expert authority. The movement has not actively networked in this context. Nevertheless, *those who have spoken in support of divestment include* Christina Figueres, former Executive Secretary of the United Nations Framework Convention on Climate Change (UNFCCC), Jim Yong Kim, World Bank President and Connie Hedegaard, European Commissioner for Climate Action in the European Commission.

The endorsement or broad support of those individuals and organisations gives the movement much-needed legitimacy,[37] without which it would be unable to exercise authoritative power or achieve its long-term goals. In contrast to private standard setting (where the weaker the standard, the greater its uptake and ensuing legitimacy), with divestment, effectiveness and legitimacy go hand in hand: the more legitimacy the movement has on the international stage, the greater the likelihood it will be taken seriously by target audiences (including divesting organisations, institutional investors and international organizations).

The support of IOs is also strategically important, since such organizations have the capacity to engage in what Finnemore and Sikkink call 'an active process of international socialization intended to induce norm breakers to become norm followers',[38] something which, they argue, is crucial to achieving broad norm acceptance. Such socialization might involve the mechanisms of diplomatic praise or censure reinforced by material sanctions and incentives. Accordingly, IOs can play an important role in norm promotion. They may also play an instrumental role, since their recommendations may prompt action by member states and empower civil society organizations, as with initiatives to develop consistent climate-related financial risk disclosure standards.

The relationship between the divestment movement and sympathetic IOs might also be thought of as one of mutual reinforcement. On the one hand, endorsement of divestment, or recognition of the risk of stranded assets, by IOs or influential figures within them provides legitimacy to the divestment movement. But it is also the case that leaders of IOs such as Christiana Figueres 'have asked movement

---

[37] Ayling (2017).
[38] Finnemore & Sikkink (1998), at 902.

leaders to push for new commitments after the Paris climate talks as a means to keep the pressure on policymakers to pass and implement effective climate legislation'.[39]

Finally, given the largely hostile, or at best muted, reaction of most nation-states to divestment initiatives, the endorsement of IOs might be seen in terms of the 'boomerang effect ... where non-state actors, faced with ... blockage at home, seek out state and non-state allies in the international arena, and in some cases are able to bring pressure to bear from above on their government to carry out domestic political change'.[40] This involves the mechanism of forum shifting from the domestic stage (where a movement is losing) to the transnational stage, where the movement has much closer allies and a much greater chance of success. But while its proponents assume that externalization depends on domestic activists networking with international allies following the failure of efforts to engage with the nation-state, the divestment movement has simply assumed that domestic pressure on national governments will rarely succeed and has accordingly focused its attention internationally. The ultimate result, nevertheless, is much the same as the boomerang effect studies would anticipate.

Having been mobilized, such organizations in turn can act as 'orchestrators' of other climate actors, 'convening, facilitating and participating in collaborative arrangements; [or] providing material and ideational support'.[41] Such orchestration, Abbott and Snidal argue, can mitigate the limitations of state-centric governance and the fragmentation of transnational governance. While there is only limited evidence of orchestration being used as a mechanism by IOs, the United Nations Environment Program, a weak agency, has engaged in a number of initiatives of this nature.

The divestment movement also engages with elements of the business community, albeit to varying degrees. One significant development is the number of financial market actors that have established arms specializing in 'sustainable finance' (also referred to as 'responsible investment' and 'impact investment'). HSBC, for example, now aims 'to provide the best analysis of climate change and its implications for economies, industries and sectors'. It does so by focusing on capital allocation to deliver the transition to a low carbon world and on climate policy in terms of decarbonizing the energy mix. Similarly, global consultants Mercer's Responsible Investment have been engaging in research and producing reports on the impact of climate change on investment returns, and advising clients on the implications thereof for asset allocation under a climate-constrained future. Some specialist investment advisors also focus on opportunities to 'do well by doing good' in the sphere of responsible investment, which by definition would screen out fossil fuel assets. As such, the previously mentioned financial market actors might be

---

[39] Divest-Invest (2015), at 11.
[40] Sikkink (2005), at 154.
[41] Abbott (2014).

seen as complementary transition agents, policy entrepreneurs who facilitate a shift towards a low-carbon economy.[42]

Self-evidently, there is considerable commonality between these themes and the agenda of the divestment movement. However, the degree of interaction between the movement and such sustainable finance initiatives is modest. Rather, a driver for the establishment of such initiatives has been the identification of a niche market opportunity, which they hope can be extended beyond 'responsible' to mainstream investors, some of whom are coming to see fossil fuel investments as 'risky business'. The fact that these are reputable financial institutions who have expert authority, and credibility with institutional investors, means that they have little to gain from overt association with the divestment movement. Nevertheless, the activities of banks, consultants and specialist investment advisors have served to complement those of the divestment movement, disseminating a similar message but reaching a different audience, and both sides benefit from discreet informal contact. Viewed in economic terms, the sustainable investment arms of banks and consultancy firms are bringing supply-side solutions to satisfy needs that the divestment movement and others have created on the demand-side. While their relationship is arms-length, it is nevertheless symbiotic.

Finally, a useful way of conceptualizing the contribution of the divestment movement – and those with whom it interacts – to climate change governance might be in terms of what Braithwaite and Drahos refer to as webs of influence. These 'comprise of many actors wielding many mechanisms'[43] that in combination play important roles in steering events on the global stage. Within such webs, the divestment movement's distinctive contribution is to act as a moral entrepreneur, exhorting the public and key actors to embrace a new norm. While most analyses assume that such movements will target a critical mass of states (norm leaders) it might equally influence international organizations, as it has demonstrably done. Other actors in the web are pulling in the same direction, albeit that there are different interactions between different actors, using different mechanisms, and with varying consequences. The strength of connections between different strands in the web is also variable. While there is a strong connection between the divestment movement and like-minded civil society organizations, the connection with others, such as IOs, is quite limited, while with the responsible investment arm of financial institutions and consultants, the connection is somewhere between weak and non-existent. Nevertheless, these organizations seem to be singing from broadly the same song sheet: there is climate risk, there will be stranded assets, taking a path towards a low-carbon economy is likely to minimize these risks *and* to be more profitable.

However, on the other side (and with very little middle ground) are stacked not only the forces of the energy status quo (i.e. the financially and institutionally

---

[42] Bouteligier (2011).
[43] Braithwaite & Drahos (2000), at 31.

powerful fossil fuel industry), but many national governments (particularly in fossil fuel–rich countries) and (as much through inertia as intent) most institutional investors as well. Facing such powerful adversaries, what hope is there for the proponents of a low-carbon economy? Is it the case that there is 'strength in weak ties', and that these might effectively enhance diffusion across a network, as Granovetter[44] would argue? Might 'webs of dialogue', which Braithwaite and Drahos show are available to the weak, provide a potentially potent means of building regimes, given their evidence that even weak strands in webs of influence often become strong by being tied to other weak strands?[45] Going further, might it be possible to nurture webs of 'collective evangelism' or even 'a winning coalition of believers whose conceptions of socially desirable activity set the terms for subsequent moral debate'?[46] Might the growth of low-carbon industries also create economic constituencies that contribute to coalitions for decarbonization that ultimately challenge and overpower the deeply entrenched power of the fossil fuel industry?[47]

These are large questions, the answers to which must await further developments in the ongoing struggle between advocates of climate change mitigation and their opponents. What can be said is that the conclusion of that struggle, notwithstanding the gross disparity in power and resources between the principal antagonists, is not preordained. Networks of weak actors do sometimes achieve victories over much stronger opponents. Networked dialogue may prove to be a potent antidote to material power, and the forces of civil society may yet succeed in a 'divide and conquer' strategy that makes allies of far-sighted institutional investors and that, over time, counters the entrenched power and inertia of their more conservative peers.[48]

11.5 SMART REGULATION AND TRANSNATIONAL GOVERNANCE

While Section 11.4 has explored the nature of the various interactions between the divestment movement and other non-state climate actors, there remains the normative question: Which combinations *should* be preferred and why?[49] This question will be addressed through the lens of Smart Regulation, revisiting the questions posed in Section 11.1 as to whether this approach to instrument and policy design (and in particular its emphasis on smart mixes) might have value at the transnational level.

---

[44] Granovetter (1973).
[45] Braithwaite & Drahos (2000), at 31.
[46] Suchman (1995), at 582.
[47] Meckling et al. (2015).
[48] Braithwaite & Drahos (2000).
[49] Of course, the answer to this question depends upon the perspective from which it is being asked. While one might argue that mixes should be designed to best achieve the public interest, identifying what that interest might be in such a contested and complex area as climate change is problematic. For present purposes, recognizing that no value-free approach can be provided, the question will be addressed from the perspective of the divestment movement and its allies.

Some of these questions are process based and would be similarly useful in both contexts. Divestment strategists, for example, might sensibly ask: What are the desired policy goals and the trade-offs necessary to achieve them? What are the distinctive characteristics of this environmental problem, and what do these suggest about potentially fruitful strategies to address it? What is the range of potential governance actors and mechanisms available to them, and how might they best be harnessed in the public interest?[50]

However, when it comes to choosing substantive instrument mixes, then translating from the domestic to the transnational level may be more problematic. This is largely because smart regulation is state-centric: it assumes that the state will play an active and influential role in public policy and that, while business and civil society can and should also have significant roles, it is the state that will usually 'call the shots'.[51] This need not be the case with transnational governance; the divestment movement, for example, was established largely *because* of dissatisfaction with the perceived unwillingness or inability of nation-states to mitigate climate change. Indeed, the case of divestment illustrates that effective governance may take place even in the absence of *any* form of government regulation, whether nationally, sub-nationally or internationally based.

The transnational level is also different to the domestic one in terms of what mechanisms might be available to various actors. Domestically, the suite of possible policy instruments includes, centrally, all the conventional tools of state regulators, from standards to economic instruments to mandating information disclosure, all of which can be underpinned by the coercive powers of the state. In contrast, a transnational activist network such as the divestment movement must settle for a narrower range of non-coercive mechanisms, such as information dissemination and persuasion in tandem with informal sanctions such as naming and shaming. The result is that the capacity to design complementary instrument mixes, to escalate up a dynamic and three-sided instrument pyramid or to engage in instrument sequencing – strategies that are central to smart regulation in a domestic context – might appear to be constrained by the absence of coercive power.

However, these constraints may be more apparent than real. If one refers back to smart regulation's three-dimensional pyramid (the three faces representing government as regulator, business as self-regulator, and third parties such as NGOs,[52] then escalation to the tip of the pyramid need not be through government coercion; it might equally be through informal sanctions such as boycotts and naming and shaming. Indeed, there may be occasions on which shaming may be more potent than conventional legal sanctions, as when it threatens the social license to operate

---

of reputation-sensitive corporations.[53] And by extension, and borrowing from John Braithwaite's seminal work on responsive regulation,[54] in the absence of government intervention (or effective intervention) a pyramid of networked escalation might be developed, harnessing various of the actors identified in Section 11.4. These actors, moreover, may seek to hold each other mutually accountable, further strengthening their interconnectedness and collective impact though multiple forms of decentred governance.

Again, it might be argued that other of smart regulation's design principles, such as 'empower participants who are in the best position to act as surrogate regulators', resonate somewhat less for transnational actors since they usually have much less capacity to empower others than do states. But on closer examination, the capacity of non-state actors to recruit, enrol and co-opt others (consistent with the spirit of the previously mentioned design principle) may be considerable. Such capacities may be crucial to achieving desired outcomes, because, at the transnational level, multiple capacities that no one actor possesses are necessary for successful social steering. For example, it might be possible to enrol financial institutions as gatekeepers, as when auditors, credit rating agencies and financial analysts, under pressure from the movement and others, come to recognize their interest in evaluating the risks that climate change poses to particular classes of investments, which in turn may shape the behaviour of shareholders and asset owners and managers themselves. Accordingly, as Drahos[55] points out, 'power in global regulatory contests is harnessed by enrolling the capacities of others into a network that works towards a common purpose'.

It is also the case, as we saw in Section 11.4, that there is a diversity of potential interactions between the divestment movement and a range of other climate actors, including complementarity, mutual reinforcement, amplification and orchestration. These are not materially different from the sorts of actor interactions that smart regulation suggests can and should be tailored in complementary fashion to achieve better regulatory/governance outcomes. Enrolling organizations with similar aspirations or values, amplifying impact through coordinated action, or seeking mutual legitimacy enhancement are examples of how transnational climate actors might seek to combine with others to maximize their impact by invoking particular mechanisms. So too is the potential for sequenced responses, as when an aggressive moral entrepreneur such as the divestment movement advocates for norm change and garners broad-based support to achieve it, but may gain the traction necessary for co-ordinated international action only when international organisations 'act as conduits for the codification, monitoring and enforcement of policies based on those norms'.[56]

---

[53] Gunningham, Kagan & Thornton (2003).
[54] Braithwaite (2011).
[55] Drahos (2017), at 258.
[56] Reich (2010), at 49.

Government interactions with non-state actors might similarly be predicted to either accelerate or constrain the capacity of the divestment movement and its allies to achieve their aspirations. For example, government-imposed constraints on rights of access to information, freedom of speech or freedom of protest are likely to constrain the divestment movement's progress, while government actions that confer such rights, or which expand the concept of fiduciary duty of asset managers to embrace climate risks, would have the converse effect. Similarly, mandating institutional asset owners to assess climate risks and disclose the results might be seen as an example of working with markets to make them more efficient (by factoring full information into decision-making), as might initiatives that facilitate the development of low-carbon stock indices.

Going forward, opportunities for positive interactions and complementary combinations to evolve are not hard to identify. These include the expansion of webs of influence that encompass not just traditional allies but also other groups with whom the divestment movement might make common cause. Broad alliances such as the Australian 'Lock the Gate' Alliance against fracking, for example, have been remarkably successful, precisely because they involve everyone from environmental activists to rural landholders, farmers associations and the conservative Country Women's Association. Engaging with international human rights groups and those concerned with international security also seems increasingly plausible. Also crucial will be the extent to which the movement is able to engage with business-dominated epistemic communities[57] and in particular with institutional investors as indicated previously. The latter can play a pivotal role in providing capital to finance the transition to a low-carbon economy, driven by the necessity to protect client assets from global climate risks. Moreover, the 'quiet revolution' in the perceptions and behaviour of institutional investors that is just beginning[58] gives the divestment movement a rare opportunity to exploit divisions among institutional investors, and to engage in a 'divide and conquer' strategy that can may be a particularly effective means by which the weak may overcome the strong.[59] But, as smart regulation emphasizes, it is not just combinations of actors, but matching actors with instruments that will have the greatest payoff, and here too, non-state actors, while unable to invoke core instruments of the regulatory state (not least coercion), may find multiple credible alternatives. For example, harnessing social media and information technology, developing mechanisms that facilitate scaling up through directed action campaigns and engaging in non-violent direct action on the model of the American civil rights movement of the 1950s and 1960s all hold considerable potential. Finally, scepticism about the possibilities of constructive engagement with the nation-state must, at least

---

[57] Braithwaite & Drahos (2000).
[58] UNEP (2015).
[59] Braithwaite & Drahos (2000).

on the evidence on the evidence of social movement theorists such as Finnemore and Sikkink,[60] give way to constructive engagement.

In summary, while the scope for identifying complementary combinations of policy mechanisms *underpinned by state action* is considerably more constrained at the transnational level than it is in the domestic sphere, that for developing complementary interactions with other actors is not. On the contrary, as the examples provided in this chapter amply illustrate, transnational climate government offers rich ground for complementary interactions between a multiplicity of non-state actors. Some of these have already been taken up, but there remains considerable opportunity for experimentation and for further and complementary interactions between actors invoking a range of different mechanisms, particularly those which are not the exclusive domain of states, such as modelling.[61] Not all of these experiments will succeed, but the precepts of smart regulation, while far from providing comprehensive answers, at the very least identify a set of procedural and substantive questions that strategists might find helpful in determining which experiments to pursue, and credible paths through which to pursue them.

To return to the fundamental insight of smart regulation, what is needed – in the transnational sphere as in domestic affairs – is not just a broader range of policy tools (mechanisms) but the willingness to match the tools with the party or parties best capable of implementing them, and with each other, thereby building on the strengths of individual actors and mechanisms while compensating for their weaknesses.

11.6 CONCLUSION

This chapter has addressed the question of whether, as well as to what extent and in what ways, approaches to designing smart mixes that evolved in the domestic context can be extended to the sphere of transnational climate governance; more specifically, whether these approaches can be applied not just to such arrangements as public-private partnerships, private standards and certification schemes, but also to 'deep green' initiatives, which variously aspire to create norms, set agendas and reframe environmental debates.

It has cautioned that not all aspects of smart regulation lend themselves to such extrapolation, not least because of the extent to which this approach assumed the existence of an interventionist state, capable of consciously harnessing second and third parties towards public interest outcomes. On the other hand, other tenets of smart regulation, which are not dependent upon any such state, appear to be equally helpful in understanding the potential of transnational climate governance.

This is not to suggest that this is the 'best' way of approaching the latter subject; quite the contrary. There has been a considerable amount of sophisticated

---

[60] Finnemore & Sikkink (1998).
[61] Braithwaite & Drahos (2000), at 581–595.

scholarship on interactions within the sphere of transnational governance, including the chapters in this volume, the work of Abbott on the climate change 'regime complex'[62] and the ambitious work on the 'interactive dynamics' of non-state governance being undertaken by the Transnational Business Governance Interactions Project.[63] The claim of this chapter is more modest: that approaching transnational climate governance by asking some of the process and substantive questions identified by smart regulation, might provide one additional source of insight, particularly in terms of what combinations of mechanisms and actors might best achieve intended outcomes. Approaching these issues through the lenses of norms, mechanisms, interactions and webs of influence may further assist such inquiry.

The analysis of the present chapter also serves to emphasise that climate governance is not only a top-down process involving states and international agreements, but also a bottom-up and polycentric process involving civil society and the corporate sector. And it is not only an instrumental process of coercing changes in behaviour and in the exercise of material power, but an expressive and symbolic one of nurturing a new norm, institutionalizing a new set of moral principles and, through this, of influencing behavioural change. However, whether moral entrepreneurs such as the divestment movement, in tandem with a cluster of other climate actors, will succeed in precipitating a 'norm cascade' and whether this, in conjunction with the higher level of international commitment evidenced by the Paris Agreement, will be enough to avert the cataclysm of dangerous climate change, remains an open question.

REFERENCES

Abbott, K. W. 2014. 'Strengthening the Transnational Regime Complex for Climate Change'. *Transnational Environmental Law* 3(1), 57–88.

Abbott, K. W. & Sindal, D. 2009. 'The Governance Triangle: Regulatory Standards Institutions and the Shadow of the State'. In Mattli, W. & Woods, N. (eds.), *The Politics of Global Regulation*. Princeton, Princeton University Press, 44–88.

Abbott, K. W., Levi-Faur, D. & Snidal, D. 2017. 'Theorizing Regulatory Intermediaries: The RIT Model'. *The Annals, American Academy of Political and Social Science* 670(1), 14–35. Available at SSRN: https://ssrn.com/abstract=2925411 or http://dx.doi.org/10.2139/ssrn.2925411.

Ayling, J. 2017. 'A Contest for Legitimacy: The Divestment Movement and the Fossil Fuel Industry'. *Law & Policy* 39(4), 349–371.

Biermann, F., Pattberg, Ph. & Zelli, F. (eds.). 2012. *Global Climate Governance Beyond 2012: Architecture, Agency and Adaptation*. Cambridge, Cambridge University Press.

Bloor, M., Datta, R., Gilinsky, Y. & Horlick-Jones, T. 2006. 'Unicorn among the Cedars: On the Possibility of Effective "Smart Regulation" of the Globalized Shipping Industry'. *Social and Legal Studies* 15(4), 534–551.

---

[62] Abbott (2014).
[63] Wood et al. (2015).

Bodansky, D. 2016. 'Climate Change: Transnational Legal Order or Disorder'. In Halliday, T. & Shaffer, G. (eds.), *Transnational Legal Orders*. Cambridge, Cambridge University Press.

Bouteligier, S. 2011. 'Exploring the Agency of Global Environmental Consultancy Firms in Earth System Governance'. *International Environmental Agreements: Politics, Law and Economics* 11(1), 43–61.

Braithwaite, J. 2011. 'The Essence of Responsive Regulation', *University of British Columbia Law Review* 44(3), 475–520.

Braithwaite, J. & Drahos, P. 2000. *Global Business Regulation*. Cambridge, Cambridge University Press.

Bulkeley, H. et al. 2014. *Transnational Climate Change Governance*. New York, Cambridge University Press.

Burris, S., Drahos, P. & Shearing, C. 2005. 'Nodal Governance'. *Australian Journal of Legal Philosophy* 30, 30–58.

Cain, M. E. 1983. 'The General Practice Lawyer and the Client: Towards a Radical Conception'. In Dingwall, R. & Lewis, P. S. C. (eds.), *The Sociology of the Professions: Lawyers, Doctors and Others*. London, MacMillan Press, 106–130.

Danish Ministry of Economic and Business Affairs, The Netherlands Regulatory Reform Group & The United Kingdom Better Regulation Executive. 2010. *Smart Regulation. A Cleaner, Fairer and More Competitive EU*, published jointly by the Ministry for Economic and Business Affairs (Denmark) the Regulatory Reform Group (the Netherlands) and the Department for Business, Innovation and Skills (UK).

Dellas, E., Pattberg, Ph. & Betsill, M. 2011. 'Agency in Earth System Governance: Refining a Research Agenda'. *International Environmental Agreements: Politics, Law and Economics* 11 (1), 85–98.

Divest-Invest Philanthropy, *Doing Good: Performing Better*, 2015, available at: www.divestin vest.org/wp-content/uploads/2017/09/2017-DIP-Briefing-Case-Studies.pdf.

Drahos, P. 2017. 'Regulatory Globalisation'. In Drahos, P. (ed.), *Regulatory Theory: Foundations and Applications*. Canberra, ANU Press, 249–265.

Finnemore, M. & Sikkink, K. 1998. 'International Norm Dynamics and Political Change'. *International Organization* 52(4), 887–917.

Fuchs, D., 'Governance by Discourse', Paper presented at the Annual Meeting of the International Studies Association, Honolulu, March 2005.

Granovetter, M. S. 1973. 'The Strength of Weak Ties'. *American Journal of Sociology* 78(6), 1360–1380.

Green, J. 2014. *Rethinking Private Authority: Agents and Entrepreneurs in Global Environmental Governance*. Princeton, Princeton University Press.

Gunningham, N. 2017. 'Building Norms from the Grassroots Up: Divestment, Expressive Politics and Climate Change'. *Law & Policy* 39(4), 305–428.

Gunningham, N. & Sinclair, D. 2002. *Leaders and Laggards: Next Generation Environmental Regulation*. Sheffield, Greenleaf Press.

Gunningham, N., Grabosky, P. N. & Sinclair, D. 1998. *Smart Regulation: Designing Environmental Policy*. Oxford, Clarendon Press.

Gunningham, N., Kagan, R. A. & Thornton, D. 2003. *Shades of Green: Business, Regulation and Environment*. Stanford, Stanford University Press.

Hansen, J. et al. 2008. 'Target Atmospheric CO_2: Where Should Humanity Aim?'. *Open Atmospheric Science Journal* 2(1), 217–231.

Keck, M. E. & Sikkink, K. 1999. 'Transnational Advocacy Networks in International and Regional Politics'. *International Social Science Journal* 51(159), 89–101.

Keohane, R. O. & Victor, D. G. 2011. 'The Regime Complex for Climate Change', *Perspectives on Politics* 9(1), 7–23.

Leese, W. 2005. 'Smart Regulation and Risk Management'. A paper prepared at the request of the Bureau du Conseil Privé/Privy Council Office Comité Consultatif Externe sur la Réglementation Intelligente (CCERI) External Advisory Committee on Smart Regulation (EACSR) Ottawa, Canada, Government of Canada.

MacLeod, M. & Park, J. 2011. 'Financial Activism and Global Climate Change: The Rise of Investor-driven Governance Networks'. *Global Environmental Politics* 11(2), 54–74.

March, J. & Olsen, J. 1989. *Rediscovering Institutions: The Organizational Basis of Politics.* New York, Free Press.

McKibben, B. 2012. 'Global Warming's Terrifying New Math'. *Rolling Stone*, 19 July 2012, available at: www.rollingstone.com/politics/news/global-warmings-terrifying-new-math-20120719.

Meckling, J., Kelsey, N., Biber, E. & Zysman, J. 2015. 'Winning Coalitions for Climate Policy'. *Science* 349(6253), 1170–1171.

Oreskes, N. & Conway, E. M. 2010. *Merchants of Doubt: How a Handful of Scientists Obscured the Truth on Issues from Tobacco Smoke to Global Warming.* New York, Bloomsbury Press.

Ostrom, E. 2009. 'A Polycentric Approach for Coping with Climate Change'. In Bierbaum, R. M., Fay, M. & Ross-Larson, B. (eds.), *World Development Report 2010: Development in a Changing Climate.* Washington, DC, World Bank Group, 34.

Pattberg, Ph. & Stripple, J. 2008. 'Beyond the Public and Private Divide: Remapping Transnational Climate Governance in the 21st Century'. *International Environmental Agreements: Politics, Law and Economics* 8(4), 367–388.

Raustiala, K. & Victor, D. G. 2004. 'The Regime Complex for Plant Genetic Resources'. *International Organization* 58(2), 277–309.

Reich, S. 2010. *Global Norms, American Sponsorship and the Emerging Patterns of World Politics.* Basingstoke, Palgrave McMillan.

Sikkink, K., 2005. 'Patterns of Dynamic Multilevel Governance and the Insider-outsider Coalition'. In Della Porta, D. & Tarrow, S. G. (eds.), *Transnational Protest and Global Activism.* Lanham, MD, Rowman & Littlefield, 151–174.

Suchman, M. C. 1995. 'Managing Legitimacy: Strategic and Institutional Approaches'. *The Academy of Management Review* 20(3), 571–610.

Sunstein, C. R. 1996. 'Social Norms and Social Roles'. *Columbia Law Review* 96(4), 903–968.

Tollefson, C., Gale, F. & Haley, D. 2008. *Setting the Standard: Certification, Governance, and the Forest Stewardship Council.* Vancouver, BC, University of British Columbia Press.

UNEP, UN Environment Programme. October 2015. *The Financial System We Need: Aligning the Financial System with Sustainable Development.* Available at: http://web.unep.org/inquiry.

Vandenbergh, M. P. & Gilligan, J. M. 2015. 'Beyond Gridlock'. *Columbia Journal of Environmental Law* 40(2), 217–303.

Vaughn, A., 'Fossil Fuel Divestment: A Brief History', *The Guardian*, 2014, available at: https://www.theguardian.com/environment/2014/oct/08/fossil-fuel-divestment-a-brief-history.

Wood, St., Abbott, K. W., Black, J., Eberlein, B. & Meidinger, E. 2015. 'The Interactive Dynamics of Transnational Business Governance: A Challenge for Transnational Legal Theory'. *Transnational Legal Theory* 6(2), 333–369.

Wood, S. & Johannson, L. 2008. 'Six Principles for Integrating Non-Governmental Environmental Standards into Smart Regulation'. *Osgoode Hall Law Journal* 46, 345–395.

12

Private Control of Public Regulation: A Smart Mix?

The Case of Greenhouse Gas Emission Reductions in the EU

Marjan Peeters and Mathias N. Müller

12.1 INTRODUCTION

This chapter focuses on a specific mix of instruments established for reducing greenhouse gas emissions in an effective way. The case that will be studied is a core element of EU climate law and concerns greenhouse gas emissions trading (also referred to as the EU ETS: the EU emissions trading scheme). This market-based instrument is established with the aim of reducing greenhouse gas emissions from industries and aircrafts in a cost-effective way.[1] However, in order to reach an effective application of emissions trading, two complementary approaches established by EU law are relevant.[2] First, an important part of the regime for checking compliance by emitters covered by the emissions trading scheme has been outsourced to private actors, the so-called verifiers. The proper functioning of the verifiers is crucial for achieving the intended reduction of emissions. Second, the fundamental right of the public to get access to environmental information, which is established as a general right in EU environmental law, is applicable to the emissions trading instrument as well.[3] This means that, in principle, civil society at

---

[1] See Table 1.1 for a typology of instruments, including emissions trading. This table does not specify EU law, but distinguishes between international law and domestic law; EU law tends to be international law. See De Witte (2017) for a discussion of whether EU law can indeed be qualified as international law.

[2] As explained in Section 1.3 of this volume, effectiveness is chosen as the primary criterion for discussing the smartness of a mix; other factors, such as coherence, efficiency, unintended effect, legitimacy and the adaptability of the instrument mixes will serve as secondary benchmarks.

[3] The right of access to environmental information is established in EU environmental law, and, importantly, can be enforced by courts, see Directive 2003/4/EC of the European Parliament and of the Council of 28 January 2003 on public access to environmental information and repealing Council Directive 90/313/EEC [2003] OJ L 41/26 (hereafter Directive 2003/4/EC) and Regulation (EC) No 1367/2006 of the European Parliament and of the Council of 6 September 2006 on the application of the provisions of the Aarhus Convention on Access

large can employ control on the proper implementation of the emissions trading instrument. It is generally assumed that 'improved access to information' enhances 'the quality and the implementation of decisions'.[4] In this sense, the right of access to environmental information is expected to contribute to the effectiveness of environmental regulation.[5] Hence, a procedural instrument, established by public regulation, is applicable as a complementary approach to the market-based instrument of emissions trading.[6] In sum, the case to be studied in this chapter focuses on a threefold instrumental approach: (1) emissions trading, being a market-based instrument for reducing greenhouse gas emissions; (2) a control-regime expecting private actors (verifiers) to check compliance behaviour of emitters; and (3) the procedural right of access to information that enables civil society to exert control on the proper implementation of regulatory approaches. This chapter takes a legal perspective to this instrument mix and examines whether the 'outsourcing' of compliance control by the government to private verifiers in the case of emissions trading constitutes a recommendable instrument mix, particularly in view of the potential consequences this may have for civil society regarding their possibilities to access the relevant information needed for checking the effective functioning of the instrument.

The structure of the chapter is as follows: Section 12.2 puts in context the choice of the EU legislator to establish a market-based mechanism complemented with verification provisions to be carried out by private actors. Section 12.3 analyses the possibility for civil society to get insights into compliance information held by the verifier. Section 12.4 sheds light on the use of information by environmental non-governmental organisations (ENGOs). Section 12.5 concludes.

12.2 PUBLIC REGULATION AND PRIVATE MONITORING: THE CASE OF THE EU ETS

12.2.1 *The EU ETS and the Broad and Evolving Regulatory Package of EU Climate Law*

The instrument of emissions trading is advocated by economic theory since its market-based feature, established by the tradability of the allowances, would lead

---

to Information, Public Participation in Decision-making and Access to Justice in Environmental Matters to Community institutions and bodies, [2006] *OJ* L 264/13.

4 See the preamble to the Convention on access to information, public participation in decision-making and access to justice in environmental matters, done at Aarhus, Denmark, on 25 June 1998 (hereafter the Aarhus Convention).

5 This seems to be supported by practical experiences, see Gunningham, Grabosky & Sinclair (1998), at 63 ff., stating that the 'community right to know' can create pressure for stricter enforcement; they refer also to potential drawbacks such as the fact that the public can misunderstand information.

6 Also procedural instruments, including access to information, are mentioned in Table 1.1.

to a cost-effective reduction of emissions.[7] Also the EU has been motivated by economic reasons for introducing emissions trading, since the EU ETS Directive explicitly aims 'to promote reductions of greenhouse gas emissions in a cost-effective and economically efficient manner'.[8] Another reason, however, is that the effectiveness – meaning that a reduction of pollution takes place – is in principle ensured. This follows from a core design element of emissions trading, which is that the maximum allowable amount of greenhouse gas emissions is fixed by establishing an EU-wide cap on pollution.[9] With this cap on the total amount of pollution, a predictable limitation of pollution is ensured, provided that full compliance takes place.[10]

The total allowable amount of pollution, hence the cap, is divided into allowances. Each allowance represents a unit of pollution, and the sum of all allowances totals the maximum amount of pollution that may be caused. Within the EU greenhouse gas emissions trading scheme, the distribution of the allowances takes place by means of two methods: 1) an auction and 2) a free allocation; the auction method is gradually replacing the free allocation method. Emitters may trade these allowances, but they have to comply with the rule that for each tonne of greenhouse gas emissions, one allowance has to be surrendered to the government.[11] Particularly in view of considering whether there is an effective approach for ensuring compliance with this rule, the instrument mix consisting of private verification and the procedural right of access to information held by the verifier will be discussed in the rest of this chapter. But interestingly, in the context of instrument mixes, it is to be noted that the EU ETS is as such part of a broad instrument mix applied by the EU in order to reduce the greenhouse gas emissions on its territory. In fact, even before the Treaty on the Functioning on the European Union introduced the mandate that Union policies shall 'promote measures at the international level to combat climate change',[12] a package of regulatory measures dealing with *inter alia* greenhouse gas

---

[7] See Dales (2002) for the original idea. For further elaboration, see also Cole (2016) and Tietenberg (2006), at 27.

[8] Directive 2003/87/EC of the European Parliament and of the Council of 13 October 2003 establishing a scheme for greenhouse gas emission allowance trading within the Community and amending Council Directive 96/61/EC [2003] *OJ* L 275/32 (hereafter the EU ETS Directive, refers to the consolidated version), Article 1.

[9] This core design element of emissions trading motivates ENGO's to prefer emissions trading above taxation. For an example of support from the ENGO community for emissions trading (in this case, the Environmental Defense Fund supporting the US acid rain emissions trading program), see Dudek & Palmisano (1988).

[10] This is different in case of an environmental tax, or in case of traditional permitting, where a fixed total of allowable pollution is not part of the design. See for a discussion of instruments for environmental law Stewart (2007).

[11] Every year, before 1 May, emissions have to be surrendered covering the emissions of the previous year, see EU ETS Directive, Article 12(2a) and (3). The competent authorities are designated by the Member States.

[12] Article 191(1) TFEU (entered into force on 1 December 2009).

emissions, renewable energy and energy efficiency was established. Among these approaches, the EU ETS is a core pillar. Its position has been strengthened in the course of updating the regulatory package in view of contributing to the aims of the Paris Agreement.[13] In fact, the EU aims to simplify its regulatory approach by focussing and improving two instruments of its climate policy: first, the EU greenhouse gas emissions trading scheme, and second the 'Effort sharing approach' through which national emission reductions for sources falling outside the scope of the emissions trading scheme will be established.[14] Both the EU ETS and the Effort sharing approach are seen by the European Council as key instruments to achieving the overall goal of the EU, which is to reduce greenhouse gas emissions to 40 per cent below 1990 levels by 2030. Nonetheless, several complementary instruments will still be used, such as a legal regime for carbon capture and storage and a regulation introducing a new governance approach for the Energy Union.[15] So, while the rest of this chapter focuses on the specific instrument mix for making EU greenhouse gas emissions trading effective by ensuring compliance, it is important to understand that the EU ETS itself is part of a broader and further evolving mix of instruments used in EU climate policy.

12.2.2 *Identifying the 'Mix of Actors' in the EU ETS in View of Compliance*

For the emissions market to work effectively, compliance by industries is crucial, particularly with regard to the basic rule that for the amount of greenhouse gas emissions caused in a given year, a corresponding number of tradable allowances has to be surrendered to the competent public authority.[16] Industries must file an annual report setting out their emissions, and for having the credibility of this report

---

[13] In order to contribute to the Paris Agreement, the EU aims to reduce greenhouse gas emissions by at least 40 per cent below 1990 levels by 2030. The EU submitted this policy aim, on behalf of itself and its member states, to the international climate change negotiations on 6 March 2015; see www4.unfccc.int/sites/submissions/INDC/Published%20Documents/Latvia/1/LV-03-06-EU%20INDC.pdf.

[14] European Council conclusions, 23 and 24 October 2014, http://data.consilium.europa.eu/doc/document/ST-169-2014-INIT/en/pdf; this instrument choice is discussed by Peeters (2016a). This chapter will concentrate on the EU ETS that is a common regulatory framework for major industries, and flight operators, in the EU; for implementing the Effort Sharing Decision, the Member States have to develop their own specific national policies themselves.

[15] Directive 2009/31/EC of the European Parliament and of the Council of 23 April 2009 on the geological storage of carbon dioxide and amending Council Directives 85/337/EC, 2004/35/EC, 2006/12/EC, 2008/1/EC and Regulation (EC) No 1013/2006 [2009] OJ L 140/114 and Proposal for a Regulation of the European Parliament and of the Council on the Governance of the Energy Union, amending Directive 94/22/EC, Directive 98/70/EC, Directive 2009/31/EC, Regulation (EC) No 663/2009, Regulation (EC) No 715/2009, Directive 2009/73/EC, Council Directive 2009/119/EC, Directive 2010/31/EU, Directive 2012/27/EU, Directive 2013/30/EC and Council Directive (EU) 2015/652 and repealing Regulation (EU) No 525/2013, [2017] COM(2016) 759 final/2.

[16] EU ETS Directive, Article 12(3) (applicable to industries).

checked, they must hire a verifier. The EU ETS Directive obliges Member States to ensure that emission reports submitted by industries are verified in accordance with a set of criteria set out in Annex V of the Directive, and any detailed provisions of the Verification Regulation adopted by the European Commission.[17] The importance of this verification activity can be illustrated by the fact that an industry is not allowed to make further transfers of allowances until the emission report has been verified as 'satisfactory'.[18] In essence, if the verifier does not approve the emission report of the emitter, the emitter is acting in breach of the emissions trading regulations, and sanctions must be imposed by the competent authority.[19] In other words, the approval by the verifier of the emission report from the emitter is crucial for the latter for being in compliance with the emissions trading regime. Moreover, the approval of the emission report by the verifier also determines the amount of allowances that must be surrendered by the emitter.[20] In this sense, the decision by the verifier whether to approve an emission report is of pivotal legal and economic importance for the EU ETS industry.

The specific design of the EU ETS as explained in Section 12.2.1 implies that the actual regulatory effect needs to be realized by various private actors. First of all, the *emitters* themselves (the industries and aviation companies covered by the EU ETS regime) need to take action in order to comply with the regulatory requirements.[21] Within the EU ETS, in essence, ample freedom for decision-making is given to emitters, since they – depending on the price of allowances and costs of their emission reduction possibilities – may choose either to cover their emissions with the tradable allowances, or to reduce their emissions themselves.[22] Second, *verifiers* play an important role since they have the task to check the emission reports developed by the emitters. Here, contractual relationships will be developed between emitters and verifiers: industries must hire an accredited verifier to get

---

[17] EU ETS Directive, Article 15, Commission Regulation (EU) No 600/2012 of 21 June 2012 on the verification of greenhouse gas emission reports and tonne-kilometre reports and the accreditation of verifiers pursuant to Directive 2003/87/EC of the European Parliament and of the Council [2012] *OJ* L 181/1 and Commission Regulation (EU) No 601/2012 of 21 June 2012 on the monitoring and reporting of greenhouse gas emissions pursuant to Directive 2003/ 87/EC of the European Parliament and of the Council [2012] *OJ* L 181/30.

[18] EU ETS Directive, Article 15.

[19] See Article 16 of the EU ETS Directive regarding the obligation of the emitter to surrender allowances equal to the total of emissions of the installation in a calendar year as stipulated in Article 6(2)(e) of the EU ETS Directive. For the precise applicable rules, including sanctions, the implementing national legislation has to be consulted.

[20] EU ETS Directive, Article 12(3).

[21] According to its official terminology, the EU ETS Directive applies to 'operators' and 'aircraft operators'; see EU ETS Directive, Article 3(f).

[22] There exists a vast amount of literature on emissions trading, from several disciplines. Recent examples include Weishaar (2016); for research overviews from a legal perspective, see Bogojević (2013) and Peeters (2016b).

approval for their annual emission report. But, in addition to emitters and verifiers, civil society – and more particularly, ENGOs and investigative journalists – may play a role by using their procedural rights in order to control, to the extent legally possible, how industries, verifiers and the responsible authorities comply with the applicable requirements. With regard to the emissions trading instrument, it is even the case that specialised ENGOs (such as *Carbon Market Watch* and *Carbon Trade Watch*) have been established with the specific aim of critically following the design and application of the specific instrument.[23] But also ENGOs with a general focus, such as Greenpeace, have shown interest for using the right of access to information for checking the emission behaviour of EU ETS industries.[24]

12.2.3 *The EU ETS: The Choice for a Double Market-Based Approach*

With the emissions trading instrument, the EU in fact applies a double market-based approach. Clearly, with introducing the possibility for polluters to trade in emission rights, a market-based regulatory approach is taken. But another market dimension is introduced by the EU's choice to make use of private verifiers in order to control the emission reports of emitters: verifiers must compete with each other to win contracts with the emitting industries for conducting the verification of emission reports. The resulting competition among verifiers may lead to a decrease in the cost of verification for the emitters. The choice of the EU legislator to task private verifiers with controlling greenhouse gas emitters can in this respect be seen as a way to let polluters pay an important part of the regulatory costs. However, the competition among verifiers may lead to concerns with regard to the integrity of the verification-regime: will the strive to deliver the verification-task at least costs be detrimental to its quality?[25]

The establishment of this second market dimension – competition in the control chain – is not strictly necessary: the control of the emission reports from industries could also be carried out under the responsibility of administrative authorities by civil servants. For instance, in the case of the Industrial Emissions Directive – a core directive of EU environmental law, aiming at the protection of the environment as a whole by means of a permit-system – no use is made of private verifiers or other types

---

23 For the specific missions of *Carbon Market Watch* and *Carbon Trade Watch*, see the following websites: http://carbonmarketwatch.org/ and http://www.carbontradewatch.org.

24 See the legal dispute decided by the Dutch Administrative Court to the Council of State from 28 October 2009, file number ECLI:NL:RVS:2009:BK1375, in which Greenpeace requested access to information included in the emission reports from 17 EU ETS industries. The requested information was refused by the Administrative Authority holding this emission report, and this refusal was considered lawful by the administrative court. However, the correctness of this court decision in view of EU law is questioned by Thurlings (2017), at 270–275, arguing that a preliminary question should have been submitted to the CJEU.

25 As far as is known to the authors, no empirical research has been employed in this respect.

of third-party action for controlling the performance of industries.[26] Nonetheless, in the field of greenhouse gas emissions trading the use of private actors for checking the performance of emitters has emerged into a large practice across the world, not only with respect to the trading provisions as established by Kyoto Protocol but also in the case of voluntary carbon trading mechanisms.[27] Furthermore, particularly for the EU, for which the establishment of the internal market is a main goal, the choice for using private verifiers fits to its market-oriented focus and seems to be part of a trend: the approach is also taken in the EU's regime for the reduction of CO_2 emissions from maritime transport, which requires companies to draft an emission report that must be verified by a private verifier.[28] Next to this, the EU also chose to involve private actors for controlling the sustainability of biofuels, which happens by means of voluntary certification regimes approved by the European Commission.[29]

12.2.4 *Access to Information Related to Compliance*

The question is, however, whether this privatisation of compliance control may have negative effects on the possibility for the public to check the performance of industries. In particular, ENGOs who specialise in controlling carbon trading may be interested in compliance information: if the emission reporting were to show deficiencies, the functioning of the emissions trading regime would be fundamentally damaged and the environmental effectiveness would be compromised. While the certification regime for biofuels has already been criticised for its lack of transparency with regard to the verification process,[30] Section 12.3 will examine to what extent ENGOs and other interested parties may face barriers in accessing

---

[26] Directive 2010/75/EU, Article 23 has this clear focus on public authorities, and does not provide any specific provision for auditing or third-party verification, except for its reference to the European Union eco-management and audit scheme (EMAS as regulated by Regulation (EC) No 1221/2009), meaning that for the systematic appraisal of the environmental risks of an installation the participation to EMAS needs to be taken into account.

[27] Ebbesson (2011), at 82; Peeters (2009); Wang et al. (2016), at 382–383 ff.; Levin et al. (2011), at 1906 and Livingston, Lee and Nguyen (2015), at 57 f. Also, the US Greenhouse Gas Reporting Program 'requires enterprises to entrust third-party verification institutions with the verification of their GHG emission list reports and to submit the same to the US Environment Protection Agency'. Wang et al. (2016), at 386.

[28] Regulation (EU) 2015/757 of the European Parliament and of the Council of 29 April 2015 on the monitoring, reporting and verification of carbon dioxide emissions from maritime transport, and amending Directive 2009/16/EC [20015] OJ L 123/55, see more specifically Article 11(1).

[29] The European Commission has the competence to authorise voluntary certification schemes – developed by and carried out by private companies – for the purposes of certifying whether biofuels comply with sustainability standards as established by the Directive 2009/28/EC of the European Parliament and of the Council of 23 April 2009 on the promotion of the use of energy from renewable sources and amending and subsequently repealing directives 2001/77/EC and 2003/30/EC, [2009] OJ L 140/16, Article 18.

[30] Romppanen (2015), at 49 and 106 ff. (pointing at the weak legislative and non-binding provisions as regards to the transparency provided by verifiers).

information used during the EU ETS verification process.[31] The practical relevance can be illustrated by observations from the European Environmental Agency, stating that 'it is not possible to conclude on how well the verification system is functioning in practice'.[32] Furthermore, Fleurke and Verschuuren have pointed at some critical issues with regard to how the monitoring of the EU ETS is carried out in practice. They observe that the control of verified emission reports varies greatly between the six Member States they examined, with the least thorough approach found in Hungary.[33] If it is the case that governmental oversight of the correctness of the emission reports and their verification is not conducted at the highest level possible, additional action by ENGOs could be helpful to identify possible shortcomings or even mistakes in the correct measurement of emissions. After all, in principle, it should not be disregarded that verifiers (and emitters) make mistakes, or, worse, commit fraud for instance by deliberately approving an emission report in which less emissions are mentioned compared to the actual data. While access to information may not be sufficient for preventing all non-compliance with law, it may be one of the important strategies for preventing such behaviour.

12.3 TRANSPARENCY WITH REGARD TO VERIFICATION

The role of civil society for strengthening (compliance with) environmental law is stressed by the Aarhus Convention, giving important procedural rights such as the right to access environmental information held by the public authorities upon request.[34] This procedural right may be helpful to some extent for checking compliance, and already the threat of being exposed in the case of non-compliance might encourage emitters to comply.[35] However, the application of this right may encounter legal problems. In light of the general observation by Liz Fisher that ' . . . transparency may be a truism in regards to public administration, its operation is

---

[31] As observed in an earlier footnote, it remains necessary to conduct further in-depth research, including empirical research, towards the applicability of the provisions on access to information in the Aarhus Convention with regard to the verification of biofuels. This falls outside the scope of this chapter. For more general discussions of the applicability of Aarhus provisions in case of privatisation, see Ebbesson (2011) and Etemire (2012).

[32] European Environmental Agency (2016), at 30.

[33] Fleurke and Verschuuren (2016), at 221–222. Furthermore, in Greece, Hungary and Poland, the authorities primarily rely on the verified reports, conducting less ex-post control compared to the UK, Germany and The Netherlands (Fleurke and Verschuuren (2016), at 220). Here, additional checking by ENGOs could be helpful in view of getting insight into the trustworthiness of the emission reports.

[34] Convention on Access to Information, Public Participation in Decision-Making and Access to Justice in Environmental Matters done at Aarhus, Denmark on 25 June 1998, 2161 UNTS 447; 38 ILM 517 (1999) entered into force on 30 October 2001, approved by the EU on 17 February 2005 (Decision 250/360/EC), (hereafter Aarhus Convention).

[35] Explanatory Memorandum of the Proposal for a Directive of the European Parliament and of the Council on public access to environmental information COM (2000) 402 final, para. 1.2.

profoundly complex',[36] this section will delve into some complexities with regard to the question whether the right of access to environmental information may also be successfully executed in the specific case of information held by a verifier operating under the EU ETS.

The EU ETS Directive illustrates that the EU legislator deems it important that the public at large is informed about compliance and non-compliance by industries. An example is the concept of naming and shaming set out in the EU ETS Directive. The names of operators who are in breach of the requirement to surrender sufficient allowances are published.[37] However, if operators are trying to cheat it is unlikely that they will do so by surrendering fewer allowances than they are supposed to since this can be easily detected. Instead, it is more likely that they will try to submit false emission reports that enable them to emit more than they are declaring, and consequently to pay less than they should. In this vein, false verification reports are also imaginable.[38] Broad public access to environmental information in the realm of the EU ETS is important as it might contribute to detecting such cases of fraud. Unfortunately, the process by which civil society can determine whether emitters and verifiers are meeting their duties as stipulated in the compliance provisions of the EU ETS directive, and subsequently bring violators to the attention of public authorities, is as yet unclear. One crucial question to be investigated is that of whether private verifiers are covered by access to information legislation, and, if yes, to what extent or in what circumstances they are obligated to disclose the information they hold.

12.3.1 *The Aarhus Convention: Access to Environmental Information*

The first pillar of the Aarhus Convention establishes the right of the public to access environmental information held by the government.[39] This stipulates that public authorities must disclose environmental information (if no grounds for refusal apply) upon request by 'a member of the public'.[40] 'The public' is defined very broadly and refers to any natural or legal person and associations thereof as defined under national law. This means that not only individuals but also the media and non-

---

[36] Fisher (2010), at 314.

[37] See Article 16(2) EU ETS Directive for the precise legal provision. A discussion of the value and applicability of this provision falls outside the scope of this chapter. For a critique of its effectiveness, see Fleurke and Verschuuren, (2016), at 224.

[38] For submitting a false verification report, collusion between the operator and the verifier is most likely the case.

[39] The second pillar provides the right to participation to governmental decision-making, and the third pillar concerns access to the court in environmental matters. This chapter concentrates on the first pillar.

[40] Aarhus Convention, Article 4(1). Article 5 covers the active right of access to environmental information. It deals with instances in which public authorities must actively disseminate environmental information without a request by the public being necessary.

governmental organisations are included in this definition.[41] It is not necessary for the member of the public to state an interest in the information requested. The public authority must disclose the information in the form requested within one month after receiving the request with the possibility of extending this deadline to two months if the volume or complexity of the information requested make this necessary.[42] The EU and all its Member States are a party to the Aarhus Convention, and the EU has adopted several measures to transpose the right to environmental information into EU secondary law.[43]

12.3.2 *Access to Information as Held by the Verifier: Specific Provisions in the EU ETS Directive*

This section and Section 12.3.3 will show the main difficulties in answering the question of whether information held by the verifier should be accessible to the public. This information could, for example, concern the way a verifier has checked a specific industry, the minutes of meetings between the verifier and the industry, including the report of a site visit, or certain agreements made between the industry and the verifier on the specific methodology for calculating the emissions.[44] Sections 12.3.2.1 and 12.3.2.2 discuss some core articles of the EU ETS Directive concerning access to information related to emission trading.[45]

12.3.2.1 Disclosure of Information upon Request: A Limited Provision in the EU ETS

Access to information within the realm of the EU ETS is in principle governed by Article 17 of the EU ETS Directive.[46] Information covered by that Article must be made available to the public upon request pursuant to Directive 2003/4/EC, which transposes the first pillar of the Aarhus Convention into EU law, thereby providing a

---

[41] Aarhus Convention, Article 2(4). Importantly, the first pillar of the Aarhus Convention does not require the public to be 'concerned' as for instance Article 6 does (participation to governmental decision-making).

[42] Aarhus Convention, Article 4(2).

[43] The most prominent ones are Directive 2003/4/EC [2003] OJ L 41/26 and Regulation (EC) No 1367/2006 [2006] OJ L 264/13. In this chapter we mainly refer to the provisions of the Aarhus Convention, as this EU secondary legislation often contains the same provisions.

[44] Case law has already shown that different interpretations on the question of which emissions are covered by the EU ETS may lead to legal conflicts; for a case in which the verifier followed another approach than then competent authority, see C-148/14 *Federal Republic of Germany v. Nordzucker AG* [2015], published in the electronic Report of Cases.

[45] Next to the general provision for access to environmental information as regulated by Directive 2003/4/EC, other environmental directives may contain specific provisions on access to information. Nonetheless, this can lead to uncertainty as to which provisions prevails, as will be discussed in Sections 12.3.2.1 and 12.3.2.2.

[46] EU ETS Directive, Article 17.

general right of access to environmental information held by public authorities within Member States. However, Article 17 of the EU ETS Directive only covers 'decisions relating to the allocation of allowances, information on project activities … and the reports of emissions required under the greenhouse gas emissions permit and held by the competent authority'. The wording of the article – information on verification and the verifier are not mentioned at all – suggests that information held by the verifier, other than emission reports, is not covered by Article 17 of the EU ETS Directive.[47] The scope of the access to environmental information provision of the EU ETS Directive is hence limited. A small-scale empirical test confirms that it was impossible to get access to verification reports which were requested from the competent authorities.[48] As a first conclusion, it appears that regarding information necessary to assess the level of compliance, Article 17 of the EU ETS Directive delineates access to environmental information only to emission reports, while there might be reasons for ENGO's or members of the public to ask for various other pieces of information, including information on how the verifier has carried out its tasks.[49]

12.3.2.2 Active Dissemination Duty: Again a Limited Provision in the EU ETS Directive

One solution for getting access to information compiled during the verification process may be found in Article 15a of the EU ETS Directive, which governs an *active* information dissemination duty for the Member States and the Commission.[50] It stipulates that 'all decisions and reports relating to … the monitoring, reporting and verification of emissions are immediately disclosed in an orderly manner ensuring non-discriminatory access'. Thus, this article asks for the active disclosure of the specified information to the public, including 'all decisions and reports' related to verification. However, it is not stipulated in this article whether the

---

[47] For a definition of 'competent authority', see Article 18 of the EU ETS Directive, with no reference to the verifiers. For the duty of industries to submit their emission report to the competent authority, see Article 14(3) of the EU ETS Directive.

[48] Müller (2016), available from the authors upon request. Requests to access individual verification reports were sent to the competent public authorities in Germany and Austria. While the German authorities did not answer at all, their Austrian counterparts refused to provide the requested information, arguing that some information contained therein was protected by the national legislation on professional secrecy. No requests for information were sent to individual verifiers (it may already be difficult to know which verifier has checked the emission report of an individual industry).

[49] As explained in Section 12.3.2; it may concern minutes or e-mails in which certain agreements – for instance, on the interpretation and application of the calculation methodology – have been noted between the industry and the verifier.

[50] This obligation hence rests on both the EU level and the national level. Such a joint obligation is very unique in EU environmental law. We do not delve in this chapter into the potential complexities that may derive from this joint obligation.

active information duty covers only information held by the national authorities, or whether it also includes information held by the verifier but not (yet) physically held by the competent authority.[51] Also, it might be that ENGOs want information other than that contained in verification reports, such as information on the way the verifier has checked a specific industry, the minutes of meetings between the verifier and the industry, including the report of a site visit, or certain agreements made between the industry and the verifier on the specific methodology for calculating the emissions. Moreover, the aforementioned study – which did a small-scale empirical test – determined that the competent Austrian and German public authorities neither published verification reports nor provided access to them upon request. Until now, there is not yet case law answering the question of whether this access should have been given or, actually, whether the authorities themselves should have disclosed this information on their own initiative.

Furthermore, the second paragraph of Article 15a explains that, by derogation from the first paragraph, information covered by 'professional secrecy' may only be disclosed to third parties in accordance with the 'applicable laws, regulations or administrative provisions'.[52] As with the first paragraph of the Article, the second one is also relatively vague. There is neither any indication as to what information is covered by 'professional secrecy', nor any specification of what 'professional secrecy' means. Furthermore, it is necessary to investigate how this matter is regulated by national legislation. '[A]pplicable laws, regulations or administrative provisions' could at least in part also refer to EU legislation, including directives that need implementation by Member States. However, it is not likely, or at least very uncertain, that Directive 2003/4/EC is meant in this reference since the term 'professional secrecy' is not mentioned in this directive.[53] Furthermore, in analogy with the *Ville de Lyon*[54] case, one should perhaps consider that the EU legislator did not have the

---

[51] See for the point of view that it would not be necessary that the public authority 'physically' holds the requested information Etemire (2013), at 372, and earlier, Ebbesson (2011), at 81.

[52] EU ETS Directive, Article 15a.

[53] Only 'tax secrecy'; see Article 4(d) of Directive 2003/4/EC. The term 'professional secrecy' is also not mentioned in the Aarhus Convention. One can wonder whether this provision of Article 15a EU ETS Directive is compatible with the Aarhus Convention. There is no case law yet on this matter; such case law could develop in view of a request for access to information that would be refused using the 'professional secrecy criterion' in connection to the 'confidentiality of commercial and industrial information' clause mentioned in Article 4 of Directive 2003/4/EC. Such a request for information may not be refused where the request relates to information on emissions, discharges or other releases into the environment (which is also a multi-interpretation term).

[54] Case C524/09 (*Ville de Lyon v Caisse de dépôts et consignation*), ECR 2010 I-14115. The case concerned access to trading data by the city of Lyon. One of the questions the Court had to answer was 'whether the reporting of trading data ... is governed by one of the derogations provided for in Article 4 of Directive 2003/4 or by the provisions of Directive 2003/87 and Regulation No 2216/2004'. The Court ruled that, according to the wording of Article 17 of the EU ETS Directive, the legislator subjected only parts of the reporting and implementation data to the regime of Directive 2003/4. However, Article 17 does not cover trading data. Instead,

intention of making access to information covered by Article 15a subject to the regime of Directive 2003/4/EC.[55] Just like in *Ville de* Lyon it could be argued that with Article 15a the EU legislator 'sought to introduce a specific, exhaustive scheme for public reporting and confidentiality of that data'.[56] In this vein, Article 15a governs as a *lex specialis* access to information concerning verification, but particularly the reference to 'professional secrecy' and the use of the word 'decisions' cause legal uncertainty as to what extent the information held by verifiers should be disclosed.[57]

12.3.3 *Observations on a Possible Request for Information in View of the Aarhus Convention*

In Section 12.3.2 we have shown that the EU ETS Directive contains a specific information regime, which entails several limits and uncertainties regarding the transparency of information from the verification process. Meanwhile, the EU ETS Directive must be consistent with the Aarhus Convention because international agreements concluded by the EU take precedence over secondary legislation.[58] Thus, EU legislation must be interpreted as far as possible in line with international agreements to which the EU is a party. Hence, in this section we examine how a request for environmental information related to the verification process should be dealt with in view of the Aarhus Convention provisions.

Taking the Aarhus Convention as the starting point for our analysis, one can see that if there is a request to access verification reports (or other information held by the verifier), two conditions must be fulfilled: First, the information must qualify as

---

Article 19 covers this kind of information. This Article does not refer to Directive 2003/4, but sets out a specific scheme that governs access to the information that falls within its scope. The fact that Article 19 sets out such a specific scheme precludes the application of Article 17, and thus Directive 2003/4/EC, for information covered by Article 19.

[55] Article 15a of the EU ETS Directive does not refer to Directive 2003/4/EC in the same way Article 17 does.

[56] Case C524/09 *Ville de Lyon*, [40].

[57] In this respect, further surveys to the following specific provisions are relevant: Article 41(3) of Commission Regulation (EU) No 600/2012 states that 'a verifier shall safeguard the confidentiality of information obtained during the verification in accordance with the harmonised standard referred to in Annex II'. Subsequently, Annex II refers to Regulation (EC) No 765/ 2008, and Article 8(4) of that Regulation states that national accreditation bodies 'shall have adequate arrangements to safeguard the confidentiality of the information obtained'. In addition to the provisions of the Regulation 'adequate arrangements to safeguard the confidentiality of information obtained' shall apply. Of course, the compatibility of these provisions with the fundamental right of access to environmental information as established by the Aarhus Convention – having become part of EU law – needs to be examined. A wide coverage or application of such 'confidentiality' provisions can be in breach of the Aarhus Convention.

[58] Case C-61/94 *Commission of the European Communities v Federal Republic of Germany* [1996] ECR I-3989.

'environmental information'; second, the verifiers must be 'public authorities' pursuant to Article 2(2) of the Aarhus Convention.

12.3.3.1 Environmental Information

Firstly, the requested information needs to fall into one of the categories set out in the definition of environmental information.[59] The term 'environmental information' is defined in Article 2(3) of the Aarhus Convention and includes information in any format on the following three areas: First, information on the state of elements of the environment, which include *inter alia* air, water and soil, the landscape, natural sites and biological diversity, as well as the interaction of those elements;[60] second, information on factors affecting or likely to affect the elements of the environment mentioned under the first subparagraph. These 'factors' can, for example, be substances or energy, but they can also be activities or measures, policies and legislation;[61] last, information on 'the state of human health and safety, conditions of human life, cultural sites and built structures'[62] insofar as the factors mentioned in subparagraph (b) have an influence on them.

The term 'emissions' is not included, but since the Convention explicitly regulates that the grounds for refusing requested information 'shall be interpreted in a restrictive way, taking into account the public interest served by disclosure and taking into account whether the information requested relates to emissions into the environment',[63] it is accepted that the definition of 'environmental information' includes information on emissions in the environment. Meanwhile, the Court of Justice of the European Union has provided in its case law that the term 'emissions into the environment' must be interpreted rather broadly, which hence strengthens the right of access to environmental information.[64]

Furthermore, the Court set out in *Mecklenburg*[65] that activities of a public authority to ensure compliance with EU legislation aiming at protecting the environment might in principle be regarded as environmental information. Furthermore, the Court pointed out that a piece of information relates to the environment if it refers to an activity that either protects or adversely affects one of the elements of the

---

[59] Article 2(3) of the Aarhus Convention, the corresponding Article of Directive 2003/4/EC is Article 2(1).

[60] Aarhus Convention, Article 2(3)(a).

[61] Aarhus Convention, Article 2(3)(b).

[62] Aarhus Convention, Article 2(3)(c).

[63] Article 4(4) final sentence.

[64] Case C-442/14 *Bayer CropScience SA-NV, Stichting de Bijenstichting v College voor de toelating van gewasbescheringsmiddelen en biociden* [2016], published in the electronic Reports of Cases, [61–67].

[65] Case C-321/96 *Wilhelm Mecklenburg v Kreis Pinneberg – Der Landrat* [1998] ECR I-038009, [20].

environment.[66] In *Glawischnig*,[67] the Court specified further that information relating to the monitoring of compliance with individual pieces of EU legislation can be regarded as environmental information only if the purpose of the legislation is to protect the environment. A look at Article 1 of the EU ETS Directive shows that its goal is in fact to protect the environment by contributing 'to the levels of reductions that are considered scientifically necessary to avoid dangerous climate change'. One may, however, wonder to what extent information that is recorded during the verification process is to be seen as 'environmental information'. While the CJEU is already taking an extensive interpretation of the definition of environmental information (and specifically of "emissions into the environment"), uncertainty may exist in practice regarding the extent of the definition.[68] However, in view of the fact that the definition of 'environmental information' is not exhaustive and that the decisive element for whether 'factors or measures' are to be considered environmental information is whether they 'have … or are likely to have … an effect on the environment',[69] it seems reasonable to say that information relevant for checking the trustworthiness of the EU ETS, and thus its effectiveness, are to be seen as environmental information. As noted earlier, if a specific verification report is wrong, for whatever reasons, the operator has possibly emitted more than reported, and, consequently, surrenders fewer allowances, thereby compromising the system as a whole. Thus, verification reports can have an effect on the environment; they can therefore be considered as environmental information under the Aarhus Convention. The same may be true for other information of the verification process such as information on the way the verifier has checked a specific industry, the minutes of meetings between the verifier and the industry, including the report of a site-visit, or certain agreements made between the industry and the verifier on the specific methodology for calculating the emissions.

12.3.3.2 Are Verifiers "Public Authorities"?

The second requirement for being able to effectuate the right to access to environmental information is that private verifiers must qualify as public authorities. Article 2(2)(a) of the Aarhus Convention defines a public authority as any governmental authority on any level of administration. Furthermore, the term 'public authority' includes any natural or legal person to whom a public authority has delegated public

---

[66] Ibid., [21]; those elements are mentioned in Article 2(1) (a) of Directive 2003/4/EC.

[67] Case C-316/01 *Eva Glawischnig v Bundesminister für soziale Sicherheit und Generationen* [2003] ECR I-05995.

[68] If a piece of information does not qualify as environmental information a request to access it cannot be made under the Aarhus Convention and Directive 2003/4/EC. Instead the general (but more limited) access to information legislation applies.

[69] Case C-316/01 *Glawischnig*, [38].

administrative tasks,[70] as well as any natural or legal person who provides a public service or has another public responsibility that relates to the environment and is under the control of an entity falling under the first two categories.[71] The distinction between these two definitions needs to be emphasised: the activities of entities carrying out public administrative tasks need not to relate to the environment while those entities carrying out non-administrative public responsibilities must do so in order for the entity to fall under the definition of public authority.

The EU ETS Directive – which has introduced the verifier – does not provide any textual explanation whether the verifier should be qualified as a public authority.[72] The Accreditation and Verification Regulation defines a verifier as either a 'legal entity carrying out verification activities pursuant to this Regulation' and who is accredited by a national accreditation body in accordance with Regulation (EC) No 765/2008 or a natural person who is certified by the national certification body.[73] Accreditation means that a national accreditation body attests that the verifier meets the standards set out in Regulation (EC) No 765/2008 and Commission Regulation (EU) No 600/2012. It is important to note that these definitions do not give public authorities the leeway to decide to carry out verification themselves.[74]

According to Article 2(2)(b) and (c) of the Aarhus Convention, for private verifiers to qualify as public authorities, they must either perform (a) public administrative functions or (b) have public responsibilities or functions, or (c) provide a public service in relation to the environment and be under the control of a public authority.[75] In *Fish Legal*,[76] the CJEU interpreted the corresponding article of Directive 2003/4/EC and noted that, for the purposes of interpreting Directive 2003/4/EC, account is to be taken of the wording and aim of the Aarhus Convention.[77] Regarding Article 2(2)(b), it explained that 'the concept of 'public

---

[70] Aarhus Convention, Article 2(2)(b).

[71] Aarhus Convention, Article 2(2)(c).

[72] There is no definition of 'verifier' in the EU ETS Directive.

[73] Commission Regulation (EU) No 600/2012 of 21 June 2012 on the verification of greenhouse gas emission reports and tonne-kilometre reports and the accreditation of verifiers pursuant to Directive 2003/87/EC of the European Parliament and of the Council (hereafter Verification Regulation).

[74] The legislation does not explicitly exclude public authorities from taking the role of a verifier. However, it seems unlikely that this was intended by the EU legislator as requiring that a public authority be accredited by a national accreditation body before performing the verification would be redundant. In any event, outsourcing verification to the private sector is widespread (according to European Environmental Agency (2016), at 30, twenty-six countries have at least one accredited verifier, and there is widespread use of verifiers from other countries).

[75] To be specific, the private verifier should be under control of a public authority as meant in Article 2(2)(a) or (b) Aarhus Convention.

[76] Case C-279/12 *Fish Legal and Emily Shirley v Information Commissioner and Others* [2013] published in the electronic Reports of Cases.

[77] Case C-279/12 *Fish Legal* 2013 [37], Article 2(2) of Directive 2003/4/EC.

administrative functions' . . . cannot vary according to the applicable law' and must therefore be uniformly applied EU wide.[78] To determine whether a private entity qualifies as a legal person performing a 'public administrative function' one must determine whether it is equipped with special powers that it normally does not have under private law.[79] In other words, in order to fall under the definition of Article 2(2)(b) of the Aarhus Convention the private entity must be a governmental authority in functional terms.[80] Thus, to qualify as public authorities pursuant to Article 2(2)(b) private verifiers must perform a service in the public interest and have special powers to perform this service. One could well argue that verifiers perform a service in the public interest since they contribute to the enforcement of environmental legislation that is intended to protect the environment, which is clearly in the public interest. Moreover, the EU ETS directive requires that the emission reports *be verified,* and that an industry may not make transfers of allowances until the report has been verified as satisfactory.[81] In other words, without this verification activity, it may be possible for industries to operate in breach of the EU ETS.[82] Furthermore, Article 7(3) of the Verification Regulation states that verification must be performed in the public interest. In sum, in case of the EU ETS, one can say that the verifier acts as a public authority and thus meets the first of two criteria that need to be fulfilled in order to classify a body as a public authority.[83]

The next question is whether the private verifier has special powers to perform the provided service that go 'beyond those which result from the normal rules applicable to relations between entities governed by private law'.[84] It can be argued that verifiers do fulfil this criterion, as they are given the mandate to audit. Furthermore, by issuing a positive or negative verification report they effectively determine whether operators can surrender allowances and subsequently continue to participate in the EU ETS.[85] Moreover, verification reports cannot be issued by anyone but

---

[78] Case C-279/12 *Fish Legal,* [45].This case concerns private companies which manage a public service relating to the environment (water and sewage services).

[79] Case C-279/12 *Fish Legal,* [56].

[80] Case C-279/12 *Fish Legal,* [52] & Ebbesson (2011), at 81.

[81] EU ETS Directive, Article 15.

[82] EU ETS Directive, Article 15, second subparagraph.

[83] However, it is imaginable that, in practice, private verifiers will try to argue that they do not qualify as public authority, or that they qualify as a public authority only under Article 2(2)(c) of the Aarhus Convention instead of Article 2(2)(b). The reason for this is that, according to the interpretation of the CJEU in Case C-279/12 *Fish Legal,* [83] under subparagraph (c), 'they are not required to provide environmental information [requested] if it is not disputed that the information does not relate to the provision of the public service in the environmental field which they provide. Governmental authorities do not have the option of making this argument.

[84] Case C-279/12 *Fish Legal,* [56].

[85] Competent authorities may have competences to control the correctness of the emission reports and the verification, see for varying practices in this respect among Member States Fleurke and Verschuuren (2016).

accredited verifiers.[86] Thus, one can conclude that private verifiers also fulfil the second criterion that must be fulfilled in order for them to fall under the definition of 'public authority' and therefore be classified as public authorities according to Article 2(2)(b) of the Aarhus Convention.[87] This means that they must provide environmental information upon request, except when a valid reason for refusing this request applies.

12.3.4 *Grounds for Refusal*

The grounds based on which a public authority may refuse a request for environmental information are set out in paragraphs 3 and 4 of Article 4 of the Aarhus Convention. This section will give particular attention to the grounds for refusal that may be relevant in the case of a request by the public for information from a verifier. For the focus of our examination, which is access to information held by the private verifier, a few grounds may be particularly relevant. Firstly, a request may be refused if the information requested relates to internal communications of public authorities (following the analysis in Section 12.3.3.2, this includes verifiers), and public authorities may refuse to provide access to environmental information if disclosure would adversely affect the 'confidentiality of proceedings of public authorities'.[88] These grounds can only be invoked in situations in which confidentiality is provided for by national law. Thus, the implementing legislation of Member States would have to be analysed. It can be remarked that if the implementation legislation were to show striking differences on this point, the accessibility of information held by the verifier would be fragmented throughout the European Union. Another situation in which access to environmental information may be refused is where releasing this information would adversely affect judicial proceedings, including the ability of any person to receive a fair trial and the ability of a public authority to conduct an enquiry of a criminal or disciplinary nature.[89] Since the function of the verifier is situated at the stage of compliance, disclosure of information to the public (for instance, where this concerns information related to potential fraud with the emissions data) may be refused on these grounds, depending on the specific facts of the case.

---

[86] Verification Regulation, Article 3(3) & (4). Furthermore, verifiers have to carry out site visits and request corrections from the operator: Verification Regulation, Articles 21 and 22(1), which are specific inspection tasks, see Ebbesson (2011), at 81.

[87] In view of the limited length of this chapter, we do not delve into Article 2(2)(c) of the Aarhus Convention, but we note that, also according to this criterion, verifiers must be qualified as public authorities. Furthermore, see for instance the German national accreditation body, which is 'entrusted by the federal government to carry out its public authority accreditation tasks'. www.dakks.de/en/content/profile (accessed 15 November 2018).

[88] Directive 2003/4/EC, Articles 4(2)(a) and (4)(2)(d).

[89] Aarhus Convention, Article 4(4)(c).

Another important reason for refusing access to environmental information is the *by law* protected confidentiality of commercial and industrial information guarding legitimate economic interests.[90] Particularly in the case of monitoring and controlling greenhouse gas emissions, one may wonder to what extent commercial and industrial information should be legitimately protected by law. In any event, the EU ETS Directive does not contain a clear provision that obliges Member States to regulate in their national implementing legislation the confidentiality of commercial and industrial information on industries that is gathered by the verifier.[91] In Section 12.3.2.2 we have already discussed the provision on "professional secrecy" in Article 15a EU ETS Directive, and the uncertainty of how to interpret this provision.

Regarding the EU ETS's specific provision that emission reports held by the competent authority must be made available on request according to Article 17 of the EU ETS Directive, industries may still claim that one of the grounds of refusal, as mentioned in Directive 2003/4/EC, applies. This attempt at preventing disclosure may not be successful in view of the fact that disclosure is obligatory in cases of 'emissions into the environment'.[92] But other pieces of information, such as misstatements addressed by the verifier, or any other information exchanged between the operator and the verifier, may be requested by the public; however, the operator may claim the need to protect sensitive business information.[93] In such a case, this argument is valid only *if* such confidentiality is provided for under the national legislation, and if the information does not concern 'emissions into the environment'. Furthermore, there is no guarantee that the public is interested in accessing information that, in the opinion of the verifier, should not be disclosed to protect its legitimate economic interest or its ability to receive a fair trial in case the verifier fears criminal prosecution (for instance, related to fraud or collusion).

Furthermore, the Aarhus Convention provides that access to information may be refused if disclosure would violate intellectual property rights[94] or infringe the confidentiality of personal data.[95] Additionally, disclosure may be refused if this would have an adverse effect on the interests of a third party from which the information originated, unless that third party has given its consent to the release of the information. All in all, although EU law clearly includes the general right for the public to obtain access to environmental information, the application of this right may face many barriers. The specific circumstances under which information related to the compliance stage, including particularly the verification activity, may

---

90 Aarhus Convention, Article 4(4)(d) (same wording is used in Article 4(2)(d) of Directive 2003/4/ EC).

91 As far as we could observe, the implementing regulations also do not have such provisions.

92 Directive 2003/4/EC, Article 4(2).

93 Verification Regulation, Article 10(1).

94 Aarhus Convention, Article 4(4)(e).

95 Aarhus Convention, Article 4(4)(f).

be held confidential, are not very clear, and trial procedures may be needed to test the enforceability of this right.

12.3.5 *Interim Conclusion: Problems in Effectuating the Right of Access to Environmental Information*

The discussion in Sections 12.3.2, 12.3.3 and 12.3.4 has shown the complexities of effectuating the right to access environmental information as held by verifiers according to the current legislative provisions. While the EU legislator has introduced the function of the verifier for controlling emission reports, it did not provide a clear legislative framework with regard to the transparency of the verification process. First of all, it has not been made clear in the regulatory provisions whether the verifier can be qualified as a public authority from which access to environmental information can be requested. Secondly, the restriction of the scope of Directive 2003/4/EC in the EU ETS Directive is particularly remarkable: verification reports and other verification information are not covered. Next to this, the applicability of the active dissemination duty of decisions and reports related to verification is unclear in view of the term 'professional secrecy'. Moreover, the extent to which Member States can lawfully provide in their implementing law for either transparency or confidentiality of information used in the verification process is unclear.

Some uncertainty, however, continues to exist with respect to the grounds for refusal. How must these grounds be interpreted and applied, and may access to information related to the verification process be limited as a consequence? Currently, there are no clear answers to these questions; however, this may change by means of future case law development.[96]

In our opinion, fine-tuned legislative provisions would be needed to clarify how and to what extent the right to access environmental information held by verifiers can be used.[97] When developing such provisions, consideration must also be given to legitimate grounds for refusing the requested information, such as the arguably needed confidentiality of information in case of enforcement proceedings. Furthermore, it would be naïve to assume that there will ever be legal provisions that provide 100 per cent certainty on when access to environmental information must be provided. Hence, the public may still be confronted with refusals and it is then up to them to start legal proceedings.[98] This brings us to the point of how to enforce the

---

[96] Krämer (2011), at 136.

[97] The support for broadening access to environmental information is yet to be determined: Schomerus and Bünger (2011), at 80 have reported that because of a fear of activism, there is resentment in Germany directed at the requirement to provide broad access to information.

[98] In this sense, further research should investigate the extent to which Member States give full implementation to the current provisions, and, in light of this, how the regulatory provisions for access to environmental information held by the verifier can be improved.

right of access to environmental information. If an ENGO has submitted a request to a verifier, and if the verifier indeed qualifies as a public authority, the applicant should be able to take the verifier to court.[99] Further research should show how Member States, in their implementing law, have regulated this opportunity, and whether these procedures are 'expeditious and either free of charge or inexpensive'.[100] If ENGOs face difficulties with accessing courts, or if the courts lay down unsatisfactory decisions, they are also entitled to file a complaint with the Aarhus Convention Compliance Committee.[101] The Committee may give further interpretations on the question of whether, and under which circumstances, the public should be given access to environmental information held by private actors in cases where they conduct a monitoring activity essential to the effectiveness of the EU ETS.[102]

12.4 ACTION BY ENGOS REGARDING INFORMATION DISCLOSURE

In our findings in Section 12.3.5, we have argued that, according to the Aarhus Convention, access to environmental information held by verifiers should in principle be possible, and that the extent to which this right can be exercised, also in view of applicable grounds of refusal, should be further clarified by means of case law and interpretations by the Aarhus Convention Compliance Committee. However, arguing that ENGOs (and other members of the public) should be able to use the right to access verification information does not mean that ENGOs will be especially eager to make use of that right.[103]

Nonetheless, the fact that some specialised greenhouse gas emissions trading ENGOs exist, as has been explained in Section 12.2.2, makes it plausible to expect that they, in their desire to check the proper functioning of the instrument, may want to get insight into the compliance performance of industries. However, as far as we can determine on the basis of available sources, we have not seen much activity yet with regard to the use of this right.[104] Moreover, Fleurke and Verschuuren found, on

---

[99] Directive 2003/4/EC, Article 6 & Aarhus Convention, Article 9(1).

[100] This criterion is mentioned only in Article 6(1) of Directive 2003/4; the Aarhus Convention contains in Article 9(1) juncto Article 9(3) and (4) different wording, more beneficial to the public.

[101] Economic Commission for Europe, 2004, [18].

[102] Such interpretations by the Compliance Committee will then be discussed by the Members to the Convention, Economic Commission for Europe, 2004, [35].

[103] Chapter 1 has already raised the issue that the involvement of NGOs cannot always be controlled (by public regulators or other actors).

[104] There is no discussion of the use of this right in literature so far, nor has any case law been developed at the EU level. We only have knowledge of one Dutch administrative court procedure in which Greenpeace asked for information included in the emissions report – this request was denied – which has been critically discussed by Thurlings (2017), at 270 f. stating that the administrative court wrongfully omitted submitting a request for a preliminary ruling to the Court of Justice of the EU: Further empirical investigations to case law in EU member

the basis of an interview with the German emissions trading authority, that ENGOs are 'more concerned with the level of emissions than on compliance issues'.[105] It may also be that reluctance to request environmental information stems from practical barriers, such as costs and the lack of sufficient legal expertise and representation to engage in court procedures for enforcing the right to environmental information. Nonetheless, it may also be that some requests for information *have* been filed, and have been positively followed up with disclosure of information.[106] In other words, a more comprehensive picture is needed on the current and future use of the right to access to environmental information in order to get insight into the compliance cycle of the EU ETS. If ENGOs (or civil society at large: individual citizens are also eligible to ask for such information) hardly make use of this right, the result will be less control of the compliance behaviour of emitters and verifiers, which may have a negative impact on the effectiveness of the emissions trading instrument. For the future, strategies of ENGOs may of course change: the car emissions fraud cases have illustrated that also with major industries, with presumably highly qualified (technical) experts, non-compliance may take place.[107] This event may cause ENGOs to concentrate more on checking compliance behaviour.

If ENGOs were to become more active in checking compliance, the resulting question would then be whether the disclosed information will be used in a correct way. As Fisher has noted, even if information is disclosed, it is not certain that the received information will be understood correctly.[108] This may lead to the public being misinformed, and misinformation may also even do reputational damage to the industry being reported on.[109]

The following text provides an example of potential misinformation by using data that are made available under the EU ETS. Since all EU ETS industries have to submit emission reports to the competent administration, and these must consequently be disclosed to the public, it is easy to identify who the biggest emitters

---

states, and to administrative practice, would be necessary to understand the use of the right of access to environmental information by ENGOs with regard to the EU ETS.

[105] Fleurke and Verschuuren (2016), at 224.

[106] Legal research often focuses on problems that emerge from case law. It would also be important, however, to get insight into how administrative procedures, like requests for information, are being dealt with.

[107] The legal procedures for holding the car producers accountable are pending, so we cannot refer to final conclusions on what exactly happened (Draft Report on the inquiry into emission measurements in the automotive sector (2016/2215 (INI)).

[108] Fisher (2010), at 294. See also Gunningham, Grabosky and Sinclair (1998) about the fact that the public can misunderstand information (at 65). For an example of where an ENGO had to correct the information it provided over the Internet with regard to an EU ETS industry, see https://sandbag.org.uk/2011/11/17/note-of-correction-to-thyssenkrupp-figures-in-sandbags-klima goldesel/ (accessed on 28 November 2017).

[109] In the example given in the previous footnote, the ENGO has apologized for any reputational damage it may have caused to the specific industry on which it was reporting, also available at: https://sandbag.org.uk/2011/11/17/note-of-correction-to-thyssenkrupp-figures-in-sandbags-klima goldesel (accessed 28 November 2017).

are.[110] In other words, through knowledge of the emission reports of all individual EU ETS installations, it is easy to determine the industries that have emitted the largest quantities of greenhouse gases. Consequently, they could be portrayed as industries acting in a bad manner, since they evidently emit the most. However, within an emissions trading system, it is logical that there are differences in emission levels among industries. This follows from the rationale that polluters are able to decide whether to use emission allowances instead of reducing emissions by taking organisational, managerial or technological reduction measures. Depending on the price of the allowances and the abatement costs, there will be, on the one hand, industries that reduce emissions and, on the other hand, industries that use allowances – and hence emit more. Particularly if many different categories of industries are included in an emissions regime, as is the case with the EU ETS, with different abatement options, the existence of high-emitting and low-emitting industries is even more logical: it fits with the deliberate choice of the legislator to use emissions trading in order to achieve lower overall costs of reducing pollution than would be obtainable through a command-and-control approach. Hence, if high-emitting industries under an emissions trading regime were to be negatively portrayed by ENGOs simply because of the fact that these are the biggest emitters, this would be a misrepresentation of the rationale of the instrument, and would amount to a rejection of this regulatory approach.[111] In this respect, the EU Commission could be keen on how information about the functioning of the instrument is used by ENGOs. If, for instance, the biggest EU ETS industries were to be identified and negatively portrayed in press releases, the Commission could step in by explaining the nature of the instrument, which accepts that, under the market-mechanism, relatively big as well as small emitters exist for reasons of efficiency.

In sum, while the argument can be made that it would be valuable if ENGOs made use of the rights established by the Aarhus Convention with the aim of checking industry compliance, as they would thereby contribute to the effectiveness of regulatory instruments such as the EU ETS, it is then also necessary to examine how ENGOs subsequently use this information.[112] In addition, the responses of the government also need to be studied; if information requested by ENGOs were to show problems with the credibility of the process of monitoring, reporting and verifying emissions, appropriate governmental control and enforcement action would be needed. If this does not occur, the effectiveness of the emissions trading instrument is threatened.

---

[110] For the required disclosure, see Article 15 of the EU ETS directive. However, some confidentiality provisions may apply, although the extent to which this is possible is not yet crystallised, and further case law is needed.

[111] Furthermore, analysis of case law can provide further insight into what means exist for industries to take legal action against – in their view – misuse of information or – also in their view – unjust shaming or blaming.

[112] The normative choice that access to information is a valuable, fundamental right has been made in the Aarhus Convention as such, and hence, by the parties adhering to it.

12.5 CONCLUSION

The aim of this chapter has been to discuss the mix of governmental and private action for regulating the reduction of greenhouse gas emissions in an effective way. As such, the emissions trading instrument in itself has an important design feature that, in principle, ensures effectiveness: no more greenhouse gases can be emitted than those which are allowed pursuant to the EU-wide cap. Consequently, no more allowances will be distributed to emitters than is possible under this cap. However, the crux for reaching real effectiveness in practice is whether compliance takes place with the rule that all emissions that are caused by emitters have to be covered by allowances. In this respect, we have explained that the EU has made a regulatory choice to rely on verifiers for controlling the emission reports from the emitters. In this sense, a twofold market-based approach is established: first of all, the market in tradable allowances, and, secondly, the markets in which verifiers offer their services to the emitters. To check the trustworthiness of this regime, we have discussed the potential role of the procedural right of access to environmental information. The general assumption that access to environmental information contributes to the proper execution of regulatory instruments has yet to be tested. In this vein, the way in which disclosed information is used by ENGOs is also a point to consider, since there are indications that disclosed information is not always sufficiently understood and/or properly used. However, before delving into studies aiming to examine the extent to which access to information may help to ensure the effective functioning of regulatory instruments, including the EU ETS, the circumstances under which this right can be successfully enforced before the court need to be clarified. Sections 12.3 and 12.4 have discussed the legal complexities that the public, and in this vein particularly ENGOs, face when trying to get insights into the verification of emission reports that must be delivered by the EU ETS emitters. Our conclusion is that, in view of the Aarhus Convention, the duty to provide access to environmental information also applies to verifiers, since they should be classified as public authorities. Nonetheless, the extent to which access to environmental information can be successfully employed is surrounded with legal uncertainty. Several grounds exist for justifying the rejection of a request for environmental information. This is, for instance, the case if the request concerns confidential business information, although this reason is by no means absolute. According EU law, 'professional secrecy' may also be used to justify the non-disclosure of environmental information, but the requirements for how this criterion is to be applied, as well as the question of whether it is compatible with the provisions of Aarhus Convention, are still uncertain. Another point that has yet to be clarified is the extent to which information that may be relevant in administrative or criminal enforcement procedures should be kept secret from the public.

In conclusion, while we see some prospect that ENGOs may contribute to achieving an effective application of the emissions trading instrument, particularly by exerting control on the correct functioning of verifiers, the opportunities and limits of the right of access to information have yet to be further examined. Hence, it cannot yet be determined whether the mix of instruments as discussed in this chapter can guarantee an effective reduction of greenhouse gas emissions. Legal actions from the public requesting access to information related to the compliance process of the EU ETS would stimulate further crystallisation of this matter.

REFERENCES

Bogojević, S. 2013. *Emissions Trading Schemes, Markets, States and Law*. Oxford/Portland Oregon, Hart Publishing.

Cole, D. H. 2016. 'Origins of Emissions Trading in Theory and Early Practice', in Weishaar, S.E. (ed.), *Research Handbook on Emissions Trading*. Cheltenham, Edward Elgar, pp. 9–26.

Dales, J. 2002. *Pollution, Property and Prices. An Essay in Policy-making and Economics*. Cheltenham, Edward Elgar (first published 1968).

Dudek, D. J. and Palmisano, J. 1988. 'Emissions Trading: Why is this Thoroughbred Hobbled?'. *Columbia Journal of Environmental Law* 13, 217–256.

Ebbesson, J. 2011. 'Public Participation and Privatisation in Environmental Matters'. *Erasmus Law Review* 4(2), 71–89.

Economic Commission for Europe. 2004. Meeting of the Parties to the Convention on Access to Information, Public Participation in Decision-Making and Access to Justice in Environmental Matters, *Report of the First Meeting of the Parties Decision I/7*, ECE/MP.PP/2/Add.8.

Etemire, U. 2013. 'Book review: The Aarhus Convention at Ten: Interactions and Tensions between Conventional International Law and EU Environmental Law, edited by Marc Pallemaerts, 2011'. *Review of European, Comparative & International Environmental Law* 22(3), 371–375.

    2012. 'Public Access to Environmental Information Held by Private Companies'. *Environmental Law Review* 14(1), 7–25.

European Environmental Agency. 2016. *Application of the EU Emissions Trading Scheme Directive – Analysis of National Responses under Article 21 of the EU ETS Directive in 2015*. Luxembourg, Publications Office of the European Union.

Fleurke, F. and Verschuuren, J. 2016. 'Enforcing the European Emissions Trading System within the EU Member States. A Procrustean Bed?'. In White, R., Spapens, T. & Huisman, W. (eds.), *Environmental Crime in Transnational Context: Global Issues in Green Enforcement and Criminology*. Abingdon, Routledge Publishing, (Green Criminology), pp. 208–230.

Fisher, E. 2010. 'Transparency and Administrative Law: A Critical Evaluation'. *Current Legal Problems* 63(1), 272–314.

Gunningham, N., Grabosky, P. & Sinclair, D. 1998. *Smart Regulation Designing Environmental Policy*. Oxford, Oxford University Press.

Krämer, L. 2011. *EU Environmental Law*. London, Sweet & Maxwell.

Levin, I., Hammer, S., Eichelmann, E. and Vogel, F. R. 2011. 'Verification of Greenhouse Gas Emission Reductions: The Prospect of Atmospheric Monitoring in Polluted Areas'. *Philosophical Transactions of the Royal Society* 369(1943), 1906–1924.

Livingston, P., Lee, C. S. J. and Nguyen, D. 2015. 'Creation and Third-party Verification of Large Governmental Greenhouse Gas Emission Report'. *Strategic Planning for Energy and the Environment* 34(3), 43–62.

Müller, M. 2016. 'The Current State of the Right to Access to Environmental Information'. Master thesis, European Law School.

Peeters, M. 2009. 'Improving Citizen Responsibility in the North and its Consequences for the South: Voluntary Carbon Offsets and Government Involvement'. In Richardson, B. J., Le Bouthillier, Y., McLeod-Kilmurray, H. and Wood, S. (eds.), *Climate Law and Developing Countries*. Cheltenham, Edward Elgar, pp. 337–362.

Peeters M. 2016a. 'An EU Law Perspective on the Paris Agreement: Will the EU Consider Strengthening its Mitigation Effort?'. *Climate Law* 6(1–2), 182–195.

    2016b. 'Greenhouse Gas Emissions Trading in the EU'. In Farber, D. A. and Peeters, M. (eds.), *Climate Change Law*. Cheltenham, Edward Elgar, pp. 377–385.

Romppanen, S. 2015. 'New Governance in Context Evaluating the EU Biofuels Regime'. Dissertation, Publications of the University of Eastern Finland. Available at: http://epublications.uef.fi/pub/urn_isbn_978–952-61–1763-8/urn_isbn_978–952-61–1763-8.pdf.

Schomerus T. and Bünger, D. 2011. 'Private Bodies as Public Authorities under International, European, English and German Environmental Information Laws'. *Journal for Environmental and Planning Law* 8(1), 62–81.

Stewart, Richard B. 2007. 'Instrument Choice'. In Bodansky, D., Brunnée, J. and Hey, E. (eds), *The Oxford Handbook of International Environmental Law*.Oxford, Oxford University Press, pp. 147–181.

Thurlings, T. 2017. *Verhandelbare emissierechten in broeikasgassen*. Deventer, Wolters Kluwer.

Tietenberg, T. 2006. *Emissions Trading: Principles and Practice*. Washington DC, RFF Press.

Weishaar, S. E. (ed.). 2016. *Research Handbook on Emissions Trading*. Cheltenham, Edward Elgar.

Wang, J., Jin, S., Bai, W. and Jin, Y. 2016. 'Comparative Analysis of the International Carbon Verification Policies and Systems'. *Journal of the International Society for the Prevention and Mitigation of Natural Hazards* 84(1), 381–397.

De Witte, B. 2017. 'EU Law: Is it International Law?'. In Barnard, C. and Peers, St. (eds.), *European Union Law*. Oxford, Oxford University Press, pp. 177–197.

13

Smart Mixes of Civil Liability Regimes for
Marine Oil Pollution

Michael Faure and Hui Wang

13.1 INTRODUCTION

Marine pollution can be caused by a variety of sources and substances. It can be released from a facility or a ship, it can occur during operational processes or through accidents; the polluting substance can be oil, chemicals or any potentially hazardous or noxious substances. Within this chapter we will focus only on the civil liability regime concerning pollution caused by oil when transported at sea (as cargo) through tankers.[1] Our analysis is limited to vessel-based marine pollution due to accidents. The pollution damage caused through offshore drilling facilities is not the focus of our discussion.[2] The compensation of damage related to substances other than oil is also not discussed in this chapter.[3]

The compensation regime for marine oil pollution may be illustrative for the study of "smart mixes" since there are a number of combinations of instruments that are employed in this domain that are quite typical and certainly deserve further attention. At the international level, for example, there are combinations of conventions related to the compensation of damage resulting from marine pollution. The international regime on civil liability for marine oil pollution constitutes an interesting mix in the sense that compensation is provided first via the (limited) liability of the tanker owner (based on the Civil Liability Convention – CLC) and as a

---

[1] Hence, the discussion in this chapter does not include the compensation regime for damage caused by oil used as fuel, which is regulated through the International Convention on Civil Liability for Bunker Oil Pollution Damage of 2001. For a discussion, see Shan (2010) as well as Han and Wang (2010).

[2] There is so far no specific international regime covering the compensation for pollution damage caused by offshore facilities. See Faure, Liu and Wang (2015), at 371–373.

[3] Other polluting substances than oil can also cause substantial damage. This is covered by the 2010 International Convention on Liability and Compensation for Damage in Connection with the Carriage of Hazardous and Noxious Substances by Sea. It might be interesting to note that this convention is structured along the same lines as the marine oil pollution compensation conventions. See Faure, Liu and Philipsen (2015), at 35–44.

285

second layer via the oil receivers (through the so-called Fund Convention). Later with the help of economic tools we will analyze whether such a mix could be considered as "smart."

In addition to the conventions, private industry mechanisms also play an important role in the international compensation mechanism. Some industry mechanisms were developed even before the advent of the international conventions. When some old industry-related compensation instruments were abolished, other new private instruments were developed again in the 2000s. These private mechanisms may take on different forms and may serve different goals, but the mix of public and private instruments constitutes an interesting feature of the whole international regime for civil liability for marine oil pollution.

Many countries have joined the international conventions. However, one exception to the international regime is the US, which has adopted its own domestic legislation, the Oil Pollution Act (OPA) of 1990. There are some similarities, as well as substantial differences, between the US and the international regimes. A comparative analysis of these two regimes may identify whether there are different "mixes" at different levels of governance and whether there are indeed "smarter" mixes in one regime or the other.

Therefore, in this chapter various mixes in the civil liability regime for marine oil pollution will be described and analyzed with respect to the question of extent to which these combinations can be considered as "smart." In order to do that, particular benchmarks will be advanced. One such benchmark is whether the (compensation) mechanism is able to provide compensation for marine pollution damage in most cases; another benchmark is whether the mechanism also contributes to prevention of accidents and thus reduces the pollution risk.

This chapter is structured as follows: after this introduction, Section 13.2 first sketches how various legal instruments are used and mixed under the international conventions dealing with the compensation regime for marine oil pollution. Next, Section 13.3 describes the private mechanisms developed by the industries to compensate for marine oil pollution damages, and hence the mix of public regulation (via conventions) and private initiatives also by industry. The domestic legislation in the US, as a major deviation from the international regime, represents different yet interesting mixes. Section 13.4 analyzes these, especially in comparison with the international regime. At the end, Section 13.5 provides possible explanations for the mixes in the civil liability regime for marine oil pollution, and some lessons for policy improvement.

13.2 COMPENSATING FOR MARINE OIL POLLUTION: MIXING CONVENTIONS

Sections 13.2.1 and 13.2.2 will first sketch how the international compensation conventions related to marine oil pollution have developed and how they have dynamically adapted to changing circumstances, more particularly more serious spills. Sections 13.2.3 and 13.2.4 will examine the main legal features of these international conventions, and then identify and analyze the mixes under these international conventions.

13.2.1 *CLC and Fund Convention*

The international regime specifically dealing with marine oil pollution compensation has been developing since the late 1960s as a reaction to some major oil spill incidents. Initially, in response to the *Torrey Canyon* spill in 1967, two international conventions were introduced to provide compensation for pollution victims. The International Maritime Organization (IMO, the IMCO at the time, being Intergovernmental Maritime Consultative Organization)[4] organized an international conference in November 1969 to discuss the liability issue arising from marine oil pollution, which led to the adoption of the International Convention on Civil Liability for Oil Pollution Damage (CLC 1969).

The CLC 1969 imposes strict liability on the registered shipowner (in other words, on the tanker owner). This liability is borne exclusively by the shipowner; this can, in principle, have the effect of shielding other parties from liability for the accident (known as the channeling of liability).[5] The shipowner was required to provide insurance or another financial guarantee to cover their exclusive liability. But the shipowner's exclusive liability was limited to a certain amount per gross tonnage of the ship.[6]

The declared aim of the 1969 conference which created the CLC – sufficient compensation of oil pollution victims – could not be reached in all cases, at least not through the CLC alone. Therefore it was decided that some form of an international compensation fund, supplementary to the CLC regime, should be considered and established.[7]

Accordingly, in 1971 the International Convention on the Establishment of an International Fund for Compensation for Oil Pollution Damage (the Fund Convention 1971) was adopted.[8] Under the Fund Convention 1971, an International Oil Pollution Compensation Fund (the IOPC Fund, or simply the Fund) was established and financed by oil-receiving entities based on the amount of oil received per year. The creation of the Fund was the result of a compromise between the oil and

---

4   The IMO is a special agency of the United Nations to deal with the shipping-related issues. It is the international organization under the auspices of which negotiations among governments take place concerning matters relating to shipping engaged in international trade, including maritime safety, efficiency of navigation, and prevention and control of marine pollution from ships. See Article 1 of the International Convention on the International Maritime Organization 1948.

5   Article III(4) of the CLC 1969 stipulated that no claims for pollution damage under the CLC or otherwise 'may be made against the servants or agents of the (ship) owner'. Claims were therefore channeled to the registered shipowner (Wang (2011), at 84–88).

6   This limitation of liability of the shipowner is debated in the literature, see Faure and Wang (2008).

7   Wang (2007), at 202.

8   The Fund Convention 1971 has established an International Oil Pollution Compensation Fund, referred to as the IOPC Fund, or the Fund.

shipping industries; since the Fund would be financed by the oil industry, this would provide a mix of compensation by both the maritime industry (via the limited liability of the tanker owner under the CLC) and the oil industry (via the 1971 Fund).[9] It was actually thanks to the prospective establishment of a compensation fund paid for by the oil industry that a strict liability on the shipowner was accepted under the CLC 1969.[10]

The combination of both conventions therefore established the basis for the international regime on marine oil pollution compensation.

13.2.2 *Adaptations*

Later catastrophic oil pollution incidents illustrated the inadequacy of the international regime as it was developed in 1969/1971. For example, the incidents with the *Amoco Cadiz* in 1978 and the *Tanio* incident in 1980, were both cases where the amounts of compensation available under the then-existing international regime were insufficient to cover the total damage. This led to initiatives in the 1980s and to draft protocols of 1984 to amend the International Convention on Civil Liability for Oil Pollution Damage. However, the protocols of 1984 could not enter into force as their entry into force relied on ratification by some major shipping nations and important oil import countries, such as the United States.[11] Prior to 1990, however, the US had no uniform federal legislation on oil pollution; what existed at that time were various state laws regulating the compensation for marine oil pollution. In many state statutes, the legislation contained principles that were different from the international regime, such as unlimited liability (in contrast to the CLC's limited liability) and "joint and several" liability (in contrast to the CLC's channeling of liability).[12] Due to dissatisfaction with the amount of compensation available under the 1984 Protocols and the inherent difficulty of reaching agreement on the pre-emption of state laws, the US government was hesitant to ratify the 1984 Protocols. When the *Exxon Valdez* incident happened off the coast of Alaska in 1989, it became clear that the US would never ratify the 1984 Protocols. This led to the

---

[9] Wang (2007), at 218–219.

[10] Wang (2011), at 96–97.

[11] The entry into force conditions of the 1984 Protocol to the CLC 1969 required ratification by at least ten states and at least six of these had a tanker fleet of over 1 million tonnes. As of June 1990, only six states had ratified the 1984 Protocol to the CLC 1969, among which only France had a tanker fleet of over 1 million tonnes. The 1984 Protocol to the Fund Convention 1971 required ratification by at least eight states with a total combined oil import level of at least 600 million tonnes. As of June 1990, this Protocol had been ratified only by two states – France and Germany – and their combined imports of oil reached 164 million tonnes. The US is one of the largest oil-consuming countries and its oil importation relies largely on transportation through tankers. The participation of the US would have facilitated the ratification of the 1984 Protocols. For more on those 1984 protocols, see Wang (2011), at 131–164.

[12] See Wang (2011), at 164, 197–205.

failure of the 1984 Protocols, as the US went its own way with the creation of the Oil Pollution Act of 1990.

However, the contents of the 1984 Protocols were incorporated into a revision of the international conventions, which took place in 1992.[13] In these 1992 Conventions the amounts of compensation available, from both the shipowner's liability and the International Oil Pollution Compensation Fund, were raised from the 1969/1971 levels to those specified in the 1984 Protocols. Moreover, the 1992 Conventions also expanded the scope of compensation.[14] Despite these changes in the 1992 Conventions, the general principles of the International Oil Pollution Compensation Regime – the sharing of liability between the shipping and the oil industry, with liability resting strictly with the tanker owner, a limitation of the shipowner's liability, compulsory insurance coverage and channeling of liability – remained the same.

Later incidents again triggered further changes to the international regime. A first important incident was the sinking of the *Erika* off the coast of Brittany (France) in 1999. This led again immediately to action by the IMO, resulting in the 2000 resolutions that increased the compensation amounts available under the CLC and the Fund Convention by 50 percent.[15] The fact that the *Erika* incident took place in European waters led to strong pressure from the European Commission, which triggered the increase in the available amounts.[16] However, only a few years later, in 2002, another incident occurred in European Waters, this time with the *Prestige* off the coast of Galicia (Spain). This time the European Commission reacted with the proposal to set up a European Compensation Fund for Oil Pollution (referred to as the COPE Fund). The idea was to provide a higher ceiling of EUR 1 billion (instead of the then-available maximum of EUR 200 million under the international conventions).[17] This European activism led to a reaction by the IMO at a diplomatic conference in London in May 2003, leading to a Protocol on the Establishment of a Supplementary Fund for Oil Pollution Damage. This Supplementary Fund of 750 million special drawing rights[18] (at the time of adoption corresponding to approximately EUR 920 million or USD 1 billion) would not replace the existing

---

[13] These conventions are the International Convention on Civil Liability for Oil Pollution Damage 1992 (CLC 1992) and the International Convention on the Establishment of an International Fund for Compensation for Oil Pollution Damage 1992 (Fund Convention 1992). These two conventions are often referred to as the 1992 Conventions.

[14] For details see Wang (2011), at 165–173.

[15] It should be added that another incident during this time (the 1997 breakup of the *Nakhodka* off the coast of Japan) equally triggered the need for another increase in the compensation amount.

[16] Wang (2011), at 173–176.

[17] Proposal for a regulation on the establishment of a fund for the compensation of oil pollution damage in European waters and related measures, *OJ* C-227 E/487 of 24 September 2002.

[18] Special drawing right, known as SDR, is defined by the International Monetary Fund (IMF) to be calculated on the basis of five major currencies: the US dollar, euro, the Chinese renminbi, the Japanese yen and the British pound sterling.

1992 Fund, but would provide a third tier of compensation. It effectively made additional compensation available to victims in the states that ratify the Supplementary Fund Protocol.[19] This Supplementary Fund Protocol entered into force on 3 March 2005. As of October 2017, 31 states have ratified or acceded to this Supplementary Fund Protocol, 137 states have ratified or acceded to the CLC 1992 and 114 states have ratified or acceded to the Fund Convention of 1992.[20]

It is worth mentioning that when the international conventions on marine oil pollution compensation were developed as reactions to some major oil pollution incidents, intense debates on safety standards, tanker design, crew training and responses to oil spills were also taking place. Therefore, a series of conventions aiming at prevention of oil pollution were adopted and revised as well.[21]

After the many adaptations just discussed, Table 13.1 provides a summary of the amounts of compensation available in the past and today:

This table clearly illustrates how the amounts of compensation available for oil pollution damage have historically developed. As ships have become larger and the scale of oil spill incidents has increased in size, the compensation protocols have had to be adapted to new situations (more particularly, incidents with higher amounts of damage). Major oil spill incidents have occurred frequently enough that they provide impetus for changes in the oil pollution compensation regime: right after a large-scale pollution incident, the legal regime on compensation is submitted to the international community for wide discussion, which has led to the development of the current compensation regime. The table also makes clear that a relatively high amount of compensation can now be generated as a result of a mix of different actors working together to provide the compensation. It consists particularly of the shipping industry (via the limited liability of the tanker owner according to the CLC 1992) and the oil industry (contributing to the IOPC Fund 1992 and to the 2003 Supplementary Fund.)

13.2.3 *Features of the International Regime*

The analysis of the legal parameters of the international regime for marine oil pollution compensation may indicate the effectiveness (or ineffectiveness) of the mix of the various conventions. This can be examined from different angles. On the one hand, the examination can be of the limits of the international regime's ability to provide compensation that accurately reflects the full extent of the damage caused by serious cases of marine pollution; on the other hand, the examination can

---

[19] Wang (2011), at 179.

[20] Official website of the IOPC Funds, www.iopcfunds.org/about-us/membership/map/#.

[21] These are mainly the International Convention for the Prevention of Pollution from Ships (MARPOL 73/78), the International Convention on the Safety of Life at Sea 1974 (SOLAS), the International Convention on Standards of Training, Certification and Watchkeeping for Seafarers 1978 (STCW Convention), and the International Convention on Oil Pollution Preparedness, Response and Cooperation 1990 (OPRC Convention).

TABLE 13.1 *Development of Available Compensation (in SDR)*

Tonnage of Ship (Gross Tonnage)	CLC 1969 (SDR)	CLC 1992 (SDR)	2000 Protocol (SDR)	
Ships ≤ 5,000	133 per ton	3 million	4.5 million	
5,000<Ships<140,000	133 per ton	3 million + 420/additional ton	4.5 million + 631/additional ton	
Ships≥140,000	14 million	59.7 million	89.8 million	
	1971 Fund (SDR)	**1992 Fund (SDR)**	**2000 Protocol (SDR)**	**2003 Supplementary Fund (SDR)**
Overall limit	60 million	135 million	203 million	750 million

employ an economic approach, which not only determines compensation but also determines an appropriate penalty and has a deterrent effect, thereby promoting prevention of environmental harm as well.[22]

First, concerning the compensation function of the international conventions, the most direct way to determine the effectiveness of these conventions would be to compare the amount of damages caused through oil spill incidents and the amount of compensation that has been paid out under the international conventions. The difficulty here might lie in finding out the accurate figure for damages related to pollution incidents. Information is available from several databases that report on the amounts of oil spilled into the sea. However, the quantity of oil is only one of the many (though crucial) factors that go into determining the damages. Other factors influencing the actual extent of damages may include, among others, the currents, weather conditions and the location where the incident occurs. In particular, when it comes to the location of the incident, the costs of damages for an area that is environmentally sensitive can be completely different from those for an area that is not. Moreover, like many environmental risks, oil pollution has a long-tail problem, which means that the full effect of the spill may not be evident until years later. An oil spill influences not only the marine environment, which is difficult to quantify in monetary terms, but also fisheries and other social and economic activities. Therefore, all these complications lead to difficulties in calculating the precise amount of damages for each accident. What often happens as a result is that different authorities may give different estimates on the damages of one particular incident.

Although it is difficult to accurately determine the actual amount of damages for oil spill incidents, there are databases that have information on the amount of oil spilled into sea. These mainly include databases from ITOPF (International Tanker Owners Pollution Federation Limited),[23] the French CEDRE (Centre de

---

[22] The economic approach to accident law is probably best represented in the work of Shavell (1987).

[23] www.itopf.com/knowledge-resources/data-statistics/statistics/.

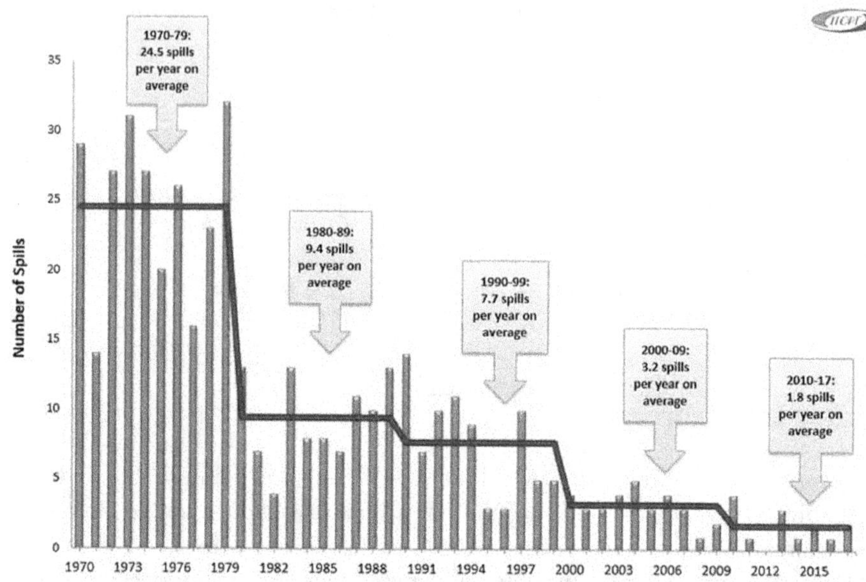

FIGURE 13.1 Number of large spills (>700 tonnes) from 1970 to 2017[24]

Documentation de Recherche et d'Expérimentations sur les Pollutions accidentelles des eaux)[25] and the US DARRP (Damage Assessment, Remediation and
Restoration Program)[26]. It might still be useful to look at the amount of oil that
has been released to the sea in relation to the compensation from the international
conventions in order to examine the effectiveness of the CLC/Fund Convention
regime as far as compensation is concerned.

Statistics show a general trend of reduction of the frequency of oil spill
incidents since the 1970s. However, nowadays when an oil spill occurs, is it
more often of a large scale. Therefore, a few very large spills are responsible for
a high percentage of oil released into sea. This is well illustrated by Figures 13.1
and 13.2.

[24] Data from the ITOPF Ltd Oil Tanker Spill Statistics 2017, see www.itopf.com/knowledge-
resources/data-statistics/statistics/.

[25] CENDRE was created in 1978 in the aftermath of the *Amoco Cadiz* accident, being dissatisfied
with the national marine pollution response system (Le Plan POLMAR, which stands for
Pollution Maritime), and to provide advice and expertise on oil spill response. See www.cedre
.fr/en/Resources/Spills.

[26] DARRP was formally created in 1992 following the *Exxon Valdez* oil spill in 1989. It is
administered by the National Oceanic and Atmospheric Administration (NOAA) and its main
tasks are to review pollution incidents (including those caused by oil from ships) and help with
cleanup actions and restorations. See the website of DARRP: darrp.noaa.gov/oil-spills.

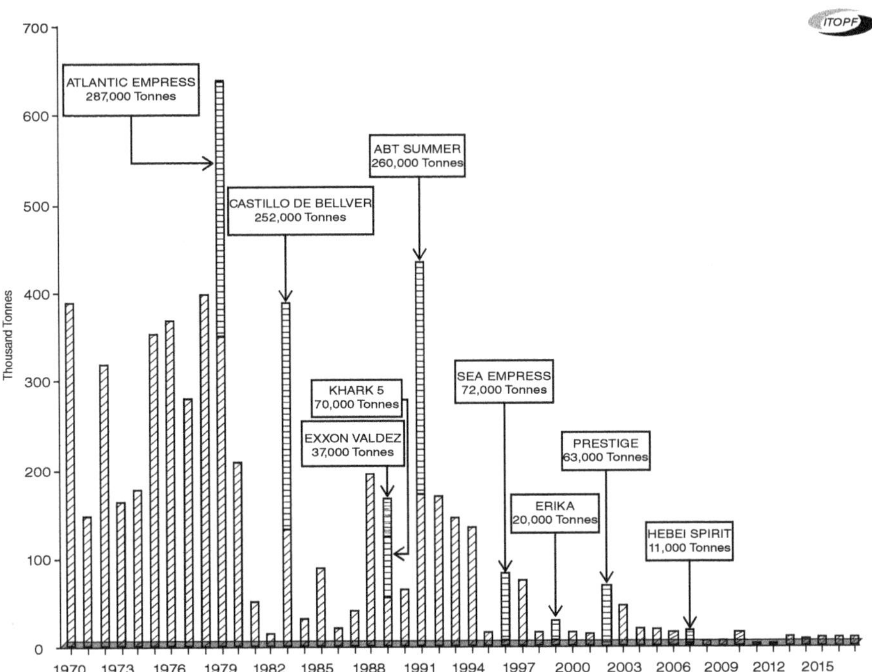

FIGURE 13.2: Quantities of oil spilt 7 tonnes and over (rounded to nearest thousand), 1970–2017[27]

Against this background, some studies were carried out that took the claimed amount of damage as an indicator to be compared with the final compensated amount.[28] These studies found that strict liability as adopted under the CLC is definitely useful in awarding compensation to the affected parties.[29] Although there are discrepancies between the amount claimed and the amount actually paid out, since the changes in 2000, the level of compensation is higher.[30] Therefore, these studies show that the CLC/Fund Convention regime does function as a compensation mechanism and since the revisions in 2000 the international regime even functions better in providing compensation.

As to the deterrence and prevention function, the basic choice in the CLC to opt for a strict liability of the tanker owner can also be supported from an economic perspective. The strict liability rule was also justified referring to the need to internalize the social costs of marine oil pollution.[31] The same can be said for the

---

[27] Data from the ITOPF Ltd Oil Tanker Spill Statistics 2017, see www.itopf.com/knowledge-resources/data-statistics/statistics/.

[28] Loureiro and Alló (2017).

[29] Ibid., at 28.

[30] Ibid.

[31] See *inter alia* Wood (1975).

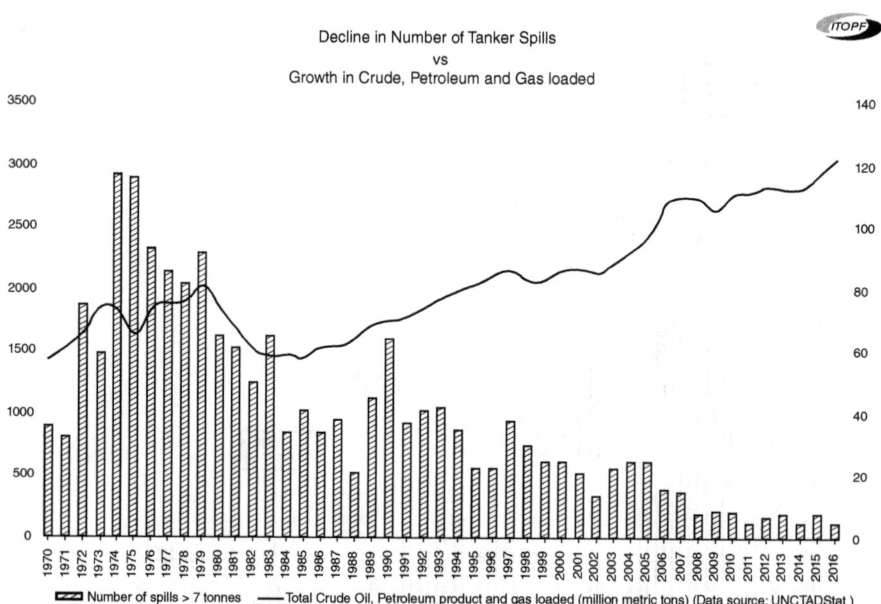

FIGURE 13.3: Decline in number of taker spills vs. growth in crude, petroleum and gas loaded[32]

fact that the strict liability in the CLC is accompanied by the liable tanker owner's obligation to seek financial coverage against any damages. Compulsory insurance can be considered as an important tool to prevent an operator (like a tanker owner) from externalizing the consequences of his actions to society, hence forcing the public to pay for the consequences of the owner's actions.[33] These features (strict liability backed up with financial security) of the international regime can hence be fully supported as providing incentives for prevention. Just as important as the prevention function, the civil liability regime should also guarantee adequate compensation. The statistics from the ITOPF also illustrate the observation that although the level of activity in the seaborne oil trade is increasing, which might imply higher risks, the number of oil spills reported shows a downward trend. This may be useful evidence of the preventive effect of the international conventions.

There are, however, several aspects of the international regime which can be considered as problematic from an economic perspective of providing incentives for prevention. The first one is the limitation of liability. Under the CLC, tanker owners are not exposed to the full weight of the burden they are potentially imposing on society in the event of an accident resulting in marine pollution. One result of this is

---

[32] Data from the ITOPF Ltd Oil Tanker Spill Statistics 2017, see www.itopf.com/knowledge-resources/data-statistics/statistics/.

[33] Faure (2016a).

that tanker owners tend to consider the severity of accidents only in terms of the maximum amount of compensation permitted by the CLC's limited liability, not in terms of the actual cost of the damages. As a result, a correspondingly lower level of prevention is chosen, thereby blunting the deterrent effect on the tanker owner.[34] Although in principle a limit on financial liability (as contained in the CLC) could be considered inefficient, in practice this should not necessarily mean that limits on financial liability will lead to a higher level of oil pollution incidents. This is because, first of all, for many (smaller) oil pollution incidents, the damage may well be lower than the limit on liability. The risk of underdeterrence, therefore, arises mainly in those (catastrophic) cases in which the actual amount of damage is higher than that permitted by the caps. Second, the prevention of oil pollution incidents is today primarily dependent upon detailed *ex ante* safety regulation, aiming at an optimal tanker design to prevent spill risks. Liability rules, therefore, have at most an additional deterrent effect to back up this regulation.

In addition to the limitation on liability, the fact that liability in the CLC is "channeled" to the tanker owner is also problematic. Article III(4) provides that no claim for compensation for pollution damage may be made against the owner unless it is in accordance with the convention, and that no claim may be made against any person other than the tanker owner. This seems inefficient since many parties other than the tanker owner may also influence the pollution risk.[35] The extent to which channeling of liability really is a serious problem, of course, depends on whether it is possible to set aside the channeling provisions in the CLC and file lawsuits against others besides the tanker owners (such as charterers and cargo owners) based on domestic law. The latter played a role in the *Erika* case. Some victims (like the French government, but also communities who had suffered environmental damage as a result of the incident) were dissatisfied with the low amounts of compensation permitted under the CLC.[36] As a result they tried to circumvent the channeling provisions in the CLC and sued Total (as owner of the cargo) *inter alia* based on French law.[37] The argument was accepted in the court of first instance[38] and the court of appeal,[39] but finally rejected in a decision of the Cour de Cassation of 25 September 2012.[40] This attempt by the victims in the *Erika* case to

---

[34] For a further analysis see Faure and Wang (2008).

[35] For a more detailed analysis see Faure and Wang (2006), at 208–211.

[36] And recall that at that time the Supplementary Fund of 2003 had not yet come into being; as a result, the lower limits under the 1992 Fund Convention applied.

[37] In fact a criminal case was brought against several parties, and according to French criminal procedure the victims could bring their civil suit in the criminal court as "civil parties." For a more detailed discussion of this procedure, see Wang (2011), at 361–366.

[38] Tribunal de Grande Instance de Paris, 16 January 2008, Judgment No. 9934895010.

[39] Cour d'Appel de Paris, 30 March 2010, Judgment No. 08/02278.

[40] Cour de Cassation, 25 September 2012, Judgment No. 10–82938. See www.legifrance.gouv.fr/affichJuriJudi.do?oldAction=rechJuriJudi&idTexte=JURITEXT000026430035&fastReqId=796592764&fastPos=1.

circumvent the channeling provisions obviously was not undertaken with the goal of promoting deterrence, but of circumventing the (low) limits of liability and channeling, and thus to obtain more compensation. However, these attempts are illustrative of the problematic nature of liability channeling, from both a compensation and a deterrence perspective.

A third problematic aspect of the international regime from a prevention perspective concerns the financing structure of the International Oil Pollution Compensation Fund. The financial duty to contribute to the Fund is determined solely on the basis of the amount of oil transported at sea through tankers. There is, however, no risk differentiation whatsoever. This means that the oil receivers shall pay a higher contribution to the Fund simply because they receive more oil through tanker transportation; they are not rewarded for choosing safer ships or punished (by, for example, being forced to pay a higher contribution) for choosing riskier ones. On the other hand, the Fund only intervenes when it comes to (mostly) large-scale oil spill incidents that cannot be fully compensated under the CLC.[41] In that respect, the lack of a risk differentiation in the financing of the Fund should not necessarily be a major problem.

Empirical evidence indicates that, since the start of the international regime, the number of oil spills in general has substantially decreased.[42] In particular, studies indicate that higher standards, increased monitoring and enforcement have led to a reduction in the number of oil spills. This can be attributed mainly to improved safety regulation (and increased enforcement), not necessarily to changes in the liability and compensation regime.[43] Specific empirical research has also been devoted to the compensation mechanisms that take effect after an oil spill. Hendrickx shows that cleanup operations have become more expensive, more particularly as a result of pressure from environmental groups and public opinion.[44]

As far as the compensation is concerned, we have shown that the international regime's need to further adapt after new spills showed the insufficiency of the then-existing amounts available for compensation. Today, with the entry into force of the Supplementary Fund Protocol, approximately USD 1 billion in compensation is available from the different sources. This amount is considered to be so large that it can provide compensation for nearly any case of marine pollution, no matter how catastrophic it may be.[45]

---

[41] See, for instance, that during the conference preceding the Fund it was opined that the Fund would only intervene in the 5 percent of large-scale oil casualties that could not be dealt with under existing rules. Official Records of the International Legal Conference on Marine Pollution Damage 1969, Document LEG/CONF/C.2/SR12, IMO, at 685–686.

[42] See the various studies by Cohen (2006), at 31–36 (his work is only focused on the US, not on the international regime).

[43] Hendrickx (2007), at 256–257.

[44] Ibid.

[45] See Jacobsson (2003), at 32.

13.2.4 *Smart Mixes in the International Regime: A Few Preliminary Lessons*

From the previous analysis it appears that the international oil pollution compensation regime is now able to provide compensation for nearly any oil pollution incident, no matter how catastrophic. Moreover, notwithstanding a few problematic design issues (such as the limitation on the liability of the tanker owner and the exclusive channeling) the international regime also has been able to have positive effects on prevention by reducing the number of oil spills. To a significant extent, this is due rather more to increased safety regulation aimed at accident prevention than to increased compensation. In any event, it is difficult to disentangle the preventive effect of regulation and liability rules. It is not unreasonable to conclude that the smart mix of both instruments has led to the reduction in the number of spills.

At the international level, there are numerous conventions dealing directly with different aspects of the safety issue. These include the International Convention for the Prevention of Pollution from Ships 1973/1978 (MARPOL), which deals with the prevention of pollution from ships due to accidents or routine operations through establishing technical standards, and the International Convention for the Safety of Life at Sea 1974 (SOLAS), which tackles the prevention aspect through setting construction, design, equipment and manning standards. These conventions constitute the major framework for the international regime dealing with marine pollution prevention. While the CLC/Fund regime does provide some impetus for the parties involved to take preventive measures, its function only complements that of the existing prevention regime. Therefore, the preventive effect of the civil liability regime should only be considered as its secondary function next to its compensation function.

In this respect, the current mix can to some extent be considered as 'smart' especially when compared with the liability and compensation regimes in other domains such as, for example, nuclear liability, where relative compensation amounts are in fact substantially lower.[46] This merits the question of how the international marine oil pollution regime managed to generate relatively high compensation amounts. In other words, what are indicators that can shed some light on the relative smartness of the mix in this particular area? One explanation lies in the nature of marine oil pollution, which inherently leads to opposing interests of states: some states with thriving shipping industries will undoubtedly strongly defend tanker owners and try to keep liability exposure limited. However, there are also many countries that, while not directly involved in maritime transport, are nonetheless potential victims of oil pollution. These are often states with long coastlines and high levels of maritime traffic. These opposing interests are illustrated quite clearly by the legal history of the CLC. When parties convened in November 1969 to discuss an international legal regime, some (such as the UK and the Netherlands)

---

[46] For a comparison between the international marine oil pollution and nuclear liability regimes, see Faure (2016b), at 172–173.

clearly defended shipping interests, while others (such as France and Spain) defended the interests of coastal states who could suffer damage as a result of marine oil pollution.[47] While the fact that the environment (via the coastal states) was given a voice at the 1969 Conference played a significant part in reaching a balanced agreement that represented not only the interests of shipowners, but also of coastal states. This factor alone does not fully explain the currently existing regime, as there will always be opposing interests.

A second feature that played a role in this particular case is that the interests involved were not only those of the shipping industry, but also those of the cargo (oil industry). This led to yet another countervailing argument whereby some shipping nations (like the Netherlands, Denmark, Greece and Norway) also pointed at the liability of the cargo interests, particularly the oil industry.[48] It was, in particular, this countervailing argument which made the acceptance of the CLC 1969 by the shipping industry possible: although the CLC 1969 had already been concluded, imposing (strict) liability on the tanker owners, it was agreed that a second convention (the Fund Convention 1971) would be concluded, with another regime financed by the oil interests.[49]

Another important feature explaining at least the latest change (particularly the creation of the Supplementary Protocol of 2003) was pressure from the regional (in this particular case, European) level. After the European Union had been confronted with serious cases of oil pollution in European waters, particularly the *Erika* (1999)[50] and the *Prestige* (2002) cases,[51] the European Commission proposed to set up a European fund with a ceiling of EUR 1 billion.[52] In its 9 February 2000 White Paper on Environmental Liability, the European Commission had already raised the issue of whether the international regime should be complemented by European measures.[53] This idea of a separate European regime was finally proposed as

---

[47] For an overview of those opposing interests during the 1969 Conference, see Wang (2011), at 62–69.

[48] See Wang (2011), at 74.

[49] As was explained in Section 13.2.1, the IOPC Fund is financed with charges levied on the oil industry.

[50] The Maltese tanker *Erika*, carrying some 31,000 tonnes of heavy fuel oil as cargo, broke in two in a severe storm on December 12, 1999, causing an oil spill of approximately 20,000 tonnes. See ERIKA, *West of France*, 1999, ITOPF, www.itopf.com/in-action/case-studies/case-study/ erika-west-of-france-1999.

[51] The *Prestige* was a tanker that sank off the coast of Galicia (Spain) in 2002, polluting thousands of kilometers of coastline in Spain, France and Portugal. For details of the spill, see www .iopcfunds.org/incidents/incident-map/#126-13-November-2002.

[52] See Wang (2011), 176–177. The legislative proposal from the European Commission concerning civil liability for oil pollution was first published on 6 December 2000 in document 2000/ 0326 (COD), proposal for a Regulation of the European Parliament and of the Council on the establishment of a fund for the compensation of oil pollution damage in European waters and related matters. Later it was amended through COM(2002) 313 final; see Official Journal of the European Union 2002/C 227 E/24, 24 September 2002, at 487–496.

[53] COM(2000) 66 final, White paper on environmental liability, presented by the European Commission, 9 February 2000, at 23.

the Fund for Compensation for Oil Pollution in European Waters, or the COPE Fund. It would complement the then-existing international regime to compensate victims of oil pollution incidents in European waters who otherwise could not obtain full compensation.[54] The European initiatives received a lot of attention at the international level. All the steps proposed by the European Commission have been closely followed by the Fund assemblies through the years.[55] The attitude of the Fund was to prefer a "uniform application of international conventions"[56] whereby a separate regional regime should be avoided. A special working group of experts was even delegated through the IOPC Fund to consider this matter.[57] These experts discussed this specific issue through several meetings between 2000 and 2001, ultimately concluding with the decision to recommend the establishment of a supplementary fund that could provide additional compensation on top of the existing CLC/Fund Convention regime. Considering the advice from the working group, in May 2003 the IMO decided to establish such a Supplementary Fund to provide a third tier of compensation.[58] Interestingly, this European initiative proposed a Fund of EUR 1 billion, and the Supplementary Fund was established at a maximum of SDR 750 million, which at the time of the adoption corresponded to approximately EUR 920 million or USD 1 billion.[59] The amount of the Supplementary Fund is quite close to the amount of the proposed COPE Fund. It might seem a coincidence that these two figures were set at a similar level; without the activism of the European Union, however, the Supplementary Fund would not have been so quickly adopted at the international level. It was therefore this regional initiative by the European Union that initiated, or at least sped up, decision-making at the international level to change the Fund convention and to eventually adopt the 2003 Supplementary Fund Protocol.[60] This constitutes an interesting example of how the fear of a regional solution can trigger a change at the international level.

[54] The proposed COPE Fund basically follows the principle of the IOPC Fund as far as compensation and contribution are concerned. However, it went further than the international fund since it added a provision which would impose financial penalties on any person involved in the transportation of oil by sea whose behavior could be shown to be grossly negligent.

[55] For example, the European White paper on environmental liability was specifically noted by the Fund in March 2000; see document 92FUND/A/ES.4/4, 10 March 2000. When the European Commission published the proposal for a regional fund 2000/0326 (COD) in December 2000, the Fund communicated on this matter in January 2001 (92 FUND/A/ES.5/ 2) and instructed the director of the Fund to take action when appropriate (92FUND/A.6/5, 24 September 2001).

[56] See document 92DUND/A.5/4, 18 August 2000, at 4–5.

[57] The proposal was published only in December 2000; however, the Fund Assembly appointed the group of experts in April 2000 as soon as they learned about the possibility of a separate European proposal. See IMO (2003), at 7–8.

[58] For the reaction at the international level to the European proposal, see document 92FUND/ A.6/5, IOPC Fund Assembly, 24 September 2001.

[59] Wang (2011), at 179.

[60] See ibid. at 207–209 (analyzing further the issue of EU activism and its influence at the European level).

13.3 INTERNATIONAL REGIME: MIXING PUBLIC AND PRIVATE

So far, we have discussed (in Section 13.2) the international conventions dealing with liability and compensation for marine oil pollution damage. In addition to those international conventions, private regimes were also created to accompany the entry into force of the international conventions. Although their importance may have been limited in practice, it is important to show how these private regimes have interacted with the international conventions. The 1969–1971 regime was accompanied by TOVALOP and CRISTAL (3.1), and the 1992 regime by STOPIA and TOPIA (3.2). The role of these private arrangements is limited in the sense that they either dissolve after serving for a particular number of years or function only within the private contracting parties. Nevertheless, these private schemes have proven to be useful since they have provided compensation to victims where there was a legal vacuum at the international level and they would establish a balance among potentially liable parties.

13.3.1 *TOVALOP and CRISTAL*

TOVALOP stands for the Tanker Owners Voluntary Agreement concerning Liability for Oil Pollution. It was a voluntary industry agreement to compensate that was signed by seven major oil companies and came into effect on 6 October 1969,[61] even before the entry into force of the CLC. One reason for the introduction of TOVALOP was the shipping industry's hope that, by offering payment of reasonable costs for preventive and cleanup measures, it could prevent the introduction of strict liability into the envisaged international liability regime.[62] TOVALOP entered into force before the CLC was created.

CRISTAL stands for the Contract Regarding an Interim Supplement to Tanker Liability for oil pollution. It came into effect in 1971.[63] Again, just as TOVALOP preceded the CLC, CRISTAL preceded the prospect of another convention involving a levy on cargo owners, which was of a large concern to oil companies. CRISTAL was a contract between cargo interests. It maintained a fund of money from which claims could be met. CRISTAL had amounts of between USD 3 and 5 million at its disposal. The prime function of CRISTAL was to supplement the pollution damage compensation available under TOVALOP. It was therefore designed as a residual scheme. The most important function of CRISTAL was to signal to the public that the oil industry could take responsibility for oil pollution damage.[64]

The principal function of TOVALOP and CRISTAL, therefore, was to provide a voluntary scheme of compensation prior to the entry into force of the CLC and the

---

[61] Wang (2007), at 200.
[62] Ibid.
[63] Wang (2007), at 202.
[64] See Gold (1985), at 26 and Doud (1973), at 541.

Fund Convention. These voluntary schemes wanted to signal that the industry was able to compensate the damage, thus preventing the need for an international convention. In that sense the voluntary schemes were a failure as they did not prevent the international conventions from being created. TOVALOP and CRISTAL adopted the principle of cost-sharing between the shipping and the oil industry, which was later copied in the CLC and Fund Convention.[65] Once the international regime established through the CLC and the Fund Convention was ratified by more states, the relevance of TOVALOP and CRISTAL was reduced. Therefore in 1995 a decision was made that these schemes had fulfilled their function as interim measures and could be abrogated on 20 February 1997.

To summarize, these schemes had an important function to provide compensation as long as the international conventions had not entered into force. This had some relevance as the entry into force of the international conventions took years (it took the CLC six years and the Fund Convention seven years to become effective) while the private agreements could become effective more quickly. It also seems that the structure of the private regime, being one of cost sharing between two involved industries, may have had some influence on the international regime, which adopted the same structure. By the time the international conventions became effective, the continuing existence of the voluntary schemes had come to be considered as a potential disincentive to joining the international regime for those states that had not yet been parties to the international conventions,[66] particularly those states that could potentially become victims (i.e. the coastal states with long coastlines and high levels of tanker traffic), if they could recover the cost of damages through private schemes, they would not be as willing to join the international conventions.

13.3.2 *STOPIA and TOPIA*

After the ending of the 1969–1971 regime, two new voluntary regimes came into place. In March 2005, at the Assembly of the 1992 Fund, the International Group of P&I Clubs (the International Group) offered to increase the limitation amount for small tankers under the 1992 CLC on a voluntary basis. This is known as STOPIA, the Small Tankers Oil Pollution Indemnification Agreement.[67] STOPIA is related to the Supplementary Fund Protocol of 2003. STOPIA entered into force on 3 March 2005, the same date on which the Supplementary Fund Protocol entered into force. The result is that, for pollution damage that occurs in a state that is a party to the Supplementary Fund Protocol, the limitation amount undertaken by the

---

[65] Wang (2007), at 203.
[66] So Jacobsson (2003), at 19–20.
[67] Wang (2007), at 210.

shipowner will be increased if the ship is insured by one of the P&I Clubs that are members of the International Group.

In October 2005, another proposal was made by the International Group to extend the scope of STOPIA to all states that are parties to the 1992 CLC and to establish a scheme called TOPIA, the Tanker Oil Pollution Indemnification Agreement. This implies that the P&I Clubs will indemnify 50 percent of the compensation paid by the Supplementary Fund.[68]

For victims, as far as the amount of compensation is concerned, these voluntary agreements are of little importance. STOPIA and TOPIA are voluntary contracts between tanker owners that are insured through the P&I Clubs. In fact, these private agreements do not affect the legal position of victims, but rather stipulate that shipowners (via the P&I Clubs) partially pay back the compensation paid by the Supplementary Fund. The result of STOPIA/TOPIA is therefore that the third tier of compensation (provided by the Supplementary Fund) is paid not only by the oil receivers (and thus by the oil industry), but also by the shipping industry (via their P&I Clubs).

To summarize, one can argue that these voluntary agreements have, especially compared to the international conventions, a relatively limited importance. TOVALOP and CRISTAL were only of importance during the period when the CLC 1969 and the Fund Convention 1971 had not yet entered into force. They are more interim than definitive measures. On the other hand, these voluntary agreements were useful in the sense that they could guarantee compensation to victims when the conventions had not yet entered into force. In fact, STOPIA and TOPIA guarantee only that the costs of the third tier of compensation (the Supplementary Fund) are borne not only by the oil industry, but (partially) also by the shipping industry (via their P&I Clubs). The idea of STOPIA and TOPIA was to try to achieve a balance in the overall contributions to the different tiers of compensation between the oil industry and the shipping industry, since the shipping industry would be liable only under the first tier (CLC) and the oil industry would contribute twice (under the Fund Convention and the Supplementary Fund Convention); therefore, to achieve a more equitable sharing of liability, the shipping industry decided to use these two voluntary to increase their level of contributions to the international regime. To the extent that the private agreements provided compensation during the period when the conventions had not yet entered into force, this could be considered as a smart mix of public (international conventions) and private schemes.

13.4 MIXING REGULATION AND LIABILITY: THE US OIL POLLUTION ACT

As mentioned in Section 13.2.2, the US did not join the international conventions. When the US was confronted with the *Exxon Valdez* incident in 1989, it decided to

[68] Ibid.

draft its own Oil Pollution Act in 1990. The most important reason for the United States' decision not to join the international conventions was their belief that the liability limits in the conventions were too low. Moreover, they rejected the principle of channeling liability to shipowners. Adherence to the conventions would also preempt state laws that included unlimited liability.[69] In response to the 1989 *Exxon Valdez* accident, the US Congress quickly passed the Oil Pollution Act of 1990. On the one hand, the US Oil Pollution Act shares a few similarities with the international regime such as strict liability, limited liability and a compulsory financial guarantee. On the other hand, OPA shows a few features that are different from the international regime: liability is not channeled; under OPA the liability limits are much higher and allow potentially responsible parties to lose their right to limited liability.In Section 13.4.1 we will first show how the financial limits depend upon the preventive efforts undertaken by tanker owners; in Sections 13.4.2 and 13.4.3, we will show that the financial caps can be set aside when regulation was breached and that, moreover, the Federal Oil Pollution Act does not set aside state laws which would provide higher liability limits or even unlimited liability.

13.4.1 *Relating the Limits to Safety*

The OPA establishes a cap on the total sum of removal costs and damages. The cap is the greater of a per-incident cap and a per-gross ton cap.[70] In 2006, the Coast Guard and Maritime Transportation Act (CGMTA) increased the cap.[71] The Coast Guard and Maritime Transportation Act (CGMTA) also established different caps for single-hull and double-hull tankers.[72] Moreover, in 2009 the Coast Guard made a further adjustment to the liability limits, which increased the limits for double-hull tankers from USD 1900 to USD 2000 per gross ton, and increased the limits for single-hull tankers from USD 3000 to USD 3200 per gross ton.[73]

The result can be summarized in the following table:

This system used by the OPA, particularly after the revision through the CGMTA in 2006, obviously provides better incentives for prevention than the (undifferentiated) international regime since the financial limits on liability are substantially lower when operators use double-hull tankers instead of the (obviously) much more dangerous single-hull tankers. This should provide incentives for phasing out single-hull tankers and replacing them with double-hull tankers.

---

[69] See Wang & Faure (2010), at 3.

[70] 33 U.S.C. § 2704 (a) (1990).

[71] Coast Guard and Maritime Transportation Act of 2006, Pub. L. No. 109–241, 120 Stat. 516 (2006), § 603.

[72] Ibid.

[73] See Consumer Price Index Adjustments of Oil Pollution Act of 1990 Limits of Liability – Vessels and Deepwater Ports, 74 Fed. Reg. 31357 (1 July 2009).

TABLE 13.2 *Comparison of Liability Limits under OPA 90 and the Coast Guard and Maritime Transportation Act of 2006*

Vessel	OPA 90 Liability Limits	2006 (2009) Liability Limits
Single-hull tanker > 3000GT	$1200/GT or $10,000,000	$3000(3200)/GT or $22,000,000 (23,496,000)
Single-hull tanker ≤ 3000 GT	$1200/GT or $2,000,000	$3000 (3200)/GT or $6,000,000 (6,408,000)
Double-hull tanker > 3000GT	$1200/GT or $10,000,000	$1900 (2100)/GT or $16,000,000 (17,088,000)
Double-hull tanker ≤ 3000 GT	$1200/GT or $2,000,000	$1900(2100)/GT or $4,000,000 (4,272,000)
Any vessel other than a tanker	$600/GT or $500,000	$950 (1000)/GT or $800,000 (854,400)

13.4.2 *Breaking the Limits*

In spite of these caps, a responsible party can lose its right to limitations on its liability if it can be proven that the immediate cause of the accident was "gross negligence or willful misconduct" or "the violation of an applicable Federal safety, construction, or operating regulation."[74] A responsible party may also face unlimited liability if it fails to report an incident, provide requested cooperation in connection with removal activities, or comply with an order of the president of the United States.[75] Hence, under the OPA the amount of the financial cap is related to the preventive measures taken by the operator (single-hull or double-hull) and the limits are easily breakable.

Furthermore, unlike the international regime established through the CLC, OPA does not channel liability to one particular party but provides for joint and several liability.[76] OPA identifies the responsible parties as being the shipowner, the operator, and the demise charterer.[77] All those features imply that OPA provides better incentives for prevention and thus presents a "smarter mix" of liability rules and safety regulation than the international regime.

13.4.3 *No Pre-emption*

Moreover, the OPA does not preempt state laws, which means that states can still impose additional liability or financial responsibility.[78] It is important to mention

---

[74] 33 U.S.C. § 2704(c)(1)(A)–(B) (1990).
[75] 33 U.S.C. § 2704(c)(2) (1990).
[76] See Wang (2011), at 209.
[77] 33 U.S.C. § 2701(32)(A) (1990).
[78] 33 U.S.C. § 2718 (1990).

that thirty of the fifty US states have a coastline and, in fact, all but six of these states have legislation on vessels liability.[79]

Some states such as Alaska, California, North Carolina and Rhode Island impose strict and unlimited civil liability for cleanup costs, natural resource damage and private losses caused by oil pollution, including pure economic losses. In some other states, unlimited liability is established only for certain categories of damage, such as in Washington (for cleanup costs and damages to persons or property), Maryland (for cleanup costs, damage to real and personal property and natural resources damages), Massachusetts (for natural resources damages) and Florida (for natural resource damages, damage to real and personal property, and losses consequential upon property damage).[80]

The result, therefore, is that the problems arising under the international regime as a result of the limitation of liability do not arise to the same extent in the US. Not only are the liability limits in the US CGMTA linked to the preventive measures taken by liable parties (double-hull rather than single-hull tankers), but in the case of a violation of federal safety, construction or operating regulations the liable party also can not call on the limits on liability. But perhaps more importantly, even when liable parties can call on a limitation of liability under federal law they can still be exposed to unlimited liability under state law, thus potentially providing higher compensation for victims and better incentives for pollution prevention.

13.5 CONCLUDING REMARKS: EXPLAINING SMARTNESS

In sum, the civil liability regime for marine oil pollution represents a mix of various elements. At the international level, it consists of a mix of public and private instruments since the international regime is embodied in the CLC/Fund Convention (public) on the one hand and the voluntary STOPIA and TOPIA (private) scheme on the other; the combination of various conventions at the international level not only guarantees compensation for the victims, but also provides incentives (in addition to those provided by safety regulation) for parties involved to take measures to prevent accidents. In this sense, such a mix may be considered smart. In the particular regime used by the US, the mix is even more interesting since the potential liability of responsible parties is associated with their compliance with regulations. Compared to the international regime, the structure of the mix in the US regime seems smarter since it not only provides better compensation to the victims, but also may generate more incentives for preventive measures.

We showed that the combination of different conventions in the international oil pollution regime is related to the involvement of different stakeholders – who sometimes have opposing interests – and different funders of the compensation.

---

[79] See Force, Davies and Force (2011), at 978–979.
[80] See ibid., at 979.

During the process of drafting these conventions, the interests of the countries representing the shipping industry were countered by those of the coastal states, thus providing more balance in the regime. The fact that more and different parties were available to provide compensation also led to a mixed system in which both the shipping industry (via the limited liability of the tanker owner under the CLC) and the oil industry (via fees paid on oil received to the IOPC Fund) contribute. The involvement of coastal states, but also the availability of different stakeholders with sometimes opposing interests, may therefore have led to relatively high amounts of compensation. At least as far as its compensation function is concerned, the international regime's mix of different instruments and conventions seems to be effective, as it is now able to provide compensation for most types of marine pollution, including catastrophic events. This finding complies to a significant extent with Becker's theory, which holds that when different interest groups compete for political influence, the outcome will not necessarily benefit only one party.[81] This would obviously not be the case if only one party (for example, representing industry interests) could lobby the government without any countervailing power.

In addition, a few other interesting conclusions can be drawn from this international regime for the compensation of oil pollution damage. The international regime shows a large amount of flexibility by dynamically adapting itself to the changing circumstances (increased demand for oil leads to more oil being transported by more and larger tanker fleets, which leads to new pollution incidents with increasingly higher amounts of damage). Statistics also show that although the seaborne oil trade has been increasing since the 1970s, the general data on oil spills seems to indicate a downward trend.[82] The fact that there is no specific international agency dealing with offshore-related damage also explains the absence of any international convention regulating liability and compensation in the case of a *Deepwater Horizon*-type incident. That once more underscores the importance of an institutional framework.

The international regime also presented various interesting interactions in a multilevel governance setting. Remember that the European Commission proposal to set up a European fund with a ceiling of EUR 1 billion led directly to the creation of the Supplementary Fund Protocol by the IMO, which provided approximately EUR 920 million or USD 1 billion of additional funding. It was this regional initiative by the European Union that initiated, or at least sped up, decision making at the international level.[83] It constitutes an interesting example of how the fear of a regional solution can trigger a change at the international level.

But interesting lessons can also be drawn from an interaction between the international and the domestic level, more particularly with the US Oil Pollution Act of 1990. Often the US is blamed for participating in the negotiation of

---

[81] Becker (1983).

[82] Statistics from ITOPF (2018), at 10.

[83] Wang (2007), at 207–209.

international treaties, but in the end not signing the international treaties. However, in this particular case of marine pollution, domestic law provides better protection than the international regime. Indeed, although we argue that the international marine oil pollution regime provides features of a smart mix, we also indicated that there are still particular features (such as the limits on liability of tanker owners and the exclusive channeling of liability) that can be criticized. For example, the way in which US law ties limits of liability to preventive efforts, and the nonexclusivity of federal law, may provide interesting lessons to improve the international regime. The same could be said about some domestic actions within EU Member States (such as those in France after the *Erika* accident) to attempt to set aside some negative effects of the international regime (like the channeling of liability) by directly suing charterers based on domestic law.

Those examples of interactions between the regional and domestic level on the one hand and the international level on the other are very interesting. The fact that other models have been and are still developed at the regional and domestic level provides scope for mutual learning and for policy improvement. In this case, the international level was clearly inspired by the European initiative to create a compensation fund for a higher amount. The nonexclusivity of the international regime for oil pollution may therefore provide scope for smart mixes with regional and domestic mechanisms, thus increasing the effectiveness of the overall regime (in terms of providing compensation and preventing environmental harm). Therefore, if there is one lesson to be learned from this international marine oil pollution regime for the debate on smart mixes, it is probably that a nonexclusivity of international conventions may provide scope for a smart multilevel governance system that may be more effective and provide better prevention and compensation for environmental harm than it currently does.

REFERENCES

Becker, G. S. 1983. 'A Theory of Competition among Pressure Groups for Political Influence'. *The Quarterly Journal of Economics* 98, 371–400.

Cohen, M. A. 2006. 'Oil Pollution Prevention and Enforcement Measures and their Effectiveness: A Survey of the Empirical Research from the US'. In Faure, M. G. and Hu, J. (eds.), *Prevention and Compensation of Marine Pollution Damage: Recent Developments in Europe, China and the US*. Alphen aan den Rijn, Kluwer Law International, pp. 25–39.

Doud, A. L. 1973. 'Compensation for Oil Pollution Damage: Further Comment of the Civil Liability and Compensation Fund Conventions'. *Journal of Maritime Law and Commerce* 4(4), 525–542.

Faure, M. G. 2016a. 'Compulsory Liability Insurance: Economic Perspectives'. In Fenyves, A., Kissling, C., Perner, S. and Rubin, D. (eds.), *Compulsory Liability Insurance from an Economic Perspective*. Vienna, De Gruyter, pp. 319–341.

2016b. 'In the Aftermath of the Disaster: Liability and Compensation Mechanisms as Tools to Reduce Disaster Risks'. *Stanford Journal of International Law* 52(1), 95–178.

Faure, M. G. and Wang, H. 2006. 'Economic Analysis of Compensation for Oil Pollution Damage'. *Journal of Maritime Law and Commerce* 37(2), 179–217.

   2008. 'Financial Caps for Oil Pollution Damage: A Historical Mistake?'. *Marine Policy* 32 (4), 592–606.

Faure, M., Liu, J. and Philipsen, N. 2015. 'Liability for Terrorism-Related Risks under International Law'. In Bergkamp, L., Faure, M. G., Hinteregger, M. and Philipsen, N. (eds.), *Civil Liability in Europe for Terrorism-Related Risk*. Cambridge, Cambridge University Press, pp. 11–55.

Faure, M., Liu, J. and Wang, H. 2015. 'A Multilayered Approach to Cover Damage Caused by Offshore Facilities'. *Virginia Environmental Law Journal* 33, 356–422.

Force, R., Davies, M. & Force, J. S. 2011. 'Deepwater Horizon: Removal Costs, Civil Damages, Crimes, Civil Penalties, and State Remedies in Oil Spill Cases'. *Tulane Law Review* 85(4), 889–982.

Gold, E. 1985. *Handbook on Marine Pollution*. Norway, Assuranceforeningen Gard.

Han, L. and Wang, D. 2010. 'Discussion on Limitation of Liability and Compulsory Insurance of Compensation for Bunker Oil Pollution Damage from Ships in China'. In Faure, M. G., Han, L. and Shan, H. (eds.), *Maritime Pollution Liability and Policy. China, Europe and the US*. Alphen aan den Rijn, Wolters Kluwer, pp. 145–158.

Hendrickx, R. 2007. 'Maritime Oil Pollution: An Empirical Analysis'. In Faure, M. & Verheij, A. (eds.), *Shifts in Compensation for Environmental Damage*. Vienna, Springer, pp. 256–257.

IMO. 2003. 'Better deal for oil pollution victims as IMO adopts third tier of compensation'. *IMO News* (3), 7–8. Available at: www.imo.org/en/MediaCentre/MaritimeNewsMaga zine/Documents/2003/IMONews_Issue2.pdf.

ITOPF – International Oil Tanker Owners Pollution Federation Limited. January 2018. *Oil Tanker Spill Statistics 2017*, pp. 1–15. Available at: www.itopf.org/fileadmin/data/Photos/ Statistics/Oil_Spill_Stats_2017_web.pdf.

Jacobsson, S. 2003. 'The International Compensation Regime 25 Years on'. In *IOPC Funds' 25 Years of Compensating Victims of Oil Pollution Incidents*. London, IOPC Funds, pp. 22–32.

Loureiro, M. L. and Alló, M. May 2017. 'The Economics of Oil Spills'. In *Oxford Research Encyclopedia of Environmental Science* (online publication). pp. 1–32. Available at: environmentalscience.oxfordre.com/view/10.1093/acrefore/9780199389414.001.0001/acre fore-9780199389414-e-501.

Shan, H. 2010. 'The Era after the Bunker Convention: Is the Gap in China's Regime for Compensating Victims of Vessel-source Oil Spills Filled?'. In Faure, M. G., Han, L. and Shan, H. (eds.), *Maritime Pollution Liability and Policy. China, Europe and the US*. Alphen aan den Rijn, Wolters Kluwer, pp. 123–143.

Shavell, S. 1987. *Economic Analysis of Accident Law*. Cambridge, Harvard University Press.

Wang, H. 2007. 'Shifts in Governance in the International Regime of Marine Oil Pollution Compensation: A Legal History Perspective'. In Faure, M. and Verheij, A. (eds.), *Shifts in Compensation for Environmental Damage*. Vienna, Springer, pp. 197–241.

   2011. *Civil Liability for Marine Oil Pollution Damage. A Comparative and Economic Study of the International, US and Chinese Compensation Regime*. Alphen aan den Rijn, Wolters Kluwer.

Wang, H. and Faure, M. 2010. 'Civil Liability and Compensation for Marine Pollution – Methods to be Learned for Offshore Oil Spills'. *Oil, Gas & Energy Law Intelligence* 8(3), 1–27.

Wood, L. D. 1975. 'An Integrated International and Domestic Approach to Civil Liability for Vessel-source Oil Pollution'. *Journal of Maritime Law and Commerce* 7, 1–68.

14

Regulatory Mixes in Governance Arrangements in (Offshore) Oil Production

Are They Smart?

Jan P.M. van Tatenhove

14.1 INTRODUCTION

Offshore and onshore oil production is a global activity that is employed in different jurisdictions and institutional settings. Oil and gas exploration and exploitation is primarily conducted within States' territorial seas and Exclusive Economic Zones (EEZs), but it is expected that in the future, as pressures on oil and gas supplies increase, companies will look to exploit reserves found on the extended parts of their continental shelves.[1] Four countries (Canada, the USA, Norway and New Zealand) have already commenced exploration activities on their extended continental shelves, while sixteen others have issued and/or are offering offshore oil and gas exploration concession licences on their extended shelves.[2] This chapter will study different governance arrangements in offshore oil and gas exploitation as a result of the interplay of states, oil companies (market) and (indigenous) communities (civil society). Each of these governance arrangements presents different mixes of regulation, as a result of different roles of states, and different levels of governance (multi-level). The aim of this chapter is to understand the smartness of the regulatory mixes, by analysing three examples of governance arrangements in oil production. Smartness is defined as the behavioural and problem-solving effectiveness (see Chapter 1 of this volume). Behavioural effectiveness refers to the ability of a mix to induce states and private actors to modify their behaviour (outcome). Problem-solving effectiveness refers to the contribution of a mix of instruments to the reversal or alleviation of environmental problems (impact).

Offshore oil production takes place in different governance arrangements, ranging from state-based and market-based to hybrid multi-level governance arrangements In this chapter the characteristics and dynamics of these different (onshore

---

[1] Mossop (2016), at 36–37.
[2] Ibid., at 38.

309

and offshore) oil governance arrangements will be described and the effectiveness of the regulatory mixes within these governance arrangements will be analysed. Section 14.3.1 analyses the development of market-based arrangements to regulate the environmental consequences of Dutch offshore oil platforms as a result of changes in Dutch national environmental policy, Section 14.3.2 the development of Social Licence to Operate arrangements in the Arctic, and Section 14.3.3 the development of benefit sharing arrangements between individual oil companies and indigenous communities in sub-Arctic Russia and Alaska. These cases give insight in the diversity and co-existence of governance arrangements governing the offshore oil production in different institutional settings. More specifically, the analysis will focus firstly on characteristics of these governance arrangements (the levels of governance, and the interplay of state, market and civil society actors), and secondly on how effective the regulatory mixes (instruments) within these specific governance arrangements are in solving social and environmental problems related to oil production.

14.2 DIFFERENT GOVERNANCE ARRANGEMENTS IN (OFFSHORE) OIL PRODUCTION

14.2.1 *Changing Role of States*

14.2.1.1 Processes of Political Modernization

The role of states and the relations between states, markets and civil societies have changed due to processes of political modernisation. Political modernisation refers to 'the shifting relationships between the state, the market and civil society in political domains of societies – within countries and beyond – as a manifestation of the "second stage of modernity" implying new conceptions and structures of governance'.[3] Structural processes of transformation such as globalisation affect the political and economic domains of societies, because the interconnectedness of all aspects of contemporary social life and economic activities are widening, deepening and speeding up.[4] In this globalised world, politics has become a mixture of conventional (inter-) state politics and politics as the result of (un-institutionalised) interactions between a plethora of actors across the international–national–subnational divide.[5] In general, the increased power of non-state actors and the increasing forces of cross-border social, political and economic processes have affected the position of national states. Contemporary world politics shows shifting relationships between territory, authority and sovereignty. Important contributions in this debate are

---

[3] Arts & Van Tatenhove (2006), at 29.
[4] Held et al. (2000).
[5] Rosenau (1997, 2006); Beck (2005).

presented by Castells;[6] Held et al.[7] and Sassen.[8] According to Held et al., 'a new "sovereignty regime" is displacing traditional conceptions of statehood as an absolute, indivisible, territorially exclusive, and zero-sum form of public power'.[9] According to Sassen, globalisation has destabilized the marriage of territory, authority and rights within the nation-state and is characterised by the emergence of different assemblages of territory, authority, and rights.[10] In this emerging complex setting of transnational networks,[11] assemblages[12] and network states[13] individual states become increasingly dependent on market parties, international organisations, Non-Governmental Organisations (NGOs), etc., to deliver policies. In this growing web of multiple actors, institutions and policies, authority is dispersed over various governance arrangements.

14.2.1.2 Webs of (Nested) Governance Arrangements

Governance is about the rules of collective decision making in settings where there are a plurality of actors or organisations and where no formal control system can dictate the terms of the relationship between these actors and organisations.[14] A governance arrangement is a temporary stabilisation of the substance and organisation of a policy domain. In a governance arrangement, different (and more or less stable) coalitions of public and private actors try to influence the activities and developments, and to design legitimate initiatives, based on shared discourses, the managing of resources and defining rules of the game (on different levels).[15] These governance arrangements institutionalise as a result of the interplay between processes of political modernisation and interactions between interdependent actors in policy practices. In these governance arrangements, authority is defined as having the right to govern and the ability to generate compliance of those actors who are governed. More specifically, authority is the ability to influence and to adopt steering mechanisms and to ensure that these steering mechanisms are implemented and enforced. Steering mechanisms are the instrument to change the behaviour of actors or to change societal or organisational processes. 'Steering mechanisms can be placed on a continuum ranging from having a voluntary to

---

[6] Castells (1996; 2009).

[7] Held et al. (2000).

[8] Sassen (2006).

[9] Held et al. (2000).

[10] Sassen (2006).

[11] Keohane (1995).

[12] Sassen (2006).

[13] A network state is 'characterised by shared sovereignty and responsibility between different states and levels of government and a flexibility of governance procedures …'. See Castells (2009), at 40.

[14] Chhotray & Stoker (2009).

[15] Van Tatenhove (2013, 2016).

compulsory nature'.[16] Steering mechanisms are the instruments to change the behaviour of actors. Each governance arrangement consists of a mix of instruments, ranging from a norm (like the precautionary principle) to legal acts. In between these formal laws/regulations and informal and soft mechanisms of steering lies an array of instruments, such as work programmes, voluntary agreements, standards and procedures. Governance arrangements co-exist and are nested within a specific institutional setting which is formed by processes of political modernisation.[17]

In this chapter, three types of governance arrangements for oil exploration and exploitation will be presented and analysed. The analysis will focus on the different roles of states, market parties and civil society actors in oil production. To enable us to understand the interplay of state, market and civil society the concepts of regime complexes, networked polity and network state will be introduced. Webs of governance arrangements develop and institutionalise within a networked polity. A networked polity or institutional setting 'is a structure of governance in which both state and societal organization is vertically and horizontally disaggregated, but linked together by cooperative exchange. The logic of governance emphasizes the bringing together of unique configurations of actors around specific projects oriented toward institutional solutions rather than dedicated programs'.[18] Governance arrangements institutionalise within this institutional setting, bringing together state and governmental actors, market parties and civil society actors in different regime complexes (for example, around oil and gas exploration and exploitation) in order to govern these activities and to find solutions for the environmental and social problems they cause. A regime complex is 'an array of partially overlapping and non-hierarchical institutions governing a particular issue area'.[19] Such a complex would fall somewhere in the middle of a continuum running from fully integrated institutional arrangements at one extreme to highly fragmented collections of arrangements at the other.[20] Examples of regime complexes in the maritime domain are shipping, fishing and oil and gas regimes. Each maritime policy domain has its own institutional dynamic, reflecting the different levels at which sectoral maritime activities are regulated. These regime complexes, defined as systems of rules and expectations, have a strong tendency to overlap and produce 'institutional orders',[21] which are co-produced over time by public and non-public actors. Finally, to help us understand the changing role of the state in the era of globalisation I make use of Castells' concept of the emerging network state. According to Castells, the sovereign nation-state is transforming to a network state, which is characterised 'by shared sovereignty and responsibility between different states and levels of government and

[16] Van Leeuwen (2010), at 37.
[17] Van Tatenhove et al. (2000); Arts & Van Tatenhove (2006).
[18] Ansell (2000), at 311.
[19] Raustiala & Victor (2004), at 279.
[20] Keohane & Victor (2011).
[21] Carter & Smith (2008).

a flexibility of governance procedures; and greater diversity of times and spaces in the relationship between governments and citizens compared to the preceding nation-states'.[22]

In sum, different governance arrangements co-exist in a web of governance arrangements. Each governance arrangement is characterised by a specific interplay of state, market and civil society, in this chapter analysed as the interplay between nation-state/network state, regime complexes and indigenous communities and NGO. The interactions between actors and coalitions take place within a trans-national and international networked polity. The main argument of this chapter is that the constellation of governance arrangements legitimises specific forms of authority and favours certain steering mechanisms and instruments over other mechanisms and instruments (regulatory mixes).

14.3 WEBS OF OIL PRODUCTION GOVERNANCE ARRANGEMENTS: THREE EXAMPLES

In this section three oil production governance arrangements will be presented and analysed: the development of market-based arrangements in nation-state setting, Social Licence to Operate arrangements in settings of diffused authority and the development of benefit sharing arrangements by multi-national companies, local indigenous communities and nation states. Each of these arrangements shows different coalitions of state, market and civil society actors, different forms of authority and different institutional settings. The first case is the environmental covenant between the Dutch government and the oil and gas industry. The covenant is an example of a market-based instrument within a nation state setting. The Social Licence to Operate (SLO) case is an example of a governance arrangement in a fragmented (inter)national institutional setting in which there are no clear accountability and authority relations between state, market and civil society organisations and in which the instrument of a SLO aims at enhancing trust and legitimacy for the activities of oil and gas companies in fragmented settings. The example of benefit sharing arrangements emphasizes the direct relation between companies and indigenous communities about the revenues of, for example, oil and gas exploitation. Benefit sharing arrangements are implemented within a national setting, but their form is influenced by transnational and international regime complexes and networks. These three cases are selected because they present different forms of oil and gas governance arrangements, and more specific they give insight in the development of specific mixes of steering mechanism and instruments in different institutional settings.

---

[22] Castells (2009), at 40.

14.3.1 *Dutch Offshore Oil-platforms: The Institutionalization of Market-based and Voluntary Instruments*[23]

The production of oil and gas on the Dutch part of the North Sea started in the early 1970s. Oil and gas production was regulated by Dutch government. The ministry of Economic Affairs was the national authority regulating oil and gas production by Dutch mining laws. The ministry was responsible for implementing the environmental standards for the offshore industry from the Paris/OSPAR (before 1992 Paris Convention) decisions and regulations. The State Supervision of Mines (inspectorate of the Ministry) controlled compliance with the mining laws. In the 1970s and 1980s the Ministry and representatives of the oil and gas industry, organised within the Netherlands Oil and Gas Exploration and production Association (NOGEPA) met in informal working groups, but oil companies and NOGEPA did not have a formal role in policy- and decision-making. This changed during the 1980s and 1990s. During these years, the role of the state in Western welfare states was debated under influence of the upcoming neo-liberal/conservative discourse expressed by Thatcher and Reagan. In Dutch environmental policy also the drawbacks of its fragmented institutionalisation became clear, such as the problem of cross-media transfer (as a consequence of developing separate legislation and regulations for each component (water, air, soil), inadequate coordination of licensing procedures and weak implementation and enforcement structures.[24] One of the strategies to overcome the drawbacks of the fragmented institutionalization of Dutch environmental policy was the development of a target group policy. Target groups were defined as more or less homogeneous groups of polluters. Dutch target group policy evolved from 'target groups as objects of management' to 'target groups as interlocutor and negotiating partner'. The steering mechanism/instruments of target group policy transferred authority from the state (Ministry of Economic Affairs) to the companies and industries, who should take responsibility for finding solutions for environmental problems, within the margins of environmental regulation. A direct outcome of this target group policy was the development of environmental covenants with companies and industrial sectors. In 1995, an environmental covenant was signed between the Ministries of Economic Affairs, Environment and Transport on the one hand, and NOGEPA as well as individual gas and oil companies active in the Netherlands on the other.[25] The covenant consisted of an Integrated Environmental Target Plan, which laid down reduction objectives (for climate change, acidification, heavy metals, benzenes, added chemicals, waste and environmental care) for the oil and gas industry. Besides the formal laws and regulations the

---

[23] The analysis in this section is based on Van Leeuwen (2010) and Van Leeuwen & Van Tatenhove (2010).

[24] Van Tatenhove & Goverde (2002).

[25] Van Leeuwen & Van Tatenhove (2010).

covenant adds its own procedures to reach the objectives of the Target Plan. The covenant also established different working groups and committees, which focused on specific issues (monitoring; waste streams), replacing the existing informal working groups. Two types of committees were installed, consisting of representatives of the ministries; NOGEPA and individual companies. The level I committee is the decision-making body of the environmental covenant, while the level II committee prepares the decisions. Part of the environmental governance process is that each company operating on the Dutch continental shelf has to develop a 4-year Company Environmental Plan that is related to company's environmental care system. In the level I and II committees these plans are discussed and have to be approved by the Ministry of Economic Affairs. After approval of all the individual company plans NOGEPA develops an aggregated Industry Environmental Plan, which evaluates the total results of all the individual plans against the agreed objectives of the environmental covenant. The evaluation of the covenant in 2001 resulted in a strengthening of this instrument in the environmental governance of Dutch offshore platforms, an agreement that developments within for example OSPAR will be discussed under the covenant, and that observers (industry representatives) were allowed at OSPAR meetings and at Working group meetings, which prepare OSPAR decisions and recommendations. In other words, the environmental covenant ensures a structured process through which the industry can give input into the OSPAR process via national governmental actors.

This case is an example of the changes within Dutch environmental policy, the changing responsibilities of public and private actors and changed regulatory mixes (instruments), which are the result of this development. The focus in Dutch environmental policy changed from sectoral legislation and governmental imposed standards to market-based and voluntary agreements, such as the environmental covenant; a standard and rules jointly agreed upon by government and the industry. The covenant does not replace environmental law and regulation, but forms with regulatory forms of governance and other market-based instruments a mix of instruments which is characteristic for the specific institutionalization of Dutch environmental policy in the 1980s and 1990s. Its effectiveness depends on the one hand on the responsibility of companies to develop environmental standards to find solutions for the environmental problems they cause, and on the capacity of governments to monitor the outcome and the impact of the efforts of the company to meet the objectives as laid down in the covenant on the other hand.

14.3.2 *Social Licence to Operate Arrangements in the Arctic*

For a century the oil and gas industry has shown high interest in the exploration and extraction of oil and gas in the Arctic. The first onshore well was drilled in Norman Wells (Canada) in 1920, while the first offshore well was drilled in the Cook Inlet in

Alaska (USA) in 1963.[26] The second half of the twentieth century has witnessed a rapid growth in activities.[27] In many regions within the circumpolar North, exploitation of oil and gas is already a major economic driver.[28] After World War II large-scale onshore exploration activities took off in Alaska (USA), Canada and Russia, followed by offshore operation in Alaska, Norway and Russia.[29] In 2013, the Arctic produced about 10 percent of the world's oil and 25 percent of its gas, of which the majority is produced in the Russian Arctic onshore. But it is expected that the region holds substantial additional resources. The United States Geological Survey estimated in 2008[30] that the Arctic contains 13 percent of the world's 'yet to find' oil, 30 percent of the world's 'yet to find' gas, and 20 percent of world's 'yet to find' natural gas liquids, totalling around 400 billion barrels of oil equivalents. About 84 percent of these estimated resources are believed to lie offshore.[31]

Environmental risks of oil pollution, the costs of exploitation in the Arctic and changed regulation will influence the development of new major projects. In general, 'hydrocarbon development in the Arctic marine area could result in potentially devastating damage to the environment, particularly were a serious accident to occur at any stage of extraction activities, although normal operational activities affect the environment as well'.[32] The main concerns related to oil and gas activities are the increasing risks of oil spills from drilling impacting the fragile Arctic marine ecosystems. The unique environmental conditions in the Arctic include extended periods of darkness, reduced visibility, ice-covered ocean areas, severe cold, high winds, and extreme storms.[33] The Arctic marine environment also has a unique seasonal shoreline and oceanographic changes. The Arctic shore consists of ice shelves, glacier margins, ice foot features, and tundra coast; the unique seasonal oceanographic and shoreline changes are due to open water, freeze-up, frozen conditions, and break-up.[34]

The regulation of oil and gas activities takes place in the nested Arctic governance setting, consisting of formal and informal institutions, such as the littoral states Russia, Norway, Denmark (Greenland), Canada and the USA (Alaska), also known as the Arctic 5, the Northern Dimension (intergovernmental platform of cooperation between the EU, Russia, Norway and Iceland), the Nordic Council and the

---

[26] Blaauw (2013).

[27] Large Arctic onshore operations started in the 1970s in Alaska Prudhoe Bay and in the Urengoy area in Russia. Offshore production from the (sub-)Arctic region started much more recently in projects such as Hibernia off Newfoundland, Canada, Sakhalin 1 and 2 in the far east of Russia and Snovit in the Barents Sea in Norway (Blaauw (2013), at 176).

[28] Hossain, Koivurova & Zojer (2014), at 161.

[29] Smits, Van Tatenhove & Van Leeuwen (2014), at 334.

[30] USGS (2008).

[31] Blaauw (2013), at 176.

[32] Hossain, Koivurova & Zojer (2014), at 164.

[33] Casper (2009).

[34] Hossain, Koivurova & Zojer (2014), at 165.

Arctic Council. This networked polity of the Arctic does not replace or suppress nation states, but states are integrated in for example the oil and gas regime complex and within the merging network state of which forms the Arctic Council forms the central regional intergovernmental form.[35] The Arctic Council (established in 1996) is a high-level intergovernmental forum to provide a means for promoting cooperation, coordination, and interaction among the Arctic States (Canada, Denmark (including Greenland and the Faroe Islands), Finland, Iceland, Norway, the Russian Federation, Sweden, and the USA), with the involvement of the Arctic indigenous communities and other Arctic inhabitants (six international organisations representing Arctic indigenous peoples are permanent participants) on common Arctic issues, in particular issues of sustainable development and environmental protection in the Arctic.[36] The observer status is open to non-Arctic States, intergovernmental and inter-parliamentary organisations with a global and/or regional constituency, and NGOs that the Council determines as potential contributors to its work.[37] In 2009 the International Association of Oil and Gas Producers (OGP) applied for an observer status.

Core of the web of nested governance arrangements in the Arctic, is the Arctic Council. Related to oil and gas development the Arctic Council has formulated several non-legally binding standards (Arctic Offshore Oil and Gas Guidelines); the Arctic countries have to decide which parts they implement. The non-binding nature of Arctic decisions[38] limits creating a level playing field for oil and gas activities across the Arctic. Within this Arctic fragmented 'networked polity' companies employ their activities, and will define the power games between public and private actors over the right to explore and to exploit Arctic oil and gas reserves. In this fragmented institutional setting the regulation of the activities of gas & oil companies is only partly regulated and approved by legal and political licences, which are based on formal judicial and political procedures.[39] This gives rise to a new instrument: the 'Social Licence to Operate' (SLO). A SLO is 'the ongoing acceptance and approval from local communities and other stakeholders'.[40] Different from the political and legal licences to operate, the social licence to operate is not granted by a public authority, but is rather a process that seeks a

---

[35] Van Tatenhove (2016), at 173.

[36] Koivurova (2013); Smits, Van Tatenhove & Van Leeuwen (2014); Van Tatenhove (2016).

[37] Ottawa Declaration, 19 September 1996.

[38] Some changes can be observed. The Arctic Council fulfils a decision-shaping function for legally binding decisions as well. Examples are the Search and Rescue (SAR) agreement of 2011 and the agreement on Cooperation on Marine Oil Pollution, Preparedness, and Response in the Arctic of 2013.

[39] The political license, granted by politicians, is 'the authority that the government gives to any other organisation to undertake a particular activity' (Morrison (2014)). The legal license applies to an activity outlined by the regulatory framework, and is granted for example as a permit by an assigned governmental authority, based on well-defined legal procedures.

[40] Parsons, Lacey & Moffat (2014); Prno (2013); Thomson & Boutilier (2011); Joyce & Thomson (2000); Smits et al. (2017).

'licence', i.e. trust and the legitimacy from for example indigenous communities, NGOs and other stakeholders.[41] Also the research of Thomson and Boutilier[42] shows that trust and legitimacy play an important role in obtaining and maintaining a social licence to operate. According to them economic legitimacy, interactional trust, institutionalised trust and socio-political legitimacy[43] influence whether a SLO is approved or withdrawn.[44] The existing literature about obtaining and maintaining a social licence to operate is mainly output oriented and lacks an in-depth understanding of the relationship with the legal and political licences.[45] According to Curran[46] the meaning of social licence remains exclusive. Her analysis of the Coal Seam Gas (CSG) case in the Northern rivers area of northern New South Wales (Australia) shows convincingly how companies, local communities and governments differently utilise the SLO concept to be either an opponent, a defender or a mediator. In this case for the company Metgasco, 'the winning of a social licence to operate relied directly on having acquitted its obligations on community consultations as required by its legal licence conditions'.[47] The community, however, embedded the SLO in a broad democracy frame. In opposing the activity, communities 'increasingly invoke social licence's democratic credentials, and support their claims with respected democratic norms . . .'.[48] Because of this politicisation of the SLO, governments are placed in the role of 'mediator' between their responsibilities to their constituents and those to their corporate clients. The role of government and its relations with companies and communities is often overlooked. While the companies define a SLO in terms of its legal licences, society and communities increasingly demand influence on the development of industrial activities by emphasising broader democratic norms and procedures.

Under what conditions could a SLO contribute to effectively finding solutions for environmental problems caused by industrial activities? In other words, how can the smartness of the mix of licences be developed and improved? A crucial condition is that a SLO on the one hand, and legal and political licences on the other, are connected to each other; the result is a mix of instruments, such as consultations, formal regulations, political decisions, CSR and so on, each with its own strengths and weaknesses. A second condition could be the empowerment of local communities

---

[41] Hall et al. (2015); Prno (2013).

[42] Thomson & Boutilier (2011).

[43] According to Thomson & Boutilier (2011), interactional trust is 'the perception that the company listens, responds, keeps promises, engages in mutual dialogue and exhibits reciprocity in its interactions'. More difficult to reach is institutionalised trust, since this requires stakeholders to perceive that the relationships between their institutions and the company are based on mutual trust with respect for each other's interests.

[44] Thomson & Boutilier (2011).

[45] Smits, Van Leeuwen & Van Tatenhove (2017).

[46] Curran (2017).

[47] Ibid., at 432.

[48] Ibid., at 433.

and NGOs, for example through the obligatory disclosure of information, reporting to the community and so on. Empowering the community presupposes a more active mediating role of governments and an active monitoring and enforcement strategy towards oil companies based on the joined objectives of the three licences.

14.3.3 *Benefit Sharing Arrangements: The Russian Arctic and Alaska*

The concept of benefit sharing originated from the Convention on Biological Diversity (CBD) in 1992 and was further developed in the 2011 Nagoya Protocol, a supplementary agreement to the CBD related to the use of biological resources.[49] In particular, benefit sharing arrangements have been developed between oil and mining companies, indigenous communities and (local and national) governments.[50] The aim of benefit sharing is to compensate indigenous people and communities for damage to their surrounding environment, livelihood and natural resources that occurs as a result of oil, gas and mining exploitation activities. Although oil companies often declare their commitment to benefit sharing on the transnational or national levels, the way they negotiate and implement these commitments in communities varies significantly.[51] The institutionalisation of benefit sharing arrangements in a multi-level governance setting is the result of the interplay of defining global standards and rules at the transnational level by governments, international organisations and oil companies and the implementation of these standards and rules in (local) sites and practices where the exploration, exploitation and transport of oil takes place. Networks of oil companies (involving operators, investment banks, equity partners, and local offices), nation-states and civil society networks (e.g. international eNGOs, representatives of indigenous communities) link the processes at the transnational and local levels. The design and implementation of benefit sharing arrangements results from the interplay of these different coalitions of actors from many different levels. We have identified four ideal types of benefit sharing arrangements.

First, the *Corporate Social Responsibility (CSR) arrangement*. In this arrangement oil companies are the dominant actor. The oil company adopts benefit sharing standards and implements them on the ground, but with limited consultation with state and community actors. The dominant discourse of the CSR benefit sharing arrangements is compensation: the oil company decides about the amount and form of compensation for damage to local communities. Decisions about compensation are taken with limited involvement and engagement with indigenous people.

Second, *the Shareholder arrangement*. This arrangement is characterised by a coalition of international, national and local actors; more specifically, joint ventures

---

[49] Secretariat of the Convention on Biological Diversity (2011).
[50] Tysiachniouk et al. (2018).
[51] Henry et al. (2016).

are formed by oil companies and indigenous Native Corporations to enhance broad opportunities for economic development (discourse). The revenues of the oil companies (resources) are shared between the companies and indigenous communities and indigenous peoples. These shareholder rules are based on international standards and rules, and are translated and implemented for specific locations and sites. Alaska is one illustration of the stakeholder arrangement; every resident receives permanent fund dividends annually. Indigenous people of the North Slope are almost always shareholders in both the Arctic Slope Regional Corporation (ASRC) and (usually) one of the village corporations, and they receive dividends from both. ASRC contracts with many oil companies and receives royalties from oil extraction on Indigenous-owned land.[52]

Third, *the Partnership arrangement*. In this arrangement the global – local interplay of actors is also high, but the implementation of benefit sharing standards takes place in tripartite coalitions. These coalitions are formed among the state, the company and indigenous people based on the rules of tripartism, e.g. state, market and civil society actors cooperate in a much looser manner (compared to corporatism) through a variety of different formal and semi-formal committees, consultative bodies and meetings.[53] Tripartite partnership agreements among the state, the company and indigenous peoples are based on international and transnational benefit sharing standards defined and negotiated at the international and transnational level, and implemented on the ground. The leading discourse in the partnership arrangements emphasizes community development, empowerment, self-sufficiency and participatory decision-making in indigenous communities. Sakhalin island (Russia) is an illustration of this arrangement. In the sub-Arctic region of Sakhalin island, two large transnational consortia (Sakhalin Energy and Exxon Neftegaz Limited) are active. Tysiachniouk et al. show that Sakhalin Energy, through loans and investments, was profoundly influenced by the global rules of international financial institutions related to the environment and Indigenous people.[54] The consortium adopted all global standards, including free prior and informed consent, and became subject to annual third-party evaluations. The benefit-sharing arrangement included the participation of Indigenous people in distributing grant funding to community groups. In contrast, Exxon Neftegaz Limited was less dependent on loans and was not significantly influenced by international financial institutions. The resulting benefit sharing arrangement was initiated by the head office of the company. This arrangement included grant funding to communities where drilling occurs, but not for other communities.

Fourth, the *Paternalistic arrangement*. In this arrangement state actors are dominant, and they define the preconditions for involvement in the arrangement. The

---

[52] Tysiachniouk (2016).
[53] Newton & Van Deth (2016), at 222.
[54] Tysiachniouk et al. (2018).

state monitors and intervenes in companies' corporate social responsibility. Because the global – local interplay of actors is minimal, rules and benefit sharing standards negotiated at the transnational and international level are not essential. The company adds to or substitutes for the state's efforts to provide in-kind support to local communities and Indigenous people. An example of the paternalistic arrangement is the Nenets Autonomous Okrug (NAO) in Russia.[55] In the NAO oil money is channelled to Indigenous communities through state-led socio-economic partnership agreements without control by Indigenous people. This arrangement is rooted in the lingering effects of the Soviet past and often results in the Indigenous people's being dependent on the oil companies. The paternalistic arrangement in the NAO resulted in housing for Indigenous reindeer herders. Until 2013 oil companies provided only in-kind support, because they did not trust Indigenous people to manage their bank accounts, while the amount of money was dependent on reindeer herders' negotiation skills. Since 2013 these agreements have been replaced by formal compensation for damage to the pasture lands.

These arrangements give insight into how different types of benefit sharing affect livelihoods, financial compensation and socio-economic support for Indigenous communities. The question is: are these benefit sharing arrangements smart? In other words, what is their behavioural and problem solving effectiveness? To enhance their effectiveness, these arrangements should take into account firstly the environmental effects on communities during all phases of resource extraction, as impacts vary during the phases of exploration, construction, production and transportation of resources, and secondly the unavoidable impacts on indigenous cultural heritage and lifestyles resulting from the forced industrial development in their communities. Characteristic of both the partnership and the stakeholder arrangement are the institutionalization of coalitions between oil companies, indigenous communities and governments. The effectiveness (outcome and impact) of these benefit sharing arrangements will increase if these coalitions are able to translate and to implement both social, economic and environmental standards and regulations (from the global and transnational level) to the sites of implementation, i.e. the geographical territories where oil exploration, extraction and transportation activities and their consequences come in contact with the traditional ways of life of indigenous communities.

14.4 SMART MIXES OF GOVERNANCE ARRANGEMENTS IN OFFSHORE OIL PRODUCTION

In Section 14.3, three different types of oil production governance arrangements were presented: market-based arrangements in a nation-state setting; social licence

---

[55] Pierk & Tysiachniouk (2016); Tysiachniouk et al. (2018).

to operate arrangements in an institutional setting of diffused authority, and benefit sharing arrangements within a multi-level (nested) governance setting.

According to Van Erp et al. (this volume) 'smart regulation proposes to combine various regulatory and governance instruments, both public and private and both international and local, into sophisticated mixes of complementary instruments and actors, tailored to the specific needs of the situation'. In their approach, the focus rests on the inter-connectedness of forms of regulation, levels of regulation and policy instruments. In this chapter, smartness of mixes refers to the behavioural and problem solving effectiveness of these mixes. In this section I will discuss the conditions under which the regulatory mixes within the governance arrangements discussed in this chapter are examples of smart mixes.

The environmental covenant to regulate Dutch offshore oil exploitation is an example of a new instrument to overcome the shortcomings of environmental regulation (implementation and enforcement) as a result of the changing role of the state. This resulted in the development of target-oriented policy and voluntary instruments. The social licence to operate arrangement is an informal and negotiated steering instrument to increase the (input and output) legitimacy of oil activities by gaining the ongoing acceptance and approval of indigenous communities, NGOs and so on. The development of a SLO, besides legal (permits) and political licenses to operate, is needed in fragmented governance setting, in which authority structures are diffused. Characteristic of Arctic governance is the co-existence of regime complexes (such as oil exploration and exploitation), and the emerging Arctic network state (which consists of actors and institutions with conflicting interests and jurisdictions, such as UNCLOS, the Arctic Council, and the Arctic Five and Eight), which lacks overall authority. In this networked polity, new relations of social accountability have to be developed. An example is the SLO as an instrument providing more direct and explicit (horizontal) accountability relations between oil companies, NGOs and public agencies. Benefit sharing arrangements link global standards (negotiated by international organisations, nation states and multi-national corporations) with compensation measurements between oil companies and indigenous communities in specific localities. The four ideal typical arrangements show different relations between the state, companies and communities, different responsibilities between these actors and different steering instruments, such as financial compensation, socio-economic contributions, in kind support and so on. The analysis of the regulatory mixes in the three governance arrangements in oil production show different conditions of smartness. The smartness in the Dutch offshore oil exploitation case depends on the self-governance responsibilities of companies and the capacity of the Dutch national government to monitor the outcome and the impact of the regulations laid down in the covenant. The conditions for smartness in SLO arrangements in the Arctic are much more complex. This has to do with the fragmented institutional governance setting and the diffuse authority structures of the Arctic. An important condition to achieving smartness is that the legal, political and social licences to operate are closely related to each other and that authorities empower communities to strengthen their position in the interactions with oil companies. This condition is also crucial to increasing the smartness of benefit sharing arrangements, i.e. when coalitions of companies, communities and governments are able to successfully implement international socio-economic and environmental standards and regulations to regulate the activities of oil companies and to reduce and compensate the negative (social-economic and environmental) consequences of these activities for Indigenous peoples and their communities.

14.5 CONCLUSIONS

This chapter analysed three examples of oil production governance arrangements. The analysis showed that the oil regime complex consists of different governance arrangements (co-existing side by side), showing different levels of governance, specific relations between states, markets and communities and different forms of authority. The reason for this is that the institutionalization of governance arrangements takes place within different institutional settings, ranging from a nation state governance setting, a global – local networked polity (benefit sharing arrangements) to a fragmented governance setting of emerging network states in the Arctic. This resulted in specific regulatory mixes within each of these governance arrangements: a mix of regulatory, market-based and voluntary policy instruments in the environmental governance of Dutch offshore oil exploitation; a mix of transnational, national and corporate standards and regulations in benefit sharing arrangements; and the interplay of social, legal and political licences in SLO arrangement. The analysis showed that the smartness of mixes, i.e. the effectiveness of regulatory mixes to change the behaviour of actors involved and to solve environmental problems in a legitimate way, is context-dependent. Based on the specific regulatory mixes in each of the governance arrangements, conditions for smartness have been formulated. These conditions are 1) the capacity of the Dutch national government to monitor the self-governance of companies, 2) connecting the political and legal licenses with the SLO and the empowerment of local communities by governments and 3) the implementation of international standards and rules in benefit sharing arrangements to improve the social-economic position of local communities and to reduce the environmental consequences of oil production activities.

REFERENCES

Ansell, C. 2000. 'The Networked Polity: Regional Development in Western Europe'. *Governance: An International Journal of Policy and Administration* 13(3), 303–333.
Arts, B. J. M. & Van Tatenhove, J. P. M. 2006. 'Political Modernisation'. In Arts, B. & Leroy, P. (eds.), *Institutional Dynamics in Environmental Governance*. Dordrecht, Springer, 21–43.

Beck, U. 2005. *Power in the Global Age: A New Global Political Economy*. Cambridge, Polity Press.

Blaauw, R. 2013. 'Oil and Gas Development and Opportunities in the Arctic Ocean'. In Berkman, P. A. & Vylegzhanin, A. N. (eds.), *Environmental Security in the Arctic Ocean*. NATO Science for Peace and Security Series C: Environmental Security. Dordrecht, Springer, 175–184.

Carter, C. & Smith, A. 2008. 'Revitalizing Public Policy Approaches to the EU: "Territorial Institutionalism", Fisheries and Wine'. *Journal of European Public Policy* 15, 263–281.

Casper, K. N. 2009. 'Oil and Gas Development in the Arctic: Softening of Ice Demands Hardening of International Law'. *Natural Resources Journal* 49(3/4), 825–882.

Castells, M. 2009. *Communication Power*. Oxford, Oxford University Press.

  1996. *The Information Age: Economy, Society and Culture, Volume I, The Rise of the Network Society*. Oxford UK/Malden USA, Blackwell.

Chhotray, V. & Stoker, G. 2009. *Governance Theory and Practice. A Cross-Disciplinary Approach*. New York, Palgrave/Macmillan.

Curran, G. 2017. 'Social License, Corporate Social Responsibility and Coal Seam Gas: Framing the New Political Dynamics of Contestation'. *Energy Policy* 101, 427–435.

Hall, N., Lacey, J., Carr-Cornish, S. & Dowd, A. M. 2015. 'Social License to Operate: Understanding how a Concept has been Translated into Practice in Energy Industries'. *Journal of Cleaner Production* 86, 301–310.

Held, D., McGrew, A., Goldblatt, D. & Perraton, J. 2000. *Global Transformations. Politics, Economics and Culture*. Cambridge, Polity.

Henry, L. A., Nisten-Haarala, S., Tulaeva, S. & Tysiachniouk, M. 2016. 'Corporate Social Responsibility and the Oil Industry in the Russian Arctic: Global Norms and Neo-Paternalism'. *Europe and Asia Studies* 68(8), 1340–1368.

Hossain, K., Koivurova, T. & Zojer G. 2014. 'Understanding Risks Associated with Offshore Hydrocarbon Development'. In Tedsen, E., Cavalieri, S. & Kraemer, R. A. (eds.), *Arctic Marine Governance. Opportunities for Transatlantic Cooperation*. Heidelberg/New York/ Dordrecht/London, Springer, 159–176.

Joyce, S. & Thomson, I. 2000. 'Earning a Social License to Operate: Social Acceptability and Resource Development in Latin America'. *Canadian Mining and Metallurgical Bulletin* 93(1037), 49–53.

Keohane, R. O. 1995. '"Hobbes" dilemma and institutional change in world politics: sovereignty in international society'. In Holm, H. H. & Sorensen, G. (eds.), *Whose World Order?*. Boulder, Westview Press, 165–186.

Keohane, R. O. & Victor, D. 2011. 'The Regime Complex for Climate Change'. *Perspectives on Politics* 9(1), 7–23.

Koivurova, T. 2013. Gaps in International Regulatory Frameworks for the Arctic Ocean'. In Berkman, P. A. & Vylegzhanin, A. N. (eds.), *Environmental Security in the Arctic Ocean*. Dordrecht, Springer, 139–155.

Morrison, J. 2014. *The Social License. How to Keep Your Organization Legitimate*. New York, Palgrave Macmillan.

Mossop, J. 2016. *The Continental Shelf beyond 200 Nautical Miles. Rights and Responsibilities*. Oxford, Oxford University Press.

Newton, K. & Van Deth, J. W. 2016. *Foundations of Comparative Politics. Democracies of the Modern World*, 3rd edn. Cambridge, Cambridge University Press.

Parsons, R., Lacey, J. & Moffat, K. 2014. 'Maintaining Legitimacy of a Contested Practice: How the Minerals Industry Understands its 'Social License to Operate'. *Resources Policy* 41, 83–90.

Pierk, S. & Tysiachniouk, M. 2016. 'Structures of Mobilization and Resistance: Confronting the Oil and Gas Industries in Russia'. *The Extractive Industries and Society* 3, 997–1009.

Prno, J. 2013. 'An Analysis of Factors Leading to the Establishment of a Social License to Operate in the Mining Industry'. *Resources Policy* 38, 577–590.

Raustiala, K. & Victor, D. 2004. 'The Regime Complex for Plant Genetic Resources'. *International Organization* 58(2), 277–309.

Rosenau, J. N. 2006. *The Study of World Politics: Globalization and Governance* (Vol. 2). London/New York, Routledge.

  1997. *Along the Domestic-Foreign Frontier; Exploring Governance in a Turbulent World.* Cambridge, Cambridge University Press.

Sassen, S. 2006. *Territory, Authority, Rights: From Medieval to Global Assemblages.* Princeton, NJ, Princeton University Press.

Secretariat of the Convention on Biological Diversity. 2011. The Nagoya Protocol on Access to Genetic Resources and the Fair and Equitable Sharing of Benefits Arising from their Utilization to the Convention on Biological Diversity: text and annex. Montreal, UNEP. . Available at: www.cbd.int/abs/.

Smits, C. C. A., Van Leeuwen, J. & Van Tatenhove, J. P. M. 2017. 'Oil and Gas Development in Greenland: A Social License to Operate, Trust and Legitimacy in Environmental Governance'. *Resources Policy* 53, 109–116.

Smits, C. C. A., Van Tatenhove, J. P. M. & Van Leeuwen, J. 2014. 'Authority in Arctic Governance: Changing Spheres of Authority in Greenlandic Offshore Oil and Gas Developments'. *International Environmental Agreements: Politics, Law and Economics* 14(4), 329–348.

Thomson, I. & Boutilier, R. G. 2011. *Modelling and Measuring the Social License to Operate: Fruits of a Dialogue between Theory and Practice.* Available at: https://socialicense.com/publications/Modelling%20and%20Measuring%20the%20SLO.pdf.

Tysiachniouk, M. 2016. 'Benefit Sharing Arrangements in the Arctic: Promoting Sustainability of Indigenous Communities in Areas of Resource Extraction', *The Many Faces of Energy in the Arctic*, Arctic and International Relations Series, Fall 2016, Issue 4, Seattle, WA, Jackson School of International Relations, University of Washington, Canadian Studies Center, 18–21. Available at: https://jsis.washington.edu/arctic/wp-content/uploads/sites/21/2016/05/AIRS-Issue-4.pdf.

Tysiachniouk, M. S., Henry, L. A., Lamers, M. & Van Tatenhove, J. P. M. 2018. 'Oil Consortia and Indigenous Peoples in Sub-Arctic Russia: Partnership for Equity in Benefit Sharing Arrangements'. *Energy Research & Social Sciences* 37, 140–152.

USGS. 2008. 'Circum-Arctic Resource Appraisal: Estimates of Undiscovered Oil and Gas North of the Arctic Circle', *U.S. Geological Survey*. Available at: https://pubs.usgs.gov/fs/2008/3049/fs2008-3049.pdf.

Van Leeuwen, J. 2010. *Who Greens the Waves? Changing Authority in the Environmental Governance of Shipping and Offshore Oil and Gas Production.* Wageningen, Wageningen Academic Publishers.

Van Leeuwen, J. & Van Tatenhove, J. P. M. 2010. 'The Triangle of Marine Governance in the Environmental Governance of Dutch Offshore Platforms'. *Marine Policy* 34(3), 590–597.

Van Tatenhove, J. P. M. 2016. 'The Environmental State at Sea'. *Environmental Politics* 25(1), 160–179.

  2013. 'How to Turn the Tide: Developing Legitimate Marine Governance Arrangements at the Level of the Regional Seas'. *Ocean & Coastal Management* 71, 296–304.

Van Tatenhove, J., Arts, B. & Leroy, P. (eds.). 2000. *Political Modernisation and the Environment: The Renewal of Environmental Policy Arrangements*. Dordrecht/Boston/London, Kluwer Academic Publishers.

Van Tatenhove, J. P. M. & Goverde, H. J. M. 2002. 'Strategies in Environmental Policy – A Historical Institutional Perspective'. In Driessen, P. P. J. & Glasbergen, P. (eds.), *Greening Society. The Paradigm Shift in Dutch Environmental Politics*. Dordrecht/Boston/London, Kluwer Academic Publishers, 47–63.

PART IV

Concluding Remarks

15

Conclusion

Smart Mixes in Relation to Transboundary Environmental Harm

Judith van Erp, Michael Faure, André Nollkaemper and Niels Philipsen

15.1 INTRODUCTION

An important part of the quest for an effective global environmental governance system consists of finding 'smart mixes' of regulatory instruments. The idea is that a combination of regulatory instruments and actors can be more effective than a single instrument. Combining instruments and regulatory actors into a mix allows taking advantage of their strengths while compensating for their weaknesses. Neil Gunningham, one of the contributors to this volume, introduced the related concept of smart regulation twenty years ago. He explains that the smart mix concept originated from the growing realisation that neither traditional command-and-control regulation nor the free market provided satisfactory answers to increasingly complex environmental problems – especially those of a transboundary nature. The smart mix concept takes into account the growing role of nonstate actors and informal regulatory strategies in policy theory, and expands the regulatory toolkit with a broader range of policy tools such as economic instruments, self-regulation and information-based strategies.

The fourteen chapters collected in this volume seek to contribute to our understanding of what makes a mix 'smart', and under which conditions smart mixes emerge. The chapters bring together theoretical analyses of instrument mixes (Part I) as well as empirical case-studies of instrument mixes in the areas of fishery, forestry, climate change and oil pollution (Parts II and III). The mixes discussed in these chapters consist of different combinations of international and domestic law and different combinations of public and private or hybrid regulation, involving a wide variety of actors.

This chapter will present our conclusions based on the theoretical and empirical contributions to this volume. We will start in Section 15.2 by elaborating on the instruments that are used in environmental governance, the actors involved in these regulatory mixes, and the interactions between actors and instruments. Section 15.3

discusses how instrument mixes develop from the dynamics in transnational environmental governance. The question of what is 'smart' about a particular regulatory mix is addressed in Section 15.4. Finally, we provide our overall conclusions as well as some conceptual reflections in Section 15.5.

15.2 ELEMENTS OF SMART MIXES

The contributions to this volume confirm that in all areas covered, the 'smart mix' toolbox consists both of instruments that can be classified as *public* regulation and of *private and/or hybrid* regulation. Within these broad categories, the chapters provide examples of substantive instruments of command-and-control regulation (e.g. private standards), economic instruments (e.g. emission trading) and suasive instruments (e.g. certification) on the one hand, and procedural instruments on the other. Such combinations were noted in several chapters. For instance, in relation to the compensation of victims of marine oil pollution, the mix of instruments consists of liability rules, public regulation and private compensation arrangements (Faure and Wang, Chapter 13). In relation to fisheries, state regulation has been supplemented by rights-based instruments and market-based controls, drawing the private sector into the mix, as well as a range of soft law and policy instruments (Karavias, Chapter 6).

The wide variety of instruments is directly linked to the wide variety of actors that are engaged in developing, applying and monitoring instruments. All chapters in this volume demonstrate that many different actors are involved in the design of (and participation in) instrument mixes, ranging from international organizations and states to various private and civil society actors. In fisheries management, the large number of actors involved, all with their own interests and regulatory agendas, results in a complex regulatory framework that encompasses international, regional and local levels of legal control (Barnes, Chapter 5; Karavias, Chapter 6). For example, Regional Fisheries Management Organisations (RFMOs, with responsibility for governing certain regional fisheries) operate next to NGOs (who play a minor role in fisheries, but often provide advice, data or subject states to pressure through the media) and the Marine Stewardship Council (MSC, with its role in certifying fisheries). Likewise, in relation to offshore oil production, different governance arrangements co-exist; each of these is characterised by a specific interplay of state, market (oil companies) and civil society (NGOs and Indigenous communities). For example, in the creation of the social licence to operate in the Arctic, the Nordic Council and the Arctic Council play an important role (Van Tatenhove, Chapter 14). This diversity of actors is obviously directly related to the diversity of instruments. In transnational private regulatory regimes, the interaction between private actors and public regulators results in regimes that are often of a hybrid public-private nature (Senden, Chapter 2).

A dominant feature that has emerged from the chapters is that the role of private or hybrid instruments is intimately linked to the limits of the power of public law in transitional situations. Often, private regulation emerged as a result of the absence of public standards (Senden, Chapter 2). Gunningham (Chapter 11) notes that the transnational level is often characterized by an absence of the possibility to exercise formal state coercion in the form of sanctions. His contribution about the fossil fuel divestment movement points to the strength of informal sanctions, such as naming and shaming and consumer boycotts, that may be equally or perhaps even more powerful. A similar finding was made in relation to forestry (Liu, Chapter 8). While treaties constitute one element of the instrument mix, due to the apparent inability of existing international law and organisations to provide satisfactory solutions for forestry problems, private and hybrid regimes like the Forest Stewardship Council (FSC) and the Programme for the Endorsement of Forest Certification (PEFC) started to develop during the 1990s to fill this gap. According to Liu (Chapter 8), public regulation (including international and domestic law) and private regulation often interact with each other in attempts to overcome the failures of the government and market. Another example of an instrument mix is discussed by Peeters and Müller (Chapter 12) in their contribution on greenhouse gas emission reductions: the involvement of private actors in the reporting and transparency obligations as well as the civil society control by environmental NGOs.

Similarly, nonstate actors fill gaps left open by states and international organisations. In relation to compensation for damages resulting from oil pollution, the work of states and the International Maritime Organization (IMO) has been supplemented by the oil industry, which participated in the creation of the International Oil Pollution Compensation Fund. In addition to the relevant conventions (Civil Liability Convention and IOPC Fund Convention) private arrangements have been worked out by the shipping and oil industry to provide additional compensation (Faure and Wang, Chapter 13).

Another notable example is the success of the fossil fuel divestment movement, which involves financial institutions such as banks, institutional investors and credit rating agencies (Gunningham, Chapter 11). This demonstrates that the increased globalisation and financialisation of production processes offer huge possibilities for the involvement of financial actors in environmental governance.

However, it appears from the case studies that it would be too simple to see private instruments such as certification only as compensation for inadequate public governance systems, which would be disconnected from such public governance. In relation to both forestry and fisheries regulation, certification has emerged as a supplement to public regulation that in many countries has been actively promoted by governments, as noted by Gulbrandsen in Chapter 10; he concludes that states, sometimes despite initial resistance, have in several cases actively supported the MSC, especially in exporting countries. He found a similar positive result for the interaction between the certification by FSC and states in relation to forestry. Likewise, in relation to climate change, the EU and UN interact with actors at the local level and with private actors. The EU-FLEGT Action Plan and

UN-REDD+ inter-governmental agreements are accompanied by a wide range of multi-lateral and bi-lateral regulatory, financing and support arrangements, national and subnational government policies and private market initiatives such as forest carbon certification schemes (McDermott, Chapter 9). Similarly, in relation to offshore oil production, the use of licences to operate in the Arctic was combined with an institutionalization of market-based and voluntary instruments with respect to Dutch offshore oil platforms (Van Tatenhove, Chapter 14). Likewise, EU Directives rely on technical specifications set by private European Standards Organisations (ESOs) (Senden, Chapter 2). And in the civil aviation area, the International Civil Aviation Organisation (ICAO) recognises the private International Air Transport Association's Dangerous Goods Regulation as the field guide to be applied.

It is important to note that not all forms of private instruments and actors can be directly explained by their relation to public authorities (either because they compensate for their weakness, or because public authorities rely on them). The idea that private actors would intentionally seek to fill gaps would be too simple. As Senden notes in Chapter 2, transnational private regulation is a heterogeneous phenomenon that is very context-dependent and can thus feature numerous international, regional and national, both public and private regimes, and can have different drivers such as risk management, harmonisation of technical standards and regulating market entry.

The toolbox of instruments (public and private, formal and informal, international and national and so on) is not constant and changes over time as a result of the invention of new tools, national and international developments and the interactions of the supranational institutions. Illustrative of this is that many European countries have adopted an increasingly wide range of so-called New Environmental Policy Instruments without abandoning traditional regulatory tools (Wurzel et al., Chapter 4). This has produced complex mixes of new and old instruments; in some cases these instruments exist side by side, while in other cases they have mutated into hybrid instruments.

All of these forms of co-existence and interactions between instruments and actors may be captured by what Pattberg and Widerberg (Chapter 3) call the 'complexity' of mixes. They explain that some global governance systems (taking as an example the global climate governance architecture) are complex systems that consist of diverse entities that interact; and in which the behaviour of the entities in the system is interdependent, in that they influence each other. Following from the insight that the whole is greater than the sum of its parts, studying individual governance arrangements and ideal types is insufficient for understanding the actual behaviour of the evolving regime complex. Instead, according to Pattberg and Widerberg, studying the actors and actions must take into account their broader environment, context, and position within an interaction network. The structural position of institutions is important when thinking of global governance in terms of complex systems and networks.

A good example of such complexity is international fisheries regulation, which Barnes (Chapter 5) ascribes partly to the complexity of the physical resource system, its location and interaction with other ocean activities. Instruments (international, regional and local levels of legal control) interact with multiple actors with asymmetrical power. These actors possess and seek to advance a range of regulatory agendas and interests, sometimes compatible, sometimes conflicting.

15.3 THE EVOLUTION OF INSTRUMENT MIXES INTO SMART MIXES

Characteristic of transboundary environmental problems is that they stretch beyond the authority of nation-states. International agreements are created to solve these problems, but within these agreements states have considerable discretion to fill in international norms. The complexity of transboundary environmental problems requires a certain degree of coordination and collaboration, but international law often does not fulfil this need as it is primarily concerned with setting broad goals and objectives and allocating authority. Within the broad frameworks of international conventions, states, international organizations, private actors and civil society actors often operate in uncoordinated ways. It has been noted by several authors in this volume that the structure of international law as a horizontal system of law governing states constrains its capacity to harness good regulatory design principles.

This may explain why the instrument mixes that have been studied in this volume have mostly not been created 'by design' but iteratively, in dynamic processes, by adding instruments, actors and layers in response to local and urgent problems. Often initiatives develop in parallel, and complement – but also compete with – each other. Nevertheless, we have seen that 'smart' mixes have emerged out of these more or less spontaneous or incremental processes. What pathways of development can be seen to emerge from the cases studied in this volume?

In some mixes that have been studied in this volume, the initiative for a mix that eventually could be qualified as a smart mix emerged at the international level through the creation of international conventions governed by international organisations. In the prevention of oil pollution, for example, we have seen the International Maritime Organization play an important initial role in coordinating the establishment of the International Convention on Civil Liability for Oil Pollution Damage (CLC 1969), which forms the basis for the multiple compensation arrangements in the oil spill compensation regime that we know today. The UN Food and Agriculture Organization (FAO) was mandated by fishery states to develop guidelines for fisheries certification in an attempt to regain control after the emergence of MSC. As Gulbrandsen describes in Chapter 10, the effect of the FAO guidelines was that MSC's position as the leading global certification programme for fisheries was strengthened – an outcome that was unintended by states, and that speaks to the important coordinating role that international organisations can play in the development of instrument mixes for sustainability.

States are also seen to play a leading role in the development of smart mixes. The US, although not ratifying the international conventions on compensation of oil pollution, has created a parallel regime by adopting the US Oil Pollution Act, which, as Faure and Wang (Chapter 13) conclude, provides better incentives for prevention and thus presents a 'smarter' mix of liability rules and safety regulation than the international regime. In the area of oil pollution, the EU also took an 'activist' position by threatening to adopt more stringent liability regulations, which triggered changes on the international plane. The Supplementary Fund of 2003 was stimulated by a European Commission initiative to set up a European fund outside of the IMO regime. In the areas of forestry and fishery, states have designed policies and regulations in reaction to the emergence of MSC certification, to either support or compete with certification. A successful example is formed by the state of the Maldives, which, after MSC certification of the private tuna fisheries, created an Action Plan to implement the points of improvement that MSC had demanded in the five years after granting the certificate. As some of these points demanded supranational coordination between the Maldives and other member states of the Indian Ocean Tuna Commission Agreement, the Maldives government successfully influenced decision-making at the supranational level to adopt the precautionary principle, one of the core conditions of MSC Certification (Yeeting and Bush, Chapter 7). Last, states are also buyers, and forest certification creates a basis for tailoring more sustainable public procurement policies by timber-consuming states, thus strengthening the position of FSC.

In addition to substantive policy-making, states can also influence the development of smart mixes through changes in the processes of regulation, by adopting more market-based forms of regulation that open up the regulatory process to private actors. Würzel et al. (Chapter 4) describe the adoption of New Environmental Policy Instruments in various European countries, such as information-based instruments, voluntary agreements, and market-based instruments such as emissions trading rights and ecotaxes. By adopting such policy instruments, states invite businesses, NGOs and citizen-consumers to enter the regulatory space, and thus create the conditions for regulatory mixes to develop.

Furthermore, we have witnessed cases in which private actors take the lead in developing regulation that supplements public regulation, thus creating a mix. In aviation, Senden's Chapter 2 describes the private organisation IATA as a frontrunner in the development of certain standards (transport of dangerous goods and of life animals). The IATA developed private regulation to fill the gap that emerged as the (public) ICAO did not take any action. In the fossil fuel divestment movement, Gunningham (Chapter 11) describes how an ad hoc coalition of private actors, primarily NGOs, created loose alliances with financial institutions, without any involvement of state actors, to influence investment decisions of institutional investors on a global level.

Although in these cases we can point to an actor taking the initiative or establishing a course of action to which other actors react, it would overstate the degree of steering if we were to describe the process of formation of instrument mixes as a process of orchestration, let alone coordination. Rather, the process is incremental, iterative and experimental, with actors reacting to each other as incentives and opportunities change or new problems arise. New instruments are added in response to shortcomings of the previous ones, and new actors enter networks, changing the power balance and forcing others to react. Sometimes, the layering of instruments creates a perceived need for coordination, but the outcomes of this coordination are not always as intended. For example, the collective adoption of private compensation schemes for oil pollution by the oil shipping industry (TOVALOP) was intended to prevent the establishment of a strict liability regime for the shipping industry, but failed to do so, as a series of very damaging oil spills induced consensus within the IMO to adopt higher liability limits. Likewise, the coordination of fishery certification by the FAO was initiated by fishery states, but resulted in the strengthening of MSC rather than the development of alternatives. Thus, more often than not, processes of adaptation are incremental and unpredictable, and attempts of actors to coordinate and steer trigger others to react in unexpected ways.

For smart mix theory, this has important implications. Most importantly, we should question the possibility of designing smart instrument mixes purposively. Policy networks with regard to transboundary environmental problems more often resemble complex webs of loosely coupled actors with different interests and powers, than close-knit groups with a coordinating party. Interactions cannot be predicted or coordinated, precisely because the networks are lacking a coordinator or common interest binding parties together. The ability of actors to influence the system in a linear-causal way is therefore limited, as feed-backs and unintended consequences (both positive and negative) undermine the linear-consequential ontology of orchestration. This does not mean that attempts to coordinate or orchestrate are entirely fruitless, but this process should be conceived of as experimentalism and 'piecemeal social engineering' rather than 'steering' or 'designing'. The emergence and development of smart mixes should not be studied as a process of straightforward institutional design, but as a dynamic process of institutions enabling parties to interact and develop binding obligations towards each other, and to coordinate problems that arise out of interactions. In other words, institutions create platforms for interaction in which actors have opportunities to influence each other and to develop common ground that may result in smart mixes. Political and economic interests play important roles, as well as processes of organizational learning, policy advocacy and coalition building. For the study of smart mixes, this entails that research should be multidisciplinary, combining political, economic and sociological perspectives with legal analysis.

The case of MSC certification of fisheries, described from various angles in this volume, forms a suitable illustration of the dynamic and interactive process of emergence of smart mixes. The fact that fisheries certification is a long-term process creates continuous and sequenced interactions between actors, which helps to engage and bind actors and develop a common framework. In Chapter 7 on MSC certification, Yeeting and Bush demonstrate that, in particular, the MSC pre-assessment phase creates an open and experimental environment that helps in creating the ambition to establish international agreements on fishing. Earlier, participating states had not managed to overcome the vested interests of their home fishing industries. By applying for MSC certification collectively, they entered a framework and process to gradually develop consensus over harvest control rules. Thus, certification works as a process of experimental rule-setting, shaping and coordinating preferences of actors, and thus changing conditions within actor networks. Because certification is a process with recurrent interactions, the incentives, interests, and strategies of network members change at different stages of the certification process. This chapter also shows how processes can be adaptive to local context, as the comparison of the MSC certification process in Regional Fisheries Management Organisations in the Indian, Pacific and Atlantic Ocean, shows that MSC influences decision making through different pathways in these three areas. Even unsuccessful attempts at certification contribute to change, because participating in the certification process creates public engagement and transparency.

Similarly, in the area of forestry, the negotiation of Voluntary Partnership Agreements (VPAs) between the EU and developing countries within the framework of FLEGT, and EU actions within the framework of REDD+ both have created new platforms for previously marginalized actors to participate in forest-related decision-making (McDermott, Chapter 9). Local participation in these platforms contribute to the safeguarding of local welfare, but McDermott warns that it is unclear how much priority VPAs or REDD+ will continue to place on local participation. She points to the risk that large firms with sufficient influence in the industry will dominate the negotiation platform and join forces with environmentalists to exclude producers who do not meet the higher standards that insiders already comply with. Thus, the organic and iterative process in which instrument mixes are developed could result in major inequalities between developed and less developed countries and between large industrial actors and smaller local producers. This brings us to a next important topic: the definition of an instrument mix as 'smart', and in particular, the potential conflict between environmental outcome and legitimacy, equality and fairness.

15.4 THE QUESTION OF 'SMARTNESS': WHAT IS SMART ABOUT SMART MIXES?

The starting point for the concept of smart mixes is that, as was explained in Chapter 1, a combination of regulatory instruments and actors is often more effective than a single instrument. Often there is no single instrument that can be considered as the silver bullet that will solve all environmental problems. The various contributions to this volume confirm the emergence of a large variety of mixes that seek to overcome the limitations of single instruments.

Yet an important question – whether it is possible to substantiate the initial assumption that creating a mix might be smarter than using a single instrument – remains. That leads to the important question on how to evaluate the smartness of instrument mixes. The chapters show some illustrations of the two approaches to effectiveness introduced in Chapter 1: the problem-solving effectiveness (the contribution of the mix to solving the environmental problem; see e.g. Senden (Chapter 2) and the behavioural effectiveness (the ability of the mix to induce actors to a behavioural change; see e.g. van Tatenhove (Chapter 14).

While several chapters do point at the usefulness of effectiveness as a criterion to determine the smartness of the mix, many also point at certain limits of the effectiveness criterion or point to the importance of a combination with other criteria. One problem with measuring the effectiveness (and efficiency) of instrument mixes in the governance of common pool resources is, as indicated by Liu in Chapter 8, that this is empirically an extremely complicated issue. She shows, in relation to the mix of public regulation and private governance and the interactions between the systems in protecting common pool resources, that there may be examples of smart (and in some cases not-so-smart) mixes of public regulation, private governance and property rights. But it remains quite difficult to pinpoint exactly which aspects of a particular mix contribute to its effectiveness.

In particular cases authors have specified the criterion of effectiveness, depending upon the particular area. For example, in the area of marine oil pollution, providing adequate compensation to victims and creating incentives for the prevention of oil pollution could be considered as specific criteria to judge the effectiveness of the mix (Faure and Wang, Chapter 13). In relation to the mix between international law and private regulation in the area of airline safety, the decreasing accident rate in air traffic could be a proxy for effectiveness (Senden, Chapter 2). Those examples show that the smartness of a mix needs to be filled in in a more detailed manner, taking into account the particular goals of the specific regime (like adequate compensation of victims in the case of marine oil pollution and reducing accident rates in the case of air traffic).

However, specification of a criterion may not always be an easy solution. McDermott (Chapter 9) stresses, like Liu in Chapter 8, that in practice policy mixes (like FLEGT or REDD+ in the area of forestry) may have such a degree of complexity that it becomes very difficult to provide an accurate evaluation of the effectiveness or smartness of the particular policy mix.

Although different chapters point out these theoretical and practical difficulties in their evaluation of effectiveness as a criterion for smartness, they nevertheless also provide some examples of instrument mixes that are considered as smart in terms of problem-solving effectiveness. Faure and Wang (Chapter 13) argue that the international regime to compensate victims of oil pollution damage can be considered as a smart mix based on the criterion of providing adequate compensation to victims. This case concerns a mix of different international conventions, providing various layers of compensation that lead to adequate victim compensation. Senden (Chapter 2) observes that the mix in the air traffic safety area between the public international law organisation (ICAO) and private regulation (IATA) fulfilled the criterion of creating a higher level of safety in civil aviation, as there was a substantial decrease in the accident rate. Other authors do not explicitly declare the mix discussed in their chapter as 'smart' (in the sense of effective) but provide criteria that may affect the smartness. Van Tatenhove (Chapter 14), discussing policy mixes in the offshore oil exploitation, provides the example of the Dutch offshore exploitation case and argues that the smartness of the policy mix will depend on the self-governance responsibilities of companies and the capacity of the Dutch national government to monitor the outcome and the impact of the regulations laid down in the particular covenant. Discussing the social licence to operate arrangement in the Arctic, he argues that an important criterion to realize the smartness of the mix relates to the authorities empowering communities to strengthen their position in the interaction with the oil companies. That example shows that the evaluation of the smartness of a mix does not just depend on the instrument as such, but that it is rather possible to identify criteria which may affect the smartness of that particular instrument. This points to a more general issue: the effectiveness of a mix cannot be evaluated in the short term. Assessing the sustainability and resilience, and more particularly the behavioural effectiveness, of a mix requires a longer-term approach towards the evaluation of the effectiveness of the mix.

Many chapters also point to the fact that [the various meanings of] effectiveness are just one way of assessing the smartness of a mix. In Chapter 1 it was indicated that other criteria, such as coherence, unintended effects, legitimacy and adaptability may be equally important in evaluating the smartness of a mix. McDermott (Chapter 9) shows that smart mixes are not per se smart in the sense of being more effective, but they can still be smarter if they provide more legitimacy with greater fairness and fewer negative side-effects. The example provided by McDermott relates to the ratcheting up of global environmental standards. It can be a means by which large influential firms gain an advantage over others, or it could even reinforce global inequities and therefore lead to environmental degradation in poor countries. In Chapter 6, Karavias argues that effectiveness may appear to be a desirable goal for fisheries management in theory, but that it is difficult to ascertain when it applies within complex natural and social systems like international fisheries. He therefore argues that it is more helpful to base an assessment of the smartness of a mix on principles of good regulatory design. To the extent that legal frameworks are consistent with those principles, they provide more measurable indices of effective regulatory outcomes. The importance of key regulatory design principles is equally stressed by Barnes in Chapter 5, which refers (in the context of international fisheries law) to criteria such as coherence, complementarity, efficiency and scalability.

Flexibility and adaptability are also stressed in several chapters as important criteria to judge the smartness. In Chapter 7, Yeeting and Bush highlight the importance of the flexibility of the certification process over time because it keeps providing incentives for improvement at different stages of the rule development in response to the strategies and interests of the RFMO members. They argue that to increase the likelihood of achieving any predefined outcome, a smart mix should therefore be able to adapt to the actors involved and the nature of their involvement. Although adaptability and flexibility are undoubtedly important, they may constitute a trade-off with predictability and robustness in a given mix. The more flexible, dynamic and adaptable the mix becomes, the greater the risk that it will also become more diffuse and less predictable, which could negatively affect its capacity to provide a behavioural change. In private certification, for example, the predictability of specific features and characteristics of a product or service are crucial in order to convince consumers to pay a premium price for ecologically sound products or services.

The various examples show that even though effectiveness may be the dominant criterion for smartness, it certainly is not the only criterion and must be complemented with other criteria such as legitimacy or (distributional) fairness.

Many chapters also note that complementarities between public regulation and private governance systems can contribute to the 'smartness' of a mix. In Chapter 2, Senden argues that a smart mix (in the sense of problem-solving effectiveness) may emerge if private interests align with public policy goals. In airline safety, for example, the huge common interest of both the private sector and the public authorities in promoting safety in civil aviation was an important driver for creating the 'smart' public-private partnership in that area.

Notwithstanding the examples provided of smart mixes, several chapters equally point at mixes that can be considered as less smart. In the discussion of international fisheries laws in Chapter 5, for example, Barnes points at important restrictions in the creation of smart mixes caused by the range of actors engaged in international regulatory activities, the limited tools available to regulators and the absence of strong reflective and adaptive governance structures. In Chapter 2 Senden, providing the example of private security companies, shows that the mix created in that particular area was only affected to a limited extent as some private actors could escape the application of the private regime and could thus free-ride. She equally argues that that was precisely the reason why free-riding was not an issue in the area of airline safety (as private and public interests were aligned), thus contributing to the success of the regime. In Chapter 12, Peeters and Müller suggest that competition between verifying actors may undermine the quality of the verification, and suggest that it would be better if administrative authorities carry out this task, as is the case in the EU Industrial Emissions Directive. The private nature of the verification reports also limits the ability of the public, journalists and NGOs to access them, which hollows out the principle of transparency and access to information that form important elements of the instrument mix in EU environmental regulation.

One conclusion of the chapters as far as evaluating the smartness of the mixes is concerned is that there is no one absolute criterion by which to judge the smartness of a mix. Rather, several criteria could be found; to a great extent, this depends on the specific policy area. Not only is there not just one criterion, there is also not just one single optimal mix of policy instruments. A smart mix of new environmental policy instruments (NEPIs and traditional regulation) is strongly context-dependent and may therefore be different for different jurisdictions (Wurzel et al., Chapter 4). This context specificity of the smartness of the mix is also stressed by Liu in Chapter 8: there is not just one 'smart mix' of instruments that may work well in all circumstances; there are only particular interactions that, under the specific conditions and depending upon the country context, may work better than others. What constitutes the most appropriate policy instrument mix is, as is also argued by Wurzel et al. in Chapter 4, therefore likely to vary, not only from jurisdiction to jurisdiction, but also over time. Different organisational structures, policy styles and policy goals explain why different mixes emerge in different contributions.

That leads to a more general point: some chapters may have pointed at 'smart' mixes, depending upon the various criteria of smartness we discussed. However, that does not necessarily imply that the particular mix identified is necessarily the best one and that other mixes (that could potentially have had similar or even better results) could not have emerged as well. There is, as was stressed in Section 15.3, no possibility to point at direct causal relationships between a particular input (for example, a mix of policy instruments) and an output (for example, improvement of environmental quality). When we refer here to the 'smartness' of a mix, this is at best meant to imply that the particular mix of instruments has likely contributed to particular policy goals, but not that there is a linear causal relationship, nor that other mixes would not have led to similar (or in some cases even better) results.

15.5 MOVING FORWARD

Building on the insights offered by the chapters in this volume, we offer four thoughts on how to move forward with the research on instrument mixes in global environmental governance. These relate respectively to the concept of smart mixes, to the collapse of traditional distinctions and to the nature of the methodological inquiry, to follow-up research, and to the limitations of smart mixes research.

First, notwithstanding the above limitations pertaining to the standard of effectiveness, the chapters in this volume do suggest that evaluating particular policy mixes in terms of their 'smartness' have an added value, for example compared to other concepts, such as transnational private regulation (TPR), discussed by Senden in Chapter 2. The concept of smart mixes is one that has both analytical and normative power.

The analytical power lies in the fact that the concept leads the focus of inquiry into a particular category (a mix) of instruments that has the ability to contribute to problem-solving or behavioural change. It thus focusses on a unit of analysis that may be complex, but that is much more relevant in describing, explaining and understanding environmental governance in all the areas discussed in this volume, than a focus on individual instruments.

The normative power lies in the fact that evaluating policy mixes in terms of their smartness adds, as is shown in Chapter 11 by Gunningham, an element of normativity to the debate. The concept of smart mixes (and, more particularly, the evaluation of the smartness of a mix) allows us not only to observe particular combinations of policy instruments in practice, but also to have a critical discussion of whether a specific mix of policy instruments is able to effectively reach specific policy goals in terms of reducing environmental degradation or improving environmental quality. The combination of the analytical and the normative perspective undoubtedly is an important contribution of smart mixes theory to environmental governance.

The second point that emerges from this volume, and that can be the starting point for further inquiries, is that thinking in terms of smart mixes involves collapsing traditional distinctions that have characterized much of environmental governance; the distinction between international and national, between public and private and between formal and informal. Obviously, we do not argue that these distinctions cannot be made or that, for particular purposes, they should not be made. Rather, we argue that if the question is one of whether a particular combination of instruments can help to solve a problem or cause a change in behaviour, these distinctions usually are not very relevant.

The combination of chapters in this volume does not yield a set of concrete suggestions as to what mixes do work and do not work. Much is context-specific, and the factors that were identified in Section 15.3 as being relevant for effectiveness, may work in one situation but not in the other. This is not to say that no generalizations are possible, but before doing so, much more systematic and in particular empirical research will have to be conducted.

However, at some level of abstraction, it can be suggested that the smart mixes in all cases reviewed in this volume, smart mixes involved a combination of the international and the national, of public and private, and of formal and informal. This is captured in part by Gulbrandsen's use in Chapter 10 of the concept of co-regulation; he notes that policy changes in the forestry and fisheries sectors are not indicative of less government involvement, but 'rather of the ambition to develop a "smart mix" of private and public policy instruments at multiple governance levels'. From the perspective of the state, rather than seeing international regulation as a threat to state sovereignty, private regulation as a challenge to government authority and informal instruments as a threat to rule of law-based governance, states have generally accepted all of these dimensions as useful, and indeed necessary, parts of governance aimed at public goals.

The third point we take forward from this volume is that studying the role of smart mixes in global environmental governance inherently calls for interdisciplinary analysis. To be sure, one can ask discrete questions of law, as several chapters in this volume have done (such as, for example, of how much legal weight is carried by a private instrument that has been endorsed by states in the interpretation of a treaty). Likewise, one can study aspects of smart mixes from the perspective of political science (e.g. how do private actors exercise power by filling gaps in public regulation?), law and economics (what type of instrument mixes are likely to create incentives for changes in behaviour?) or normative theory (on what grounds can mixes contribute to the legitimacy of environmental governance?). But to understand when, how and why particular mixes lead or do not lead to changes in outcomes or changes in behaviour, all of these types of questions need to be considered.

Of course, this is not unique for smart mixes – any proper study of environmental governance calls for interdisciplinary exercises. But we would suggest that the very complexity of smart mixes makes such interdisciplinarity all the more important. For instance, if one asks the question of how two instruments are legally connected, one should understand why actors have developed these instruments and whether or not they have done so with a view to achieving similar goals. The wide variety of relationships between instruments that emerges from complexity studies sets the agenda for legal analysis, but it is difficult to engage in the latter without having done the former. The fact that the concept straddles the analytical, the descriptive, the normative and the empirical dimensions of environmental governance, means that the question of whether a particular combination of instruments is a mix that can be qualified as smart calls for interdisciplinarity.

Our fourth point relates to the limitations of smart mixes research. The concept of smart mixes embodies a paradox. If we assume that there is not one single 'silver bullet' that helps to realize objectives of environmental governance, but that we have to explore combinations of instruments and actors, there is no logical limit to what type of instruments will be relevant to the mix. Climate change, depletion of fish stocks, or deforestation result from a wide variety of causes, and reversal of the trends likewise results from multiple causes. It may be tempting to limit the inquiry to those instruments (and their interrelationship) that intentionally are designed to combat these problems. But this limitation may neglect other factors that may be equally relevant.

Gunningham's contribution to this volume (Chapter 11) indeed observes that the smart mixes concept is perhaps disproportionately focused on collaboration, cooperation and partnership, and may neglect, for instance, the counterweight of industrial powers. Gunningham raises the question of whether the instruments that are commonly treated as part of smart mixes are 'disruptive enough' to substantially transform the industrial practices that cause massive and sometimes irreversible environmental harm. If this is true, fundamental transformation may require a change of norms, perceptions, and discourses that requires actors to act as moral entrepreneurs in more powerful ways than the smart mix metaphor seems to suggest.

The paradox is that such deeper processes are either left out (in effect somewhat limiting the power of the smart mixes concept) or taken on board (what is more smart than a mix that includes all relevant causes?), but then the concept would be so overbroad as to defeat its purposes – there then would be little to distinguish smart mixes from wider political and social processes.

Studying smart mixes inevitably will have to find a pragmatic middle way: identifying particular combinations of instruments that can be separated and, to some degree, isolated from wider processes and background factors, and at the same time recognizing that the relative contributions of smart mixes to the solution of environmental problems, and their effect on behavioural change, cannot be understood apart from this background.

CPSIA information can be obtained
at www.ICGtesting.com
Printed in the USA
LVHW051520110522
718483LV00008B/271